CARBON NANOTUBES
SCIENCE AND APPLICATIONS

CARBON NANOTUBES
SCIENCE AND APPLICATIONS

EDITED BY **M. Meyyappan**

NASA Ames Research Center
Moffett Field, CA

CRC PRESS

Boca Raton London New York Washington, D.C.

Front cover: Top row, left: structure of a carbon nanotube; right: schematic of a nanotube-based biosensor (courtesy of Jun Li). Middle row, left: hydrogen functionalized single-walled nanotube (courtesy of Charles Bauschlicher); right: patterned growth of carbon nanofibers using dc plasma chemical vapor deposition (courtesy of Ken Teo, Cambridge University). Bottom row, left: vertically aligned carbon nanotubes (courtesy of Alan Cassell); right: atomic force microscope image of Red Dune sand which is a Mars simulant, generated using a single-walled nanotube tip (courtesy of Ramsey Stevens).

Back cover: This atomic force microscopy image showing the words "carbon nanotube" was created by scanning probe lithography using a carbon nanotube tip. The letters are 20 nm wide SiO_2 anodized on H-passivated Si surface. Credit: Keith Jones and Bruce Wallace of Veeco and Cattien V. Nguyen of NASA Ames Center for Nanotechnology.

Library of Congress Cataloging-in-Publication Data

Carbon nanotubes : science and applications / edited by M. Meyyappan.
    p. cm.
    Includes bibliographical references and index.
    ISBN 0-8493-2111-5 (alk. paper)
  1. Carbon. 2. Nanostructured materials. 3. Tubes. I. Meyyappan, M. II.
Title.

    TA455.C3C375 2004
    620.1'93—dc22

                  2004049389

Visit the CRC Press Web site at www.crcpress.com

© 2005 by CRC Press LLC

No claim to original U.S. Government works
International Standard Book Number 0-8493-2111-5
Library of Congress Card Number 2004049389

Printed on acid-free paper

Preface

The extraordinary mechanical properties and unique electrical properties of carbon nanotubes (CNTs) have stimulated extensive research activities across the world since their discovery by Sumio Iijima of the NEC Corporation in the early 1990s. Although early research focused on growth and characterization, these interesting properties have led to an increase in the number of investigations focused on application development in the past 5 years. The breadth of applications for carbon nanotubes is indeed wide ranging: nanoelectronics, quantum wire interconnects, field emission devices, composites, chemical sensors, biosensors, detectors, etc. There are no CNT-based products currently on the market with mass market appeal, but some are in the making. In one sense, that is not surprising because time-to-market from discovery typically takes a decade or so. Given that typical time scale, most current endeavors are not even halfway down that path. The community is beginning to move beyond the wonderful properties that interested them in CNTs and are beginning to tackle real issues associated with converting a material into a device, a device into a system, and so on. At this point in the development phase of CNT-based applications, this book attempts to capture a snap shot of where we are now and what the future holds.

Chapter 1 describes the structure and properties of carbon nanotubes — though well known and described in previous textbooks — both as an introduction and for the sake of completeness in a book like this one. In understanding the properties, the modeling efforts have been trailblazing and have uncovered many interesting properties, which were later verified by hard characterization experiments. For this reason, modeling and simulation are introduced early in Chapter 2. Chapter 3 is devoted to the two early techniques that produced single-walled nanotubes, namely, arc synthesis and laser ablation. Chemical vapor deposition (CVD) and related techniques (Chapter 4) emerged later as a viable alternative for patterned growth, though CVD was widely used in early fiber development efforts in the 1970s and 1980s. These chapters on growth are followed by a chapter devoted to a variety of imaging techniques and characterization (Chapter 5). Important techniques such as Raman spectroscopy are covered in this chapter.

The focus on applications starts with the use of single-walled and multiwalled carbon nanotubes in scanning probe microscopy in Chapter 6. In addition to imaging metallic, semiconducting, dielectric, and biological surfaces, these probes also find applications in semiconductor metrology such as profilometry and scanning probe lithography. Chapter 7 summarizes efforts to date on making CNT-based diodes and transistors and attempts to explain the behavior of these devices based on well-known semiconductor device physics theories explained in undergraduate and graduate textbooks. It is commonly forecast that silicon CMOS device scaling based on Moore's law may very well end in 10 or 15 years. The industry has been solving the technical problems in CMOS scaling impressively even as we embark on molecular electronics, as has been the case with the semiconductor industry in the past 3 decades. Therefore, for those pursuing alternatives such as CNT electronics and molecular electronics, the silicon electronics is a moving target and the message is clear: replacing silicon-conducting channel simply with a CNT-conducting channel in a CMOS may not be of much value — alternative architectures;

different state variable (such as spin)-based systems; and coupling functions such as computing, memory, and sensing are what can set the challengers apart from the incumbent. Unfortunately, at the writing of this book, there is very little effort in any of these directions, and it is hoped that such alternatives emerge, succeed, and flourish.

Field emission by carbon nanotubes is very attractive for applications such as flat panel displays, x-ray tubes, etc. The potential for commercial markets in television and computer monitors, cell phones, and other such displays is so enormous that this application has attracted not only much academic research but also substantial industrial investment. Chapter 8 discusses principles of field emission, processes to fabricate the emitters, and applications. One application in particular, making an x-ray tube, is covered in great detail from principles and fabrication to testing and characterization.

With every atom residing on the surface in a single-walled carbon nanotube, a very small change in the ambient conditions can change the properties (for example, conductivity) of the nanotube. This change can be exploited in developing chemical sensors. The nanotubes are amenable to functionalization by attaching chemical groups, DNA, or proteins either on the end or sidewall. This also allows developing novel sensors using nanotubes. Chapter 9 discusses principles and development of chemical and physical sensors. Likewise, Chapter 10 describes biosensor development.

The mechanical, thermal, and physical properties of carbon nanotubes have resulted in numerous studies on conducting polymer films, composites, and other structural applications. Chapter 11 captures these developments. Finally, all other applications that elude the above prime categories are summarized in Chapter 12.

This is an edited volume, and various authors who practice the craft of carbon nanotubes day to day have contributed to this volume. I have made an effort to make this edited volume into a cohesive text. I hope that the readers — students and other researchers getting into this field, industry, and even the established experts — find this a valuable addition to the literature in carbon nanotubes. I would like to thank Nora Konopka of the CRC Press for her support throughout this work. Finally, this book would not have been possible without the help and skills of my assistant Amara de Keczer. I would like to thank her also for the cover design of the book.

M. Meyyappan
Moffett Field, California
January 2004

The Editor

M. Meyyappan, Ph.D., is Director of the NASA Ames Research Center for Nanotechnology, Moffett Field, California. After receiving his Ph.D. in chemical engineering from Clarkson University in 1984, Dr. Meyyappan worked as a research scientist at Scientific Research Associates in Glastonbury, Connecticut where his focus was semiconductor heterostructure device physics, plasma processing, and device and process modeling. He moved to NASA Ames in 1996 to establish the nanotechnology group. His current research interests include growth of carbon nanotubes and inorganic nanowires, biosensor and chemical sensor development and nanodevices. Dr. Meyyappan is a member of IEEE, AVS, ECS, MRS, AIChE, and ASME and a Fellow of the IEEE. He is the IEEE Distinguished Lecturer on nanotechnology and ASME's Distinguished Lecturer on nanotechnology. For his contributions to nanotechnology, he has received the Arthur Fleming Award and NASA's Outstanding Leadership Medal. He has published over 130 peer-reviewed articles in various science and engineering journals. Dr. Meyyappan serves on numerous nanotechnology advisory committees across the world and has given over 200 Invited/Plenary talks and invited seminars.

Contributors

Enrique Barrera
Rice University
Mechanical Engineering and Material Science
Houston, Texas

E. L. Corral
Rice University
Mechanical Engineering and Material Science
Houston, Texas

Jie Han
Center for Nanotechnology
NASA Ames Research Center
Moffett Field, California

Jing Li
Center for Nanotechnology
NASA Ames Research Center
Moffett Field, California

Jun Li
Center for Nanotechnology
NASA Ames Research Center
Moffett Field, California

Raouf O. Loutfy
MER Corporation
Tucson, Arizona

K. McGuire
University of North Carolina
Department of Physics
Chapel Hill, North Carolina

M. Meyyappan
Center for Nanotechnology
NASA Ames Research Center
Moffett Field, California

Alexander P. Moravsky
MER Corporation
Tucson, Arizona

Cattien V. Nguyen
ELORET Corporation
Center for Nanotechnology
NASA Ames Research Center
Moffett Field, California

Apparao M. Rao
Department of Physics
Clemson University
Clemson, South Carolina

Philippe Sarrazin
Center for Nanotechnology
NASA Ames Research Center
Moffett Field, California

M. L. Shofner
Rensselaer Polytechnic Institute
Troy, New York

Deepak Srivastava
Center for Nanotechnology
NASA Ames Research Center
Moffett Field, California

Eugene M. Wexler
MER Corporation
Tucson, Arizona

Toshishige Yamada
Center for Nanotechnology
NASA Ames Research Center
Moffett Field, California

Contents

5 Characterization Techniques in Carbon Nanotube Research
K. McGuire and A. M. Rao

6 Applications in Scanning Probe Microscopy   *Cattien V. Nguyen*

7 Nanoelectronics Applications   *Toshishige Yamada*

8 Field Emission   *Philippe Sarrazin*

1

Structures and Properties of Carbon Nanotubes

Jie Han
NASA Ames Research Center

Since the discovery of carbon nanotubes (CNTs) by Iijima in 1991 [1], great progress has been made toward many applications, including, for example:

- Materials
    - Chemical and biological separation, purification, and catalysis
    - Energy storage such as hydrogen storage, fuel cells, and the lithium battery
    - Composites for coating, filling, and structural materials
- Devices
    - Probes, sensors, and actuators for molecular imaging, sensing, and manipulation
    - Transistors, memories, logic devices, and other nanoelectronic devices
    - Field emission devices for x-ray instruments, flat panel display, and other vacuum nanoelectronic applications

The advantages of these applications have been demonstrated, including their small size, low power, low weight, and high performance, and will be discussed in the following chapters. These applications and advantages can be understood by the unique structure and properties of nanotubes, as outlined below:

- Structures (Sections 1.1–1.3)
    - *Bonding:* sp^2 hybrid orbital allows carbon atoms to form hexagons and occasionally pentagons and pentagon units by in-plane σ bonding and out-of-plane π bonding.
    - *Defect-free nanotubes:* these are tubular structures of hexagonal network with a diameter as small as 0.4 nm. Tube curvature results in σ–π rehybridization or mixing.

- *Defective nanotubes:* occasionally pentagons and heptagons are incorporated into a hexagonal network to form bent, branched, coroidal, helical, or capped nanotubes.
- Properties (Sections 1.4–1.9)
  - *Electrical:* electron confinement along the tube circumference makes a defect-free nanotube either semiconducting or metallic with quantized conductance whereas pentagons and heptagons will generate localized states.
  - *Optical and optoelectronic:* direct band gap and one-dimensional band structure make nanotubes ideal for optical applications with wavelength ranging possibly from 300 to 3000 nm.
  - *Mechanical and electromechanical:* σ–π rehybridization gives nanotubes the highest Young's modulus of over 1 TPa and tensile strength of over 100 GPa and remarkable electronic response to strain and metal-insulator transition.
  - *Magnetic and electromagnetic:* electron orbits circulating around a nanotube give rise to many interesting phenomena such as quantum oscillation and metal-insulator transition.
  - *Chemical and electrochemical:* high specific surface and σ–π rehybridization facilitate molecular adsorption, doping, and charge transfer on nanotubes, which, in turn, modulates electronic properties.
  - *Thermal and thermoelectric:* inherited from graphite, nanotubes display the highest thermal conductivity while the quantum effect shows up at low temperature.

1.1   Bonding of Carbon Atoms

To understand the structure and properties of nanotubes, the bonding structure and properties of carbon atoms are discussed first. A carbon atom has six electrons with two of them filling the $1s$ orbital. The remaining four electrons fill the sp^3 or sp^2 as well as the sp hybrid orbital, responsible for bonding structures of diamond, graphite, nanotubes, or fullerenes, as shown in Figure 1.1.

In diamond [2], the four valence electrons of each carbon occupy the sp^3 hybrid orbital and create four equivalent σ covalent bonds to connect four other carbons in the four tetrahedral directions. This three-dimensional interlocking structure makes diamond the hardest known material. Because the electrons in diamond form covalent σ bonds and no delocalized π bonds, diamond is electrically insulating. The electrons within diamond are tightly held within the bonds among the carbon atoms. These electrons absorb light in the ultraviolet region but not in the visible or infrared region, so pure diamond appears clear to human eyes. Diamond also has a high index of refraction, which makes large diamond single crystals gems. Diamond has unusually high thermal conductivity.

In graphite [3], three outer-shell electrons of each carbon atom occupy the planar sp^2 hybrid orbital to form three in-plane σ bonds with an out-of-plane π orbital (bond). This makes a planar hexagonal network. van der Waals force holds sheets of hexagonal networks parallel with each other with a spacing of 0.34 nm. The σ bond is 0.14 nm long and 420 kcal/mol strong in sp^2 orbital and is 0.15 nm and 360 kcal/mol in sp^3 configuration. Therefore, graphite is stronger in-plane than diamond. In addition, an out-of-plane π orbital or electron is distributed over a graphite plane and makes it more thermally and electrically conductive. The interaction of the loose π electron with light causes graphite to appear black. The weak van der Waals interaction among graphite sheets makes graphite soft and hence ideal as a lubricant because the sheets are easy to glide relative to each other.

A CNT can be viewed as a hollow cylinder formed by rolling graphite sheets. Bonding in nanotubes is essentially sp^2. However, the circular curvature will cause quantum confinement and σ–π rehybridization in which three σ bonds are slightly out of plane; for compensation, the π orbital is more delocalized outside the tube. This makes nanotubes mechanically stronger, electrically and thermally more conductive, and chemically and biologically more active than graphite. In addition, they allow topological defects such as pentagons and heptagons to be incorporated into the hexagonal network to form capped, bent, toroidal, and helical nanotubes whereas electrons will be localized in pentagons and heptagons because of redistribution of π electrons. For convention, we call a nanotube defect free if it is of only hexagonal

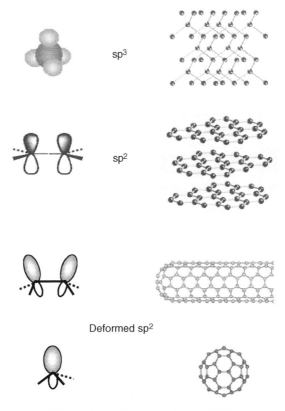

FIGURE 1.1 Bonding structures of diamond, graphite, nanotubes, and fullerenes: when a graphite sheet is rolled over to form a nanotube, the sp^2 hybrid orbital is deformed for rehybridization of sp^2 toward sp^3 orbital or σ–π bond mixing. This rehybridization structural feature, together with π electron confinement, gives nanotubes unique, extraordinary electronic, mechanical, chemical, thermal, magnetic, and optical properties.

network and defective if it also contains topological defects such as pentagon and heptagon or other chemical and structural defects.

Fullerenes (C_{60}) are made of 20 hexagons and 12 pentagons [4]. The bonding is also sp^2, although once again mixed with sp^3 character because of high curvature. The special bonded structures in fullerene molecules have provided several surprises such as metal–insulator transition, unusual magnetic correlations, very rich electronic and optical band structures and properties, chemical functionalizations, and molecular packing. Because of these properties, fullerenes have been widely exploited for electronic, magnetic, optical, chemical, biological, and medical applications.

1.2 Defect-Free Nanotube

There has been a tremendous amount of work studying defect-free nanotubes, including single or multiwalled nanotubes (SWNTs or MWNTs). A SWNT is a hollow cylinder of a graphite sheet whereas a MWNT is a group of coaxial SWNTs. SWNT was discovered in 1993 [5,6], 2 years after the discovery of MWNT [1]. They are often seen as straight or elastic bending structures individually or in ropes [7] by transmission electron microscopy (TEM), scanning electron microscopy (SEM), atomic force microscopy (AFM), and scanning tunneling microscopy (STM). In addition, electron diffraction (EDR), x-ray diffraction (XRD), Raman, and other optical spectroscopy can be also used to study structural features of nanotubes. These characterization techniques will be discussed in detail in Chapter 5. Figure 1.2 shows

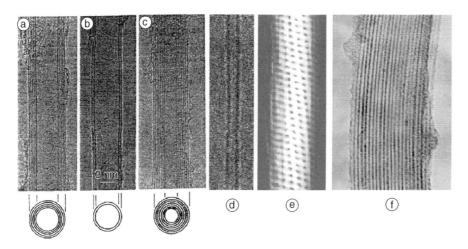

FIGURE 1.2 Homogeneous nanotubes of hexagonal network: TEM images (a), (b), and (c) for three multiwalled nanotubes (MWNTs) first discovered by Iijima in 1991 [1]; TEM image (d) for a single-wall nanotube (SWNT) first discovered by Iijima et al. in 1993 [5,6], an atomic resolution STM image (e) for a SWNT; and a TEM image (f) for a SWNT rope first reported in 1996 by Thess et al. [7]. (Figures 1.2a and 1.2b are from Iijima, S., Nature, 354.56, 1991; Figure 1.2d is from Iijima, and Ichihashi, Nature, 363, 603, 1991; Figure 1.2f is from Thess et al., Science, 273, 483, 1996.)

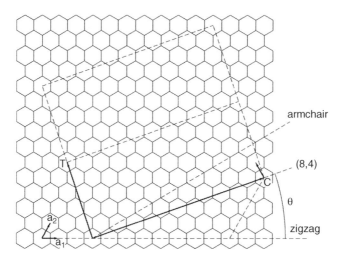

FIGURE 1.3 A nanotube (n,m) is formed by rolling a graphite sheet along the chiral vector $\mathbf{C} = n\mathbf{a}_1 + m\mathbf{a}_2$ on the graphite where $\mathbf{a}_1$ and $\mathbf{a}_2$ are graphite lattice vector. The nanotube can also be characterized by the diameter $|\mathbf{C}|$ and the chiral angle θ is with respect to the zigzag axis, $\theta = 0°$. The diagram is constructed for a (8,4) nanotube.

an STM image with atomic resolution of a single SWNT from which one can see the hexagonal structural feature and TEM images of a SWNT rope and a few MWNTs.

A SWNT can be visualized as a hollow cylinder, formed by rolling over a graphite sheet. It can be uniquely characterized by a vector $\mathbf{C}$ in terms of a set of two integers (n,m) corresponding to graphite vectors $\mathbf{a}_1$ and $\mathbf{a}_2$ (Figure 1.3) [8],

$$\mathbf{C} = n\mathbf{a}_1 + m\mathbf{a}_2 \tag{1.1}$$

Thus, the SWNT is constructed by rolling up the sheet such that the two end-points of the vector $\mathbf{C}$ are superimposed. This tube is denoted as (n,m) tube with diameter given by

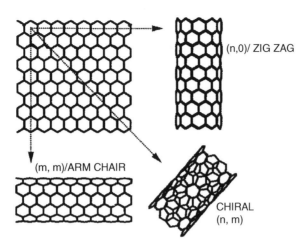

FIGURE 1.4 By rolling a graphite sheet in different directions, two typical nanotubes can be obtained: zigzag (*n*, 0), armchair (*m, m*) and chiral (*n,m*) where *n>m>0* by definition. In the specific example, they are (10,0), (6,6), and (8,4) nanotubes.

$$D = |C|/\pi = a \, (n^2 + nm + m^2)^{1/2}/\pi \qquad (1.2)$$

where $a = |\mathbf{a}_1| = |\mathbf{a}_2|$ is lattice constant of graphite. The tubes with $m = n$ are commonly referred to as armchair tubes and $m = 0$ as zigzag tubes. Others are called chiral tubes in general with the chiral angle, θ, defined as that between the vector **C** and the zigzag direction $\mathbf{a}_1$,

$$\theta = \tan^{-1} [3^{1/2}m/(m + 2n)] \qquad (1.3)$$

θ ranges from 0 for zigzag ($m = 0$) and 30° for armchair ($m = n$) tubes. Note that $n \geq m$ is used for convention.

The lattice constant and intertube spacing are required to generate a SWNT, SWNT bundle, and MWNT. These two parameters vary with tube diameter or in radial direction. Most experimental measurements and theoretical calculations agree that, on average, the C–C bond length $d_{cc} = 0.142$ nm or $a = |\mathbf{a}_1| = |\mathbf{a}_2| = 0.246$ nm and intertube spacing $d_{tt}, = 0.34$ nm [8]. Thus, Equations (1.1) to (1.3) can be used to model various tube structures and interpret experimental observations. Figure 1.4 illustrates examples of nanotube models.

We now consider the energetics or stability of nanotubes. Strain energy caused by forming a SWNT from a graphite sheet is proportional to 1/D per tube or $1/D^2$ per atom [9]. It is suggested [9–11] that a SWNT should be at least 0.4 nm large to afford strain energy and at most about 3.0 nm large to maintain tubular structure and prevent collapsing. Typical experimentally observed SWNT diameter is between 0.6 to 2.0 nm while smaller (0.4 nm) or larger (3.0 nm) SWNTs have been reported [12]. A larger SWNT tends to collapse unless it is supported by other force or surrounded by neighboring tubes, for example, as in a MWNT. The smallest innermost tube in a MWNT was found to be as small as 0.4 nm whereas the outermost tube in a MWNT can be as large as hundreds of nm. But, typically, MWNT diameter is larger than 2 nm inside and smaller than 100 nm outside. A SWNT rope is formed usually through a self-organization process in which van der Waals force holds individual SWNTs together to form a triangle lattice with lattice constant of 0.34 nm.

The structural model is of special interest to derive the tube chirality (*n,m*) from simple structural relation or experimentally measurable geometry (*D*, θ). This is because important properties of a nanotube are function of tube chirality, as will be discussed. For example, we may exclude the presence of all zigzag tubes in a MWNT from the structural relations. The spacing between any two coaxial neighboring zigzag tubes (*n*, 0) and (*m*, 0) is $\Delta D/2 = (0.123/\pi) \, (n\text{-}m)$ from Equation (1.2) and $a = 0.246$ nm.

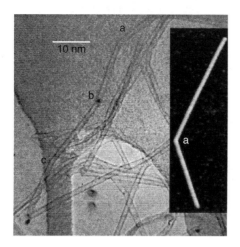
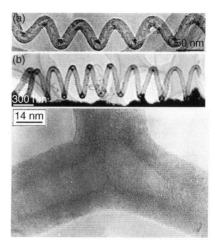

FIGURE 1.5 Representative TEM and AFM (insert) images of the individual SWNT bends. (a), (b) and (c) denote three typical bend angles of 34°, 26°, and 18°[18], MWNT coils [17], and Y branches [16]. (Figure 1.5a from Zhang et al., Appl. Phys. Lett., 83, 423, 2003; Figure 1.5c from Satishkumar et al., Appl. Phys. Lett., 77, 2530, 2000.)

This, however, cannot be close to 0.34 nm spacing required to form a MWNT regardless of values of integers n and m. However, a MWNT can be made of all armchair tubes $(5\,m, 5\,m)$ where $m = 1, 2, 3$, etc. The interspacing for all armchair MWNTs is $\Delta D/2 = (0.123/\pi)(3)^{1/2}(5) = 0.334$ nm, very close to 0.34 nm. An experimentally observed MWNT can be interpreted with other models as well. For example, a MWNT can also be viewed as a scrolled graphite sheet or a spiral graphite sheet, or mixture of scrolled structure and concentric shells [13,14], rather than coaxial SWNTs. These models, however, have not been accepted in general. But it is still likely that they present some of experimentally observed carbon nanostructures or even reported MWNTs because graphite does show diverse structures such as graphite whiskers and carbon fibers [3].

The significance of the tube chirality (n,m) is its direct relation with the electronic properties of a nanotube. STM can be used to measure tube geometry (d, θ), which, in turn, can be used to derive (n,m). In the following sections, we will see a direct correlation of (n,m) with electronic, optical, magnetic, and other properties of a nanotube.

1.3   Defective Nanotubes

In addition to defect-free nanotubes, experimentally observed structures also include the capped and bent [15], branched (L, Y, and T) [16], and helical [17] MWNTs, and the bent [18], capped [19], and toroidal [20] SWNTs. Figure 1.5 shows TEM images of some of these structures. Most of these structures are believed to have topological defects such as pentagons and heptagons incorporated in the nanotube of hexagonal network. In addition, the reported MWNTs also include nontubular structures such as multiwalled carbon nanofibers and bamboo structures, as illustrated in Figure 1.6. A bamboo structure can be viewed as many capped short nanotubes. In general, most SWNTs are defect-free whereas MWNTs are relatively more defective, containing either topological defects (pentagon-heptagon) or structural defects (discontinuous or cone-shaped walls or bamboo structure).

Many approaches have been developed to model nanotubes containing topological defects because these structures present intratube heterojunction nanoelectronic devices [21–24]. Han et al. have developed a generic approach and a computer program to generate and model configurations of bent [18], branched [25,26], toroidal [27], and capped nanotubes [28]. In this approach, a single bend or each bend in a branched, toroidal, or helical nanotube is considered to connect two types of nanotubes with the topological defects (pentagon-heptagon pairs). The bend angle between two connected nanotubes follows a simple topological relation [18]:

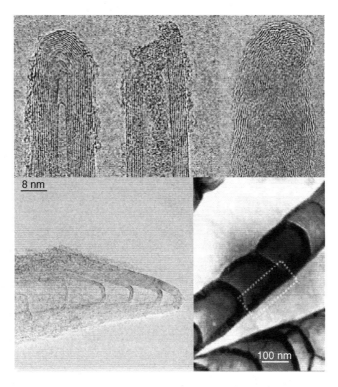

FIGURE 1.6 Capped MWNTs and MWNT variations including carbon fibers (CNF) and bamboo structures.

$$\Phi = |\ \theta_1 \pm \theta_2\ | \qquad (1.4)$$

where θ_1 and θ_2 are defined in Equation (1.3). Figure 1.7 illustrates the approach to construct and generate the model structure. Han et al. have modeled the experimentally observed 2-, 3-, and 4-terminal; toroidal; and helical nanotubes using molecular dynamics simulations of the model structures. The experimentally measured diameter of each tube and bend angle are used to derive possible tube chirality. They found that a set of chiralities could be matched to fit the same experimental parameters. For example, a 30° sharp bend can be connected by two nanotubes satisfying:

$$m_2 = n_2\ (m_1 + 2n_1)/(m_1 - n_1) \qquad (1.5)$$

If $n_1 = 0$, then $m_2 = n_2$. This indicates any zigzag tube $(n_1, 0)$ can be connected with any armchair tube (m_2, n_2) for a 30° bend. This bend can be, for example, (17,0)-(10,10), (17,1)-(11,9), (16,2)-(12,8), and (15,4)-(13.6). These isomers slightly differ energetically.

Structural modeling and simulations allow determination of number and position of the defects in defective nanotubes. Figure 1.8 shows possible structural models, which match the bend angle and diameter of the observed SWNT bends in Figure 1.5. Topologically, a 0° and a 30° bend need only a pair of pentagon-heptagons. In the 0° bend structure, this pair is fused together. In the 30° bend, the pentagon and heptagon reach the maximum separation along the tube circumference. Between these two energy-minimized configurations, as bend angle decreases, the number of pentagon-heptagon pairs increase. For example, the three and five pairs of pentagon-heptagons are required to form 26° and 18° bends, respectively.

It is a simple matter to construct branched, toroidal, and helical nanotubes from bent nanotubes through topological operation of fusion, rotation, and connection. When two or more bends are fused and connected to form branched structures, pentagons may be eliminated with only heptagons required for negative curvature. By Euler's topological theorem, an *n*-branched structure follows $n = [$(number

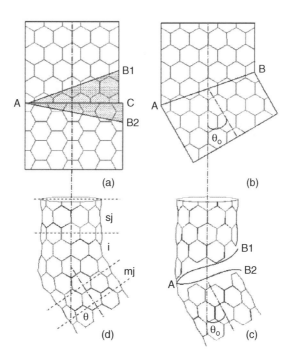

FIGURE 1.7 Construction of a SWNT bend junction (10,0)-(6,6). (a) and (b), two graphite sheets representing (10,0) and (6,6) nanotubes are connected to form a 30° planar bend; (b) and (c), the planar bend is rolled over to form a 30° tube bend; and (c) and (d), the 30° bend is relaxed to a 36° bend via a molecular dynamics simulation. The sj, mj, and I between four broken lines represent the unit cells of two tubes and junction interface [18].

of heptagons – number of pentagons) + 12]/6. Thus, to obtain 3- or 4-branched structure, the minimum number of topological defects is 6 or 12 heptagons. In addition, any number of pentagon-heptagon pairs is allowed, but this may cause extra energy.

In contrast, 6 or 12 pentagons are required to cap two or one end of a nanotube. For example, fullerenes (C_{60}) that contain 12 can be cut half to cap a (5,5) or (9,0) nanotube. However, for larger tubes, especially for MWNTs, pentagon-heptagon pairs may be required to shrink a nanotube to a smaller size before capping, as illustrated in Figure 1.6.

The current interest in nanotube junctions is largely from a theoretical point of view. Theoretically, one can structure a variety of models to study their intriguing structures and properties. However, experimentally all these structures are observed only occasionally.

In the following sections, we will mainly discuss the properties of defect-free nanotubes including (a) an individual SWNT, (b) an individual MWNT, and sometimes (c) a SWNT rope. There has been a great deal of work on defective, filmed, bundled, or arrayed SWNT or MWNT samples. However, the measured properties, for example, in electrical and thermal conductivity and elastic modulus can vary by several orders of magnitude from sample to sample. This is mainly because defective structures in a MWNT and random orientation of various nanotubes in film or bulk samples have yet to be characterized or specified and correlated with the properties of interest, which are mostly one-dimensional. These measurements, however, are still of practical interest in applications and will be discussed in detail in the following application chapters.

1.4   Electronic Properties

Electronic properties of nanotubes have received the greatest attention in nanotube research and applications. Extremely small size and the highly symmetric structure allow for remarkable quantum effects and electronic, magnetic, and lattice properties of the nanotubes. Earlier theoretical calculations [29–31] and

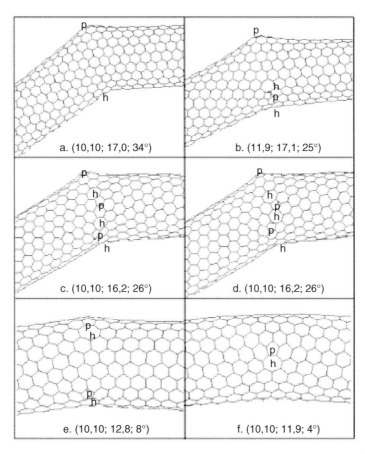

FIGURE 1.8 Examples of SWNT bend junctions. (a), a 34° bend has one pentagon and one heptagon in the opposite sites of the joint; (b) to (d), a 26° bend has three pentagon-heptagon defects with one in opposite site and the other two (fused) in different arrangements; (e), an 8° bend has two fused defects; and (f), a 4° bend has only one fused defect [18].

later experimental measurements [32–37] have confirmed many extraordinary electronic properties, for example, the quantum wire feature of a SWNT, SWNT bundle, and MWNT and the metallic and semi-conducting characteristics of a SWNT.

In the simplest model [29–31], the electronic properties of a nanotube derived from the dispersion relation of a graphite sheet with the wave vectors (k_x, k_y)

$$E(k_x, k_y) = \pm \gamma \{1 + 4\cos(\frac{\sqrt{3}k_x a}{2})\cos(\frac{k_y a}{2}) + 4\cos^2(\frac{k_y a}{2})\}^{1/2} \tag{1.6}$$

where γ is the nearest neighbor-hopping parameter and a is lattice constant. $\gamma = 2.5 - 3.2$ eV from different sources [29–37] and $a = 0.246$ nm.

When the graphite is rolled over to form a nanotube, a periodic boundary condition is imposed along the tube circumference or the **C** direction. This condition quantizes the two-dimensional wave vector $\mathbf{k} = (k_x, k_y)$ along this direction. The $\mathbf{k}$ satisfying $\mathbf{k} \cdot \mathbf{C} = 2\pi q$ is allowed where q is an integer. This leads to the following condition at which metallic conductance occurs:

$$(n - m) = 3q \tag{1.7}$$

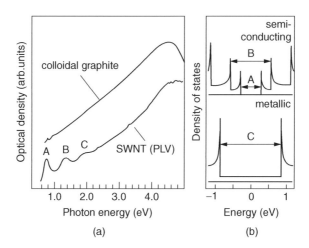

FIGURE 1.9 UV-VIS-NIR spectra from different SWNT sample types. (a) Spectrum of SWNT rope material shown for comparison together with the spectrum of colloidal graphite (offset for clarity). (b) A, B, and C features can be attributed to symmetric transitions between the lowest subbands in semiconducting (A, B) and metallic (C) tubes in representative density of states (DOS). (From A. Hagen and T. Hertel, Nano Lett., 3, 383, 2003.)

This suggests that one third of the tubes are metallic and two thirds are semiconducting. The band gap for a semiconducting tube is give by

$$E_g = 2d_{cc}\gamma/D \qquad (1.8)$$

Thus, the band gap of a 1-nm wide semiconducting tube is roughly 0.70 eV to 0.9 eV. This relation is in good agreement with STM experimental measurement for single SWNTs [35,36]. The STM measurements also confirm the density of state (DOS) or band structure predicted from the dispersion relation of graphite imposed with tubular periodic boundary condition. The DOS of SWNT will be discussed in the following section.

The derivation from graphite does not consider the curvature effect or σ–π rehybridization. This effect has been investigated using various approaches, including first principle *ab-initio* calculations [38–42,65]. It is found that σ–π rehybridization can open up a small band gap (~0.02 eV) for smaller (<1.5 nm) nonarmchair metallic tubes, as can be seen from Figure 1.13. A STM study indeed confirms such a small gap for $n - m = 3q$ SWNT [41]. However, this effect is found to be very rapidly disappearing with the tube diameter. In principle, only armchair tubes are intrinsically metallic. However, for most discussions the metallic condition $(n-m) = 3q$ and the band gap and structures predicted from only the simplest π-orbital model have been accepted.

Intertube coupling needs to be considered when the results of a SWNT are used for a SNWT rope or a MWNT. Calculations reveal interesting intertube coupling properties. The intertube coupling induces a small band gap for certain metallic tubes [42] but a reduced band gap by 40% for semiconducting tubes [43] in a SWNT rope. Similar observations can be expected for a MWNT as well, but the intertube coupling is relatively smaller because of bigger diameter in a MWNT. For example, it is predicted that two metallic tubes (5,5) and (10,10) in a coaxial MWNT can both open a small bang gap [43], but (10,10) and (15,15) tubes in a MWNT are found to remain metallic because of less intertube coupling for larger tubes. All semiconducting tubes in a MWNT tend to be semi-metallic just like graphite because of reduced band gap for large tubes and hole-electron pairing for multiwall coupling. More experiments on individual MWNT samples indeed show the dominating metallic or semimetallic nature of a MWNT while small band gap was reported and attributed to presence of defects or an electric contact barrier.

It has been experimentally confirmed that a single SWNT [32], a SWNT rope [33], or a MWNT [34] behave like a quantum wire intrinsically because of the confinement effect on the tube circumference. A

MWNT or a SWNT rope can be viewed as a parallel assembly of single SWNTs. The conductance for a SWNT, a SWNT rope, or MWNT is given by

$$G = G_o M = (2e^2/h) M \tag{1.9}$$

where $G_o = (2e^2/h) = (12.9 \text{ k}\Omega)^{-1}$ is quantized conductance. M is an apparent number of conducting channels including electron-electron coupling and intertube coupling effects in addition to intrinsic channels. $M = 2$ for a perfect SWMT. M, however, is determined not only by the intrinsic properties of a nanotube itself, but also by the intertube coupling as discussed above and the scatters such as defects, impurities, structural distortions, coupling with substrate, and contacts. Therefore, the experimentally measured conductance is much lower than the quantized value. The measured resistance for a single SWNT is ~10 kΩ [44], as compared with the perfect value of 12.9/2 or 6.45 kΩ.

Transport properties of nanotubes and physics of the devices fabricated using these structures will be discussed in Chapter 7. Here a brief discussion is made to compare the resistivity or conductivity of graphite and nanotubes. There have been a number of reports on the measured resistance of nanotubes, but most of them cannot be compared or cited here because the reported resistance, not resistivity, lacked the specification for the sample quality and measurement conditions and other details. The resistivity of graphite varies remarkably depending on sample quality. As temperature increases, it can decrease for disordered structures or increase for highly ordered structures such as a single crystal. The room temperature in-plane resistivity of the highest quality graphite is about 0.4 $\mu\Omega m$ [3]. In many measurements of SWNT ropes and MWNTs, the resistivity is found to decrease with temperature, and the room temperature values are much higher than 0.4 $\mu\Omega m$. This is mainly because nanotubes are randomly oriented in the sample. When the measurement is carried out for the purified SWNT ropes or MWNTs aligned across four electrodes, the result is consistently comparable with or lower than 0.4 $\mu\Omega m$ [44,45]. The π electron is more delocalized in a defect-free nanotube because of σ–π rehybridization and thus should give rise to higher conductivity than that of graphite. As shown in Figure 1.1, when curved, π orbitals become rich or more delocalized outside the tube, which leads to increased conductivity. Many measurements show decreasing behavior of resistivity with temperature. This, however, is not due to semiconducting nanotubes but the contact, intertube coupling, defects, tube alignment, or other issues [44]. The nanotube is a one-dimensional conductor and has to be aligned between two electrodes for transport measurement.

More theoretical attention has been paid to the electronic properties of heterogeneous nanotubes, especially bent and branched structures. There are three main features for these structures [18,21–25, 46,47]. First, these structures are molecular mimics of 2- or 3-terminal heterojunctions that connect two or three different nanotubes in the form of A-B or A-B-C in which A, B, or C can be a metallic or semiconducting tube. Second, localized states appear in the junction interface containing pentagons and heptagons. Out of this region, each tube retains its own band structure or density of state. Third, the interface may or may not be conducting, depending on how tubes are connected. For example, a (9,0)-(6,3) tube junction is not conducting for symmetric match but conducting for asymmetric match. The symmetric match retains straight tube geometry, which is difficult to be observed experimentally. Asymmetric match leads to a bend structure. Experimental observations indeed confirmed theoretical predictions for SWNT bent junctions [18,36]. MWNT heterojunctions are observed more frequently. However, there has not been any controlled way to experimentally fabricate them especially from SWNTs.

Localized states are originated from pentagons and heptagons incorporated in a hexagonal network. They are also observed in capped SWNT and MWNT ends. These localized states are responsible for enhanced field emission and interface states at nanotube junctions.

The novel electronic properties of nanotubes have attracted great interest in applications of nanotubes in nanoelectronics. Much of the effort to date has been made in using individual semiconductor SWNTs for transistors, memories, and logic devices. The striking feature of these nanoelectronic devices is higher mobility and stronger field effect. In addition, nanotube junctions such as sharp bends and T and Y branches have been studied as nanoelectronics devices [48,49]. Furthermore, the electronic properties

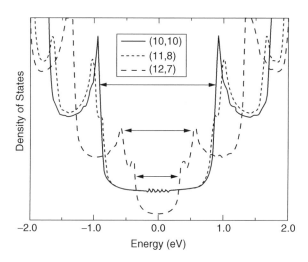

FIGURE 1.10 Calculated electronic DOS of metallic (10,10) and (11,8) and semiconducting (12,7) tubes using tight-binding calculations with the Fermi level positioned at zero energy.

have been correlated with mechanical, chemical, biological, thermal, and magnetic interactions with nanotubes. As a result, the extended electromechanical, electrochemical, thermal electronic, and electromagnetic properties are associated with applications of CNTs in sensors, actuators, field emission, batteries, fuel cells, capacitors, and many others.

1.5   Optical and Optoelectronic Properties

Defect-free nanotubes, especially SWNTs, offer direct band gap and well-defined band and subband structure, which is ideal for optical and optoelectronic applications. Optical spectra have been established for individual SWNTs and ropes using resonant Raman [50], fluorescence [51–53], and ultraviolet to the near infrared (UV-VIS-NIR) spectroscopies [54]. In addition, electrically induced optical emission [55] and photoconductivity [56] have been studied for individual SNWTs. A typical optical spectrum measured for a SWNT rope is shown in Figure 1.10 with that for a graphite sample for comparison [54]. Three peaks for the SWNT ropes cannot be observed for the graphite and attributed to symmetric transitions between the lowest subbands in semiconducting (A and B) and metallic (C) tubes. Usually, as-grown nanotube samples are a mixture of semiconducting and metallic tubes, as mentioned before. The measured peak position and intensity are correlated with electronic structures or tube chirality (n,m) or (D,θ). Therefore, optical spectra have been extensively used to determine the detailed composition of SWNT samples. This will be further detailed in Chapter 5. Here we discuss only the fundamental optical and optoelectronic properties of nanotubes.

Optical and optoelectronic properties can be understood from the band structure or DOS of a SWNT. The one-dimensional DOS of a SWNT can be derived from that for graphite with the expression as follows:

$$\rho(\varepsilon) = \frac{4}{l}\frac{2}{\sqrt{3}\gamma a}\sum_{m=-\infty}^{\infty} g(\varepsilon, \varepsilon_m) \tag{1.10}$$

where

$$g(\varepsilon, \varepsilon_m) = |\varepsilon| / \sqrt{\varepsilon^2 - \varepsilon_m^2} \quad \text{for} \quad |\varepsilon| > |\varepsilon_m|$$

$$g(\varepsilon, \varepsilon_m) = 0 \quad \text{for} \quad |\varepsilon| < |\varepsilon_m|$$

$$|\varepsilon_m| = \frac{|3q - n + m|\gamma a}{\sqrt{3}D}$$

As an example, Figure 1.10 shows calculated electronic DOS of metallic (10,10), (11,8) and semiconducting (12,7) tubes. The Fermi level is positioned to zero. The left and right side to the Fermi level define valence and conductance band, respectively. The peak of DOS is called van Hove singularity (VHS). The optical transition occurs when electrons or holes are excited from one energy level to another, denoted by E_{pq}. The selection rules, p-$q = 0$, for interband transitions that are symmetric with respect to the Fermi level require polarized light parallel to the tube axis, as shown by A, B, and C absorption features of Figure 1.9. The other selection rules that require perpendicular light to tube axis, however, have not been observed in optical spectra probably because of too-weak transitions. The energy corresponding to the symmetric transition $p = q$ for semiconducting (S) and metallic tubes (M) follows the relations with one p-orbital approximation:

$$E_{pp,S} = 2pd_{cc}\gamma/D \quad \text{and} \quad E_{pp,M} = 6pd_{cc}\gamma/D \tag{1.11}$$

The number p ($p = 1, 2...$) is used to denote the order of the valence and conduction bands symmetrically located with respect to the Fermi energy. Note that $p = 1$ defines the band gap of a semiconducting tube. Thus, a map can be made taking possible values of p and D for metallic and semiconducting tubes, as shown in Figure 1.12.

Figure 1.11 includes tube curvature-induced *s-p* rehybridization effect with which only armchair tubes ($n=m$) are truly metallic whereas others satisfying $n-m = 3q$ are semi-metallic with small band gap. The energy unit in Figure 1.11 is γ eV. Taking $\gamma = 2.5$ (low bound) and 3.0 eV (high bound), the wavelength of a semiconducting tube (= hc/E) can vary from 300 to 3000 nm. This suggests potential applications of semiconducting nanotubes in optical and optoelectronic devices from blue lasers to IR detectors.

For example, IR laser-excited photoconductivity was observed for a semiconducting SWNT within an ambipolar field effect transistor device [56]. This suggests that a semiconducting SWNT can be used for a polarized IR photo detector in which the photocurrent is nearly a linear function of IR intensity. In contrast, the same device can be also used for optoelectronic devices such as a light emitter in which emission of wavelength of 1500 nm is induced electronically [55]. Unlike conventional solid state optoelectronics, the semiconducting SWNT can emit light from injecting electrons and holes from two contact electrodes, instead of doping. Electrical control of the light emission of individual SWNTs allows detailed characterization of the optical properties. Another experiment is the observation of light emission by injecting electrons through a STM into MWNTs. The emitted photon wavelength is in the range of 600 to 1000 nm [57]. However, the emission is associated with localized states in the nanotube tips. In addition to STM ejection of electrons, light emission was also observed when coupled with electron (field) emission [58]. A peak was clearly seen close to 1.8 eV. Again, this is attributed to localized states at the nanotube tips.

It is still very challenging to study the optical and optoelectronic properties of a single nanotube. Extensive work has been carried out to establish the structure-assigned optical spectra for identification of Raman-active, infrared-active photon modes from samples containing different diameters and chiralities of nanotubes. In principle, the assignment can be readily established based on the unique VHSs in the electronic density of states. For example, one can assign tube chirality (n,m) from the measured photonic energy and image-measured diameter using Figure 1.12. Simulation and experimental studies

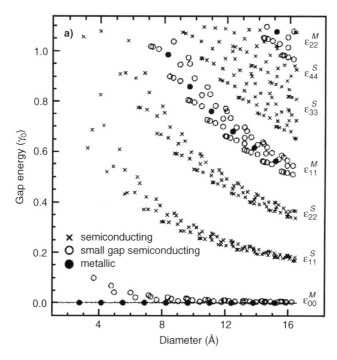

FIGURE 1.11 Energies for symmetric interband transitions in SWNTs as a function of their diameter. (From A. Hagen and T. Hertel, Nano Lett., 3, 383, 2003.)

have found much different absorption coefficient spectra for nanotubes and strong dependence of nonlinear optical properties on the diameter and symmetry of the tubes [51–55]. One can expect that these dependences become more complicated for MWNTs. In addition, the electronic and optical properties of nanotubes are strongly coupled with mechanical, chemical (environmental), thermal, and magnetic (radiation etc.) properties, as will be discussed in the following sections. This will further complicate characterization of the nanotube structure and properties.

1.6 Mechanical and Electromechanical Properties

σ bonding is the strongest in nature, and thus a nanotube that is structured with all σ bonding is regarded as the ultimate fiber with the strength in its tube axis. Both experimental measurements and theoretical calculations agree that a nanotube is as stiff as or stiffer than diamond with the highest Young's modulus and tensile strength. Most theoretical calculations are carried out for perfect structures and give very consistent results [27,59,60]. Table 1.1 summarizes calculated Young's modulus (tube axis elastic constant) and tensile strength for (10,10) SWNT and bundle and MWNT with comparison with other materials. The calculation is in agreement with experiments on average [61–64]. Experimental results show broad discrepancy, especially for MWNTs, because MWNTs contain different amount of defects from different growth approaches.

In general, various types of defect-free nanotubes are stronger than graphite. This is mainly because the axial component of σ bonding is greatly increased when a graphite sheet is rolled over to form a seamless cylinderical structure or a SWNT. Young's modulus is independent of tube chirality, but dependent on tube diameter. The highest value is from tube diameter between 1 and 2 nm, about 1 TPa. Large tube is approaching graphite and smaller one is less mechanically stable. When different diameters of SWNTs consist in a coaxial MWNT, the Young's modulus will take the highest value of a SWNT plus contributions from coaxial intertube coupling or van der Waals force. Thus, the Young's modulus for MWNT is higher than a SWNT, typically 1.1 to 1.3 TPa, as determined both experimentally and

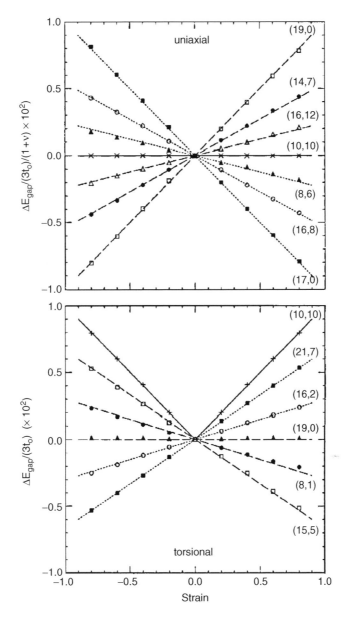

FIGURE 1.12 Band gap change of SWNTs under uniaxial strain (>0 for tension and <0 for compression) and torsional strain (>0) for net bond stretching and <0 for net bond compression), predicted by Equation (1.10) [65].

TABLE 1.1 Mechanical Properties of Nanotubes

	Young's modulus (GPa)	Tensile Strength (GPa)	Density (g/cm³)
MWNT	1200	~150	2.6
SWNT	1054	75	1.3
SWNT bundle	563	~150	1.3
Graphite (in-plane)	350	2.5	2.6
Steel	208	0.4	7.8

Source: J. Lu and J. Han, *Int. J. High Speed Electron. Sys.* 9, 101 (1998). With permission.

theoretically. On the other hand, when many SWNTs are held together in a bundle or a rope, the weak van der Waal force induces a strong shearing among the packed SWNTs. This does not increase but decreases the Young's modulus. It is shown experimentally that the Young's modulus decreases from 1 TPa to 100 GPa when the diameter of a SWNT bundle increases from 3 nm (about 7 (10,10) SWNTs) to 20 nm [64].

The elastic response of a nanotube to deformation is also very remarkable. Most hard materials fail with a strain of 1% or less due to propagation of dislocations and defects. Both theory and experiment show that CNTs can sustain up to 15% tensile strain before fracture [27]. Thus the tensile strength of individual nanotube can be as high as 150 GPa, assuming 1 TPa for Young's modulus. Such a high strain is attributed to an elastic buckling through which high stress is released. Elastic buckling also exists in twisting and bending deformation of nanotubes. All elastic deformation including tensile (stretching and compression), twisting, and bending in a nanotube is nonlinear, featured by elastic buckling up to ~15% or even higher strain. This is another unique property of nanotubes, and such a high elastic strain for several deformation modes is originated from sp^2 rehybridization in nanotubes through which the high strain gets released.

However, sp^2 rehybridization will lead to change in electronic properties of a nanotube. A position vector in a deformed nanotube or graphite sheet can be written as $r = r_o + \Delta r$ where r can be deformed lattice vector **a** or chiral vector **C** described in Section 1.4. Using a similar approach to deriving electronic properties of a nanotube from graphite, the following relations are obtained [65]:

$$E_g = E_{go} + sgn\,(2\,p + 1)\,3\gamma\,[(1 + \upsilon)\,(\cos 3\theta)\,\varepsilon_l + (\sin 3\theta)\,\varepsilon_r\,] \qquad (1.12)$$

In this relation, E_{go} is zero strain band gap given by Equation (1.6); θ is nanotube chiral angle defined by Equation (1.3); ε_l and ε_r are tensile and torsion strain, respectively; and υ is Poisson's ratio. Parameter p is defined by $(n - m) = 3q + p$ such that $p = 0$ for metallic tube; $p = 1$ for type I semiconductor tube, for example, (10,0); and $p = -1$ for type II semiconductor tube, for example, (8,0). Thus, function $sgn(2p +1) = 1$, 1 *and* –1, respectively, for these three types of tubes. Equation (1.12) predicts that all chiral or asymmetric tubes $(0 < \theta < 30°)$ will experience change in electronic properties for either tensile or torsional strain whereas symmetric armchair or zigzag tubes may or may not change their electronic properties. In asymmetric tubes, either strain will cause asymmetric $\sigma - \pi$ rehybridization and therefore change in electronic properties. However, effect of strain on a symmetric tube is not so straightforward. A detailed explanation is given in the original publications together with analysis of DOS. Figure 1.12 shows band gap change for different nanotubes.

The most interesting is the predicted metal-insulator transition. The armchair tube is intrinsically metallic but will open a band gap under torsional strain. The zigzag $(3q, 0)$ metallic tube, instead, will open a band gap under tensile strain, not torsional strain. The chiral metallic tube, for example, (9,3), will open band gap in either case. The above theory can also be extended to tube bending. For pure bending where bond stretching and compression cancel each other along tube circumference without torsion deformation, the band gap should not change from that predicted from Equation (1.12). However, there are exceptions. For example, when a SWNT with two ends fixed by two electrode leads is subject to a bending deformation applied by an AFM tip, the deformation also generates a net tensile stretching strain. Metal-insulator (semiconductor) transition and a decrease in conductance with strain are indeed observed experimentally in this case. Experiments have confirmed the predicted remarkable electromechanical properties of nanotubes or electronic response to mechanical deformation [66,67].

Equation (1.12) does not include sp^2 hybridization effect. Again, the effect is found very small, similar to that for the tube under no deformation, as shown in Figure 1.3. Figure 1.3 shows striking features of electromechanical properties of a nanotube such as splitting and merging of VHS peaks including band gap opening and closing under mechanical deformation.

There has not been much effort studying the electromechanical properties of SWNT bundles and MWNTs. Intertube coupling may play a larger role in electromechanical properties as it does for Young's modulus and tensile strength.

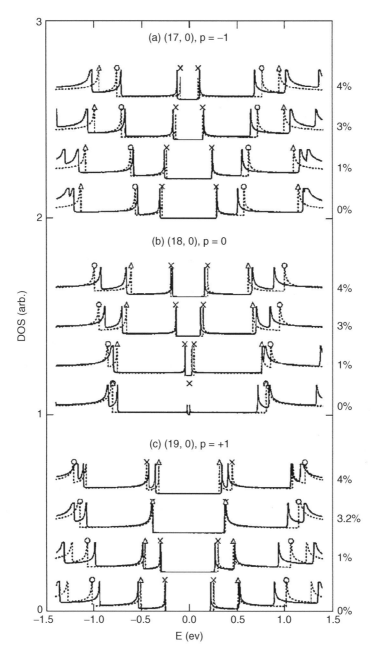

FIGURE 1.13 DOS with (dash line) and without (solid line) consideration of s-p rehybridization for three typical SWNTS. Values on the right side are stains. A small band gap of 0.02 eV at zero strain, caused by tube curvature or rehybridization, is seen for tube (18,0). Striking features of electromechanical properties include splitting and merging of VHS peaks including band gap opening and closing [65].

1.7 Magnetic and Electromagnetic Properties

Similar to mechanical and electromechanical properties, magnetic and electromagnetic properties of CNTs are also of great interest. The magnetic properties are studied with electron spin resonance (ESR), which is very important in understanding electronic properties, for example, for graphite and conjugated materials. Once again, there is a large discrepancy from different experimental measurements, especially in transport properties, because of sample quality and alignment whereas qualitatively they agree with theoretical calculations.

Magnetic properties such as anisotropic g-factor and susceptibility of nanotubes are expected to be similar to those for graphite while some unusual properties may exist for nanotubes. Indeed, it is found from ESR [68] that the average observed g-value of 2.012 and spin susceptibility of 7×10^{-9} emu/g in MWNTs are only slightly lower than 2.018 and 2×10^{-8} emu/g in graphite. Some interesting properties are also found from ESR studies of Pauli behavior. For example, aligned MWNTs are metallic or semi-metallic. The measured susceptibility gives the density of state at the Fermi level of 1.5×10^{-3} states/eV/atom, also comparable with that for in-plane graphite. The carrier concentration is about 10^{19}cm^{-3}, as compared with an upper limit of 10^{19} cm^{-3} from Hall measurement. However, similar observations have not been made for SWNTs and bundles. The possible reason is sample alignment difficulties and strong electron correlation, which may block nonconduction ESR signal.

It can also be expected that CNTs would have interesting electrical response to a magnetic field. Indeed, both experiment and theory confirm the metal-insulator transition and band gap change whereas transport again is an intriguing issue. A similar approach to those in previous sections can be adopted to predict Landau band structure of nanotubes in magnetic field. However, the hopping parameter in Equation (1.5) will be multiplied by magnetic flux, $\phi = (h/e)(\beta/3)$ or $\beta = 3\phi/(h/e)$ where (h/e) is the magnetic flux quantum. Thus, the band gap of nanotube under uniform magnetic filed parallel to the tube axis is given by [69]:

For metallic tubes of $n - m = 3q$

$$E_g = E_{go} \beta, \qquad\qquad 0 < \beta < 3/2$$
$$E_g = E_{go} |3 - \beta|, \qquad 3/2 < \beta < 3$$

For semiconducting tubes

$$E_g = E_{go} |1 - \beta|, \qquad 0 < \beta < 3/2$$
$$E_g = E_{go} |2 - \beta|, \qquad 3/2 < \beta < 3$$

These relations predict a metal-insulator transition and band gap change for semiconductor tubes under magnetic field parallel to tube axis. This is similar to electrical response of nanotubes to mechanical deformation. Qualitatively the electronic response of three types of nanotubes under either magnetic field or strain field parallel tube axis can be schematically shown in Figure 1.15. Driven by magnetic or strain field, the Fermi level will move away from the original position, and this results in the band gap change pattern in Figure 1.15. This is explained in Figure 1.14 and quantified in Yang and Han [65]. Similar response can also be observed when magnetic field or strain field is perpendicular to tube axis.

A major feature from the theory is that the band gap change is oscillatory and that the semiconducting and metallic nature of nanotubes can be altered by applying a magnetic field or strain field. This is called Aharonov-Bohm effect in magnetic field case. The oscillatory behavior is confirmed by experimental measurements of resistance change in MWNTs with tube axis parallel to magnetic field [70]. In the experiments, only the outermost tube in the MWNT is considered to contact with electrode leads and to contribute to the measured resistance.

1.8   Chemical and Electrochemical Properties

Small radius, large specific surfaces and $\sigma{-}\pi$ rehybridization make CNTs very attractive in chemical and biological applications because of their strong sensitivity to chemical or environmental interactions. These, however, also present challenges in characterization and understanding of other properties. The chemical properties of interest include opening, wetting, filling, adsorption, charge transfer, doping, intercalation, etc. Applications include chemical and biological separation, purification, sensing and detection, energy storage, and electronics.

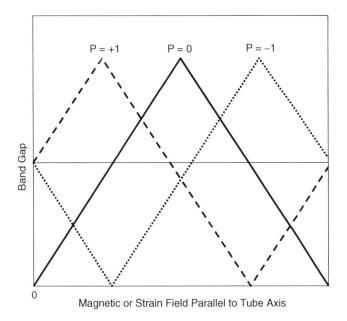

FIGURE 1.14 Band gap change pattern of three types of nanotubes (n,m) under either magnetic field or mechanical strain field. Integer $p = n - m - 3q = -1$, 0, and +1.

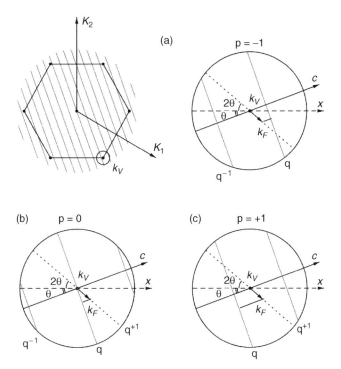

FIGURE 1.15 Driven by magnetic or strain field, the Fermi wave-vector K_F moves away from K_V at zero field, with the band gap measured by a distance between K_F and q line measure the bang gap. The band gap increases for $p = -1$ but increases for $p = 0$ and +1 tubes. The K_F move across q line for $p = -1$ but q+1 line to reach zero gap. This moving pattern results in the band gap change pattern shown in Figure 1.14 [65].

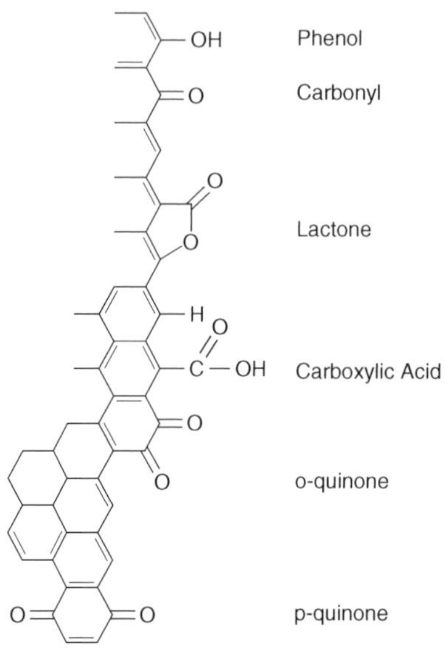

FIGURE 1.16 Possible chemical groups at opened nanotube ends.

Opening [71–73]

The nanotube end is more reactive than the sidewall because of the presence of pentagons or metallic catalysts sitting on the opened ends and greater curvature. Many approaches have been used to open nanotube ends, including, for example, vapor phase oxidation, plasma etching, and chemical reaction using acids such as HNO_3. The opened end is terminated with different functional groups such as carboxyl, etc., as shown in Figure 1.16. The opening is required for many applications as described below.

Wetting and Filling [74–76]

Nanotubes are hydrophobic and do not show wetting behavior for most aqueous solvents. It is reported that various organic solvents, HNO_3, S, Cs, Rb, Se, and various oxides such as Pb and Bi_2O_2 can wet nanotubes. A nanotube provides a capillary pressure proportional to $(1/D)$. Therefore, these wetting agents can be driven to fill inside the nanotube by the capillary pressure. It is also likely to fill nonwetting agents inside a nanotube by applying a pressure that is higher than the capillary pressure. An effective alternative is to use wetting agents such as HNO_3 to assist filling of nonwetting agents inside the nanotube.

Adsorption and Charge Transfer [77–79]

Enhanced molecular adsorption and charge transfer can be expected for nanotubes. Strong adsorption and charge transfer of oxygen to CNTs have been experimentally observed at room temperature. Extensive calculations have been carried out for various gas molecules using first principles approaches. The gas adsorption and charge transfer capability are functions of sites and gas molecules. The site on which a gas molecule can adsorb includes interstitial in tube bundles, groove above the gap between two neighboring tubes, nanopore inside a tube, and surface of a single tube. The adsorption and charge transfer capability is found to follow a decreasing order:

Sites: Interstitial, groove, nanopore, and surface

Gas: $C_8N_2O_2Cl_2$, O_2, C_6H_{12}, C_6H_6, NO_2, H_2O, NH_3, CH_4, CO_2, N_2, H_2, and Ar

The calculated equilibrium distance between the gas molecules and the nearest nanotube ranges from 0.193 nm for NO_2 to 0.332 nm for Ar, adsorption energy from 30.6 kJ/mol for $C_8N_2O_2Cl_2$ to ~1 kJ/mol for Ar, and partial charge from 0.212 for $C_8N_2O_2Cl_2$ to 0.01 for N_2. These values are between those for conventional physical and chemical adsorption. $C_8N_2O_2Cl_2$, O_2 and NO_2 display an acceptor nature with negative charge obtained from the nanotube while others show donor nature with positive charge transferred to the nanotube.

The calculations further find significant electronic modulation of certain adsorption to the nanotube, including change in the Fermi level and density of state of the nanotube for molecules from $C_8N_2O_2Cl_2$ to NO_2. There will be a conductance change when these molecules are introduced to contact with nanotubes. Electronic response of nanotubes to chemical adsorption through charge transfer has been of great interest in designing and understanding electronic devices for chemical sensor applications. Indeed, O_2 has been found to be dominantly adsorbed on as-grown nanotube surface, leading to p-type behavior whereas it has been argued that water is major reason [79]. In addition, nanotube electronic devices have been used for room temperature detection of NO_2, $C_6H_5NO_2$, C_6H_6, NH_3, CH_4, etc. Such sensing applications will be discussed in Chapter 10.

Chemical Doping, Intercalation, and Modification [80–84]

The substitutional doping with B and N dopants was pursued to make nanotubes p- and n-types. However, molecular adsorption as discussed above provides a simple, noncovalent doping approach to turn nanotubes into p-type with oxygen or water adsorption or n-type with, for example, C_6H_{12}. On the other hand, intercalation of the alkali metals with nanotubes is used for enhanced metallic conductivity or halogens with nanotubes for charge- or energy-storage applications. Experimental observation and theoretical calculations show that these intercalating agents mainly enter intertube spaces or defects on nanotubes for enhanced electrochemical capability for charge transfer and storage.

Indeed, nanotubes as electrode materials show enhanced electrochemical capability. The reduction and oxidation reactions that occur at the electrodes produce a flow of electrons that generate and store energy. In battery applications, conventional graphite, or other electrodes can reversibly store one lithium ion for every six carbon atoms. Experiments reveal an electrical storage capacity approximately double that of graphite. Theoretical studies show that the tubes' open ends facilitate the diffusion of lithium atoms into interstitial sites.

The reduction and oxidation that occur at the electrodes produce a flow of electrons that generate a signal for chemical and biological detection. The spatial and temporal resolution of sensitivity and speed for detection depend on the size of the electrodes. Carbon nanotube electrodes basically inherit from graphite electrodes several advantages such as broad window of reduction and oxidation, chemical inertness or corrosion resistance, and biological compatibility. However, their nanoscale dimension provides unique electrochemical properties in greatly improved sensitivity and speed in chemical and biological sensor applications. This topic will be discussed in detail in Chapters 9 and 10.

1.9   Thermal and Thermoelectric Properties

Graphite and diamond show extraordinary heat capacity and thermal conductivity. It can be expected that nanotubes have similar thermal properties at room and elevated temperatures but unusual behavior at low temperatures because of the effects of phonon quantization. Both theory and experiment show that intertube coupling in SWNT bundles and MWNTs is weak in temperature region of >100 K [85]. The specific heat of MWNTs has not been examined theoretically in detail. Experimental results on MWNTs show a temperature-dependent specific heat, which is consistent with weak interlayer coupling, although different measurements show slightly different temperature dependencies.

When $T > 100$ K, an SWNT, SWNT bundle, and MWNT all follow or are close to specific heat relation of graphite, about 700 mJ/gK. However, at lower temperatures, CNTs show quantum confinement effects.

For example, the heat capacity (mJ/gK) is 0.3 for a (10,10) SWNT, ~0 for SWNT bundle and graphite, and 2 to 10 for a MWNT or bundle [85,86].

The thermal conductivity of both SWNTs and MWNTs should reflect the on-tube phonon structure, regardless of intertube coupling. Measurements of the thermal conductivity of bulk samples show graphite-like behavior for MWNTs but quite different behavior for SWNTs, specifically a linear temperature dependence at low T, which is consistent with one-dimensional phonons. Thermal conductivity is one-dimensional for nanotubes like electrical conductivity. Therefore, the measurements give a broad range of 200 to 6000 W/mK, again, showing a strong dependence on the sample quality and alignment. Theoretical calculations and experimental measurements showed that the thermal conductivity for a SWNT ropes and MWNTs at room temperature could vary between 1800 and 6000 W/mK [86,87] whereas more than 3000 W/mK is firmly confirmed [88] from the measurement of a single MWNT.

The thermoelectric power, defined by TEP = $\Delta V / \Delta T$ in which V is thermoelectric voltage and T is temperature, is of great interest in understanding transport due to its extreme sensitivity to the change of electronic structure at the Fermi level. When a bias is applied across a single tube, the temperature gradient will be built up along the tube axis through Joule heating. TEP for a single metallic or semiconducting tube follows linear temperature dependence with positive and negative slope, respectively, for p- and n-doped tube. Its room temperature value is around 280 μV/K for a semiconducting SWNT [87] and 80 μV/K for a MWNT [86]. Thermoelectric properties vary significantly from sample to sample for filmed and bundled SNWTs and MWNTs.

Summary

Both theory and experiment show extraordinary structures and properties of carbon nanotubes. The small dimensions, strength and the remarkable chemical and physical properties of these structures enable a broad range of promising applications.

A SWNT can be metallic and semiconducting, dependent on its chirality. Semiconducting SWNTs have been used to fabricate transistors, memory and logic devices, and optoelectronic devices. SWNT nanoelectronics can be further used for chemical and biological sensors, optical and optoelectronic devices, energy storage, and filed emissions. However, it is currently not possible to selectively control the tube chirality or obtain either metallic or semiconducting SNWTs. These constraints in addition to problems of nanoscale contacts and interconnects stand in the way of large-scale fabrications and integration and applications of CNT electronics.

A MWNT basically behaves like a metal or semimetal because of the dominating larger outermost tube. Therefore, MWNTs are suitable for nanoelectrodes, field emission, and energy storage applications. In these applications, the tube chirality control is not critical. But MWNTs allow incorporation of diverse defects, which significantly affect electrical and mechanic properties. Exploiting the diverse structure and properties discussed here for a variety of applications will be the subjects of several chapters later in this book.

References

1. S. Iijima, Nature, 354, 56 (1991).
2. M.A. Prelas, G. Popovici, and L.K. Bigelow, Eds. Handbook of Industrial Diamonds and Diamond Films, Marcel Dekker, New York (1997).
3. B.T. Kelly, Physics of Graphite, Applied Science, London (1981).
4. H.W. Kroto et al., Nature, 318, 162 (1985).
5. S. Iijima and T. Ichihashi, Nature, 363, 603 (1993).
6. D.S. Bethune et al., Nature, 363, 605 (1993).
7. A. Thess et al., Science, 273, 483 (1996).

8. M. Dresselhaus, G. Dresselhaus, and P. Eklund, Science of Fullerenes and Carbon Nanotubes, Academic Press, San Diego (1996).

9. D.H. Roberson, D.W. Brenner, and J.W. Mintmire, Phys. Rev. B, 45, 592 (1992).

10. S. Sawada and N. Hamada, Solid State Comm., 83, 917 (1992).

11. A.A. Lucas, P. Lambin, and R. E. Smalley, J. Phys. Chem. Solids, 54, 587 (1993).

12. N. Wang, Z.K. Tang, G.D. Li, and J.S. Chen, Nature, 408, 50 (2000).

13. O. Zhou et al., Science, 263, 1744 (1994).

14. S. Amelinckx et al., Science, 267, 1334 (1995).

15. S. Iijima, T. Ichihashi, and Y. Ando, Nature, 356, 776 (1992).

16. B.C. Satishkumar et al., Appl. Phys. Lett., 77, 2530 (2000).

17. Z.Y. Zhong, S. Liu, and E.G. Wang, Appl. Phys. Lett., 83, 423 (2003).

18. J. Han et al., Phys. Rev. B, 57, 14983 (1998).

19. P. Kim et al., Phys. Rev. Lett., 82, 1225 (1999).

20. J. Liu et al., Nature, 385, 780 (1997).

21. B.I. Dunlap, Phys. Rev. B, 49, 5463 (1994).

22. A. Fonseca et al., Carbon, 33, 1759 (1995).

23. P. Lambin et al., Chem. Phys. Lett., 245, 85 (1995).

24. L. Chico et al., Phys. Rev. Lett., 76, 971 (1996).

25. J. Han, Electrochemical Society Proceedings, Vol. 98-8, pp 875–884 (1998).

26. J. Han, Chem. Phys. Lett., 282, 187 (1998).

27. J. Lu and J. Han, Int. J. High Speed Elec. Sys., 9, 101 (1998).

28. J. Han and R. Jaffe, J. Chem. Phys., 108, 2817 (1998).

29. R. Saito et al., Phys. Rev. B, 46, 1804 (1992).

30. N. Hamada, S. Sawada, and A. Oshiyama, Phys. Rev. Lett., 68, 1579 (1992).

31. J.M. Mintmire, B. I. Dunlap, and C. T. White, Phys. Rev. Lett., 68, 631 (1992).

32. M. Bockrath et al., Science, 275, 1922 (1997).

33. S.J. Tans et al., Nature, 386, 474 (1997).

34. S. Frank et al., Science, 280, 1744 (1998)

35. J.W.G. Wildoer et al., Nature, 391, 59 (1998).

36. T.W. Odom et al., Nature, 391, 62 (1998).

37. Z. Yao et al., Nature 402, 274 (1999).

38. C. White, D. Robertson, and J. Mintmire, Phys. Rev. B, 47, 5485 (1993).

39. X. Blasé et al., Phys. Rev. Lett., 72, 1879 (1994).

40. H. Yorikawa and S. Maramatsu, Phys. Rev. B, 50, 12203 (1996).

41. M. Ouyang et al., Science, 292, 702 (2001).

42. P. Delaney et al., Nature, 391, 466 (1998); Phys. Rev. B 60, 7899 (1999).

43. P. Lambin et al., Comp. Mat. Sci., 2, 350 (1994).

44. H.T. Soh et al., Appl. Phys. Lett., 75, 627 (1999).

45. T.W. Ebbesen et al., Science, 272, 523 (1996).

46. J.E. Fisher et al., Phys. Rev. B, 55, R4921 (1997).

47. N. Mingo and J. Han, Phys. Rev. B, 64, 201401 (2001).

48. M. Menon and D. Srivastava, Phys. Rev. Lett., 79, 4453 (1997).

49. J.P. Lu, Phys. Rev. Lett., 79, 1297 (1997).

50. A.M. Rao et al., Science, 275, 187 (1997).

51. S.M. Bachilo et al., Science, 298, 2361 (2002).

52. M.J. O'Connell et al., Science, 297, 593 (2002).

53. A. Hartschuh et al., Science, 301, 1354 (2003).

54. A. Hagen and T. Hertel, Nano Lett., 3, 383 (2003).

55. J.M. Misewich et al., Science, 300, 783 (2003).

56. M. Freitag et al., Nano Lett., 3, 1067 (2003).

57. W.A. de Heer et al., Science, 268, 845 (1995).

58. J.M. Bonard et al., Phys. Rev. Lett., 81, 1441 (1998).

59. J.P. Liu, Phys. Rev. Lett., 79, 1297 (1997).

60. A. Garg, J. Han, and S. B. Sinnott, Phys. Rev. Lett., 81, 2260 (1998).

61. T. Ebbesen et al., Nature, 382, 54 (1996).

62. E.Wong, P Sheehan, and C. Lieber, Science, 277, 1971 (1997).

63. P. Poncharal et al., Science, 283, 1513 (1999).

64. J.P. Salvetat, Appl. Phys. A, 69, 255 (1999).

65. L. Yang and J. Han, Phys. Rev. Lett., 85, 154 (2000).

66. J. Cao, Q. Wang, and H. Dai, Phys. Rev. Lett., 90, 157601 (2003).

67. S. Paulson et al., Appl. Phys. Lett., 75, 2936 (1999).

68. G. Baumgartner et al., Phys. Rev. B, 55, 6704 (1997).

69. J.P. Lu, Phys. Rev. Lett., 74, 1123 (1995).

70. C. Schonenberger et al., Appl. Phys. A, 69, 283 (1999).

71. S.C. Tsang et al., Nature, 372, 159 (1994).

72. P.M. Ajayan et al., Nature, 362, 522 (1993).

73. H. Hirura et al., Adv. Mater., 7, 275 (1995).

74. P.M. Ajayan et al., Nature, 375, 564 (1995).

75. E. Dujardin et al., Science, 265, 1850 (1994).

76. D. Ugarte et al., Science, 274, 1897 (1996).

77. J. Zhao, A. Buldum, J. Han, and J.P. Lu, Nanotechnology, 13, 195 (2002).

78. J. Zhao, J.P. Lu, J. Han, and C. Yang, Appl. Phys. Lett., 82, 3746 (2003).

79. M. Kruger et al., N. J. of Phys., 5, 138 (2003).

80. J. Zhao, A. Buldum, J. Han, and J. P. Lu, Phys. Rev. Lett., 85, 1706 (2000).

81. X. Liu, C. Lee, C. Zhou, and J. Han, Appl. Phys. Lett., 79, 3329 (2001).

82. E. Frackowiak et al., Carbon, 37, 61 (1999).

83. G.T. Wu et al., J. Electrochem. Soc., 146, 1696 (1999).

84. B. Bao et al., Chem. Phys. Lett., 307, 153 (1999).

85. J. Hone et al., Science, 289, 1730 (2000).

86. W. Li et al., Phys. Rev. B, 59, R9015 (1999).

87. J. Hone et al., Phys. Rev. B, 59, R2514 (1999).

88. P. Kim et al., Phys. Rev. Lett., 87, 215502 (2001).

89. J.P. Small et al., Phys. Rev. Lett., 91, 256801 (2003).

2

Computational Nanotechnology of Carbon Nanotubes

Deepak Srivastava
NASA Ames Research Center

2.1 Introduction

The science and technology of nanoscale materials, devices, and their applications in functionally graded materials, molecular-electronics, nanocomputers, sensors, actuators, and molecular machines form the realm of nanotechnology. The prefix "nano" corresponds to a basic unit on a length scale, meaning 10^{-9} meters, which is a hundred to a thousand times smaller than a typical biological cell or bacterium. At few nanometer length scale the devices and systems sizes begin to reach the limit of 10 to 100s of atoms, where even new physical and chemical effects are observed and form the basis for the next generation of cutting-edge products based on the ultimate miniaturization where extended atomic or molecular structures form the basic building blocks. The earliest impetus to the scientific and technological possibility of coaxing individual atomic and molecular building blocks into the making of useful materials, devices, and applications was given by the late Nobel prize–winning physicist Richard Feynman in a landmark lecture "There's Plenty of Room at the Bottom," delivered at the American Physical Society

(APS) meeting at Cal Tech in 1959, in which he said, "*The problems of chemistry and biology can be greatly helped if our ability to see what we are doing, and to do things on an atomic level, is ultimately developed — a development which I think cannot be avoided.*" Indeed, scanning probe microscopes (SPMs) and other related techniques, in recent years, have given us these abilities in limited domains and spurred a vigorous growth in the pursuit of nanotechnology during the past decade.

The real progress in nanotechnology, however, has also been spurred by the discovery of atomically precise nanoscale materials such as fullerenes in the mid-1980s and carbon nanotubes in the early 1990s. CNTs, as described in Chapter 1, can be thought of as sheets of carbon rolled up atoms into a tubular structure such that atoms at the seams are connected in a flawless manner according to rules of graphitic (sp^2) type chemical bonding. A single sheet of carbon atoms rolled up into a tubular structure is called a single-wall carbon nanotube (SWNT), and a rolled up stack of sheets results in multiwall carbon nanotubes (MWNTs). Since the discovery of MWNTs in 1991 by Iijima [1], and subsequent synthesis of SWNTs by Iijima [2] and Bethune [3], there are numerous experimental and theoretical studies of their electronic, chemical, and mechanical properties [4,5]. CNTs with very good mechanical strength/stiffness and elasticity characteristics, electronic properties ranging from semiconductors to metals, high electronic sensitivity to chemical adsorbates and mechanical strains, and very large aspect and surface-to-volume ratios have been proposed for applications as reinforcing fibers in functionally graded light-weight composite materials, components of molecular electronic devices, chemical and mechanical sensors and actuators, metrology probe tips, and gas and energy storage materials, respectively [6]. Investigation of the experimental- and theory/simulation-based characterization and conceptualization of novel applications in the above areas are pursued vigorously. The theory and computational modeling of the characterization and application design of CNTs thus has played a significant role in leading the developments from the very beginning.

At nano-meter (10^{-9} m) length scale it is possible to describe the structural, mechanical, thermal, and electronic behavior fairly accurately through computational nanotechnology, i.e., physics- and chemistry-based modeling and simulations of nanomaterials, devices, and applications. This is perhaps because the devices and systems sizes have shrunk sufficiently on the one hand, and computing power has continued to increase on the other. In many cases, the quantum mechanics–based simulation technologies have become also predictive in nature, and many novel concepts and designs were first proposed through modeling and simulations and were then followed by their realization or verification in experiments [7]. Computational nanotechnology is emerging as a fundamental engineering analysis tool for novel nano materials, devices, and systems design in a similar way that the continuum finite element analysis (FEA) was and has been used for design and analysis of micro- to macroscale engineering systems such as integrated circuits (ICs) and micro electro-mechanical systems (MEMS) devices in the submillimeter length scale regime and automobiles, airplanes, ships, etc., in large-scale engineering structures.

The computational nanotechnology-based modeling and simulations of CNT nanomechanics, functionally graded composite materials, electronic devices, sensors and actuators, and molecular machines essentially cover the entire range in the multilength and time-scale simulation techniques [7]. For example, simulations of electronic characteristics and sensor applications are very well within the high-accuracy quantum regime, whereas simulations for the processing and characterization of CNT-reinforced functional composite materials would typically require mesoscale simulation techniques, which are not yet very well developed. In between there are atomistic simulations for nanomechanics, reactivity, and molecular machines of individual CNTs and tight-binding (TB) quantum mechanical approaches for the same but at more accurate level. Many of the applications come about through response behavior of the nanoscale materials system to external electromagnetic or thermal fields where the equilibrium response or transport characteristics are simulated for advance prototyping of applications.

A broad interest in computational nanotechnology of CNTs thus derives from the possibilities of multiscale simulations for characterization of CNTs for nanostructured materials, nanoscale electronics, sensors and actuators, energy and gas storage devices, and molecular machines. Additionally, in general, it is noted that carbon-based materials are ideally suitable for molecular-level building blocks for nanoscale systems design and applications. This is because carbon is the only element that exists in a variety of topological shapes such as bulk (diamond), stack of layers (graphite), spherical (fullerenes), tubular (CNTs), conical (nanocones and nanohorns), and toroidal, etc., with varying physical and chemical properties. All basic shapes and forms needed to build any complex molecular-scale architecture are available and possible with carbon. By coating any carbon-based nanoscale devices and applications with biological lipid layers or protein molecules, it may also be possible to extend into the rapidly expanding area of biocompatible or bionanotechnology. In this chapter, however, our focus will be mainly on computational nanotechnology of CNTs with emphasis on either characterization of the basic properties or characterization in different application scenarios.

In Section 2.2 the multiscale simulation techniques needed to study CNTs via modeling- and simulation-based approaches are discussed. Section 2.3 briefly reviews the structure and symmetry properties of individual CNTs. In Sections 2.4 to 2.7, the simulation-based characterization of individual CNTs and CNTs in different applications scenarios such as thermomechanics and composite materials, chemical functionalization and gas-storage, nanoelectronics and sensors, and molecular machines are described succinctly.

2.2 Multiscale Simulation Techniques for Computational Nanotechnology

The importance of computational nanotechnology–based simulations in advancing the frontiers for the next generation of nanostructured materials, devices, and applications is based on three reasons. First, the length and time scales of important nanoscale systems and phenomenon have shrunk to the level where they can be directly addressed with high-fidelity computer simulations and theoretical modeling. Second, the accuracy in the atomistic and quantum-mechanical methods has increased to the extent that, in many cases, simulations have become predictive. Third, the raw CPU power available for routine simulation and analysis continues to increase so that it is regularly feasible to introduce more and more "reality" in the simulation-based characterization and application design.

The relevant problems for modeling- and simulation-based investigations of CNTs are truly multiple length- and time-scale in nature. At the atomistic level there are accurate semiclassical and quantum simulation methods that feed into the large-scale classical molecular dynamics (MD) simulations with 10s of millions of atoms, which then can be coupled to mesoscopic (few hundreds of a nanometer length scale) devices and systems. There have been many attempts to develop an integrated "grand-simulation tool"–based approach that cuts across different length scales on one hand and attempts to achieve a seamless integration across the interfaces on the other [8,9]. In reality, most of these integration schemes are geared toward and work only for very specific materials and devices but do not work otherwise. In the past few years, attempts, however, have also been made on attacking the "right" type of problems with appropriate "right" type of techniques [7] across a wide range of CNT characterization and application simulations. In this section, we briefly summarize the main simulation approaches that have been used in investigating CNTs across many different length and time scales.

2.2.1 Quantum Electronic Structure and Dynamics

Starting from the bottom-up, a few tens to hundreds of atoms are very accurately simulated with the *ab initio* (or first principles) quantum mechanics–based methods, wherein simulations are aimed toward the solution of the complex quantum many-body Schroedinger equation of the atomic system (including nuclei and electrons), using numerical algorithms [10]. The atoms are described as a collection of quantum mechanical particles, nuclei and electrons, governed by the Schrodinger equation, $H \Phi[\{\mathbf{R}_I, \mathbf{r}\}] = E_{tot} \Phi[\{\mathbf{R}_I, \mathbf{r}\}]$, with the full quantum many-body Hamiltonian operator $H = \mathbf{P}_I^2/2M_I + Z_I Z_J e^2/R_{IJ} + \mathbf{p}^2/2m_e + e^2/r - Z_I e^2/|\mathbf{R}_I - \mathbf{r}|$, where $\mathbf{R}_I$ and $\mathbf{r}$ are nuclei and electron coordinates. Using the Born-Oppenheimer approximation, the electronic degrees of freedom are assumed to follow adiabatically the corresponding nuclear positions, and the nuclei coordinates become classical variables. With this approximation, the full quantum many-body problem is reduced to a quantum many-electron problem $H[\mathbf{R}_\mathrm{I}]\,\Psi[\mathbf{r}] = E_\mathrm{el}\,\Psi[\mathbf{r}]$, where $H = \Sigma \mathbf{P}_\mathrm{I}^2/2M_\mathrm{I} + H[\mathbf{R}_\mathrm{I}]$, $H[\mathbf{R}_\mathrm{I}]$ is electronic Hamiltonian Operator.

Current *ab-initio* simulation methods are based on a rigorous mathematical foundation of the density functional theory (DFT) [11,12]. This is derived from the fact that the ground state total electronic energy is a functional of the density of the system. Kohn and Sham [11,12] have shown that the DFT can be reformulated as single-electron problem with self-consistent effective potential including all the exchange-correlation effects of electronic interactions:

$$H_1 = \mathbf{p}^2/2\,m_\mathrm{e} + V_\mathrm{H}(\mathbf{r}) + V_\mathrm{XC}[\rho(\mathbf{r})] + V_\mathrm{ion\text{-}el}(\mathbf{r}),$$

$$H_1\,\psi(\mathbf{r}) = \varepsilon\,\psi(\mathbf{r}), \text{ for all atoms}$$

$$\rho(\mathbf{r}) = \Sigma|\psi(\mathbf{r})|^2.$$

This single-electron Schrodinger equation is known as Kohn-Sham equation, and the local density approximation (LDA) has been introduced to approximate the unknown effective exchange-correlation potential. This DFT-LDA method has been very successful in predicting materials properties without using any experimental inputs other than the identity of the constituent atoms.

The DFT-based *ab-initio* methods have proven successful in a variety of simulations involving structural, chemical, and electronic characterization of nanostructured materials such as clusters, fullerenes, nanotubes, and nanowires. For practical applications, the DFT-LDA method is implemented with a pseudopotential approximation and a plane wave (PW) basis expansion of single-electron wave functions [10]. These approximations reduce the electronic-structure problem to a self-consistent matrix diagonalization problem. At the end of the iterative matrix diagonalization procedure, the resulting eigen values correspond to the quantum-mechanically possible electronic energy states of the system, and the eigen functions contain information about the electronic density distribution in the computed space. One of the popular DFT simulation programs is the Vienna *ab-initio* simulation package (VASP), which is available through a license agreement [13].

In this regime, there are also the quantum-chemistry–based methods, which are generally more suitable to molecular systems of smaller size, but the advantage is that the electronically excited states can be accommodated in these methods. The applications in optoelectronics, quantum computing, or quantum information processing need the transitions among the excited states that are not covered in the above-mentioned DFT-based simulation techniques. Coupled with the electronic-structure solution of the many-atoms system is also the possibility of simulating the dynamic evolution as a function of time. Several accurate quantum molecular dynamics schemes have been developed, in which the forces among atoms are computed at each time step via quantum mechanical calculations within the Born-Oppenheimer approximation. The dynamic motion for ionic positions is still governed by the Newtonian or Hamiltonian mechanics, and described by molecular dynamics. The most widely known and accurate scheme is the Car-Parrinello (CP) molecular dynamic method [14], in which the electronic states and atomic forces are described using the *ab-initio* DFT method.

In the intermediate length scale regime of few hundred- to thousand-atom systems, tight-binding [15] or semiempirical quantum mechanics–based methods provide an important bridge between the *ab-initio* quantum mechanics–based approaches described above and the classical atomistic force field–based methods that will be described below. The computational efficiencies of the tight-binding or semiempirical methods derive from the fact that the quantum Hamiltonian of the system can be parameterized. Furthermore, the electronic-structure information can also be easily extracted from the tight-binding Hamiltonian, which also contains the effects of angular interatomic forces in a natural way. The tight-binding structure and molecular dynamics methods have been parameterized for a variety of molecular and materials systems. For structural problems, diagonalization of large matrices are needed only a few times, and hence relatively large systems (up to 10,000 of atoms) have been simulated with order-N tight-binding methods implemented on parallel computers, but these have not been applied to

CNTs. For dynamic problems, nonorthogonal tight-binding molecular dynamics schemes are more accurate but are not easily converted to order-N type and can typically handle only systems with up to a few thousand atoms [16].

In the tight-binding model [15], an approximation is made to further simplify the quantum many-electron problem. It is assumed that the crystal potential is strong, which is the same as assuming that the ionic potentials are strong so that when an electron is captured by an ion during its motion through the lattice, the electron remains at that site for a long time before leaking, or tunneling, to the next ion site. During the capture interval, the electron orbits primarily around a single ion uninfluenced by other atoms so that its state function is essentially that of an atomic orbital. Most of the time the electron is tightly bound to its own atom. In other words, the atomic orbital is modified only slightly by the other atoms in the solid. The tight-binding wave function is, therefore, constructed by taking a linear combination of localized atomic orbitals, modulated by a Bloch wave function phase factor for a periodic lattice. This ensures that an electron in a tight-binding level will be found, with equal probability, in any cell of the crystal, because its wave function changes only by the phase factor as it moves from one cell to another. The computational efficiency of the tight-binding method derives from the fact that the electronic Hamiltonian $H[\mathbf{R}_I]$ can be parameterized. Furthermore, the electronic structure information can be easily extracted from the tight-binding Hamiltonian, which in addition also contains the effects of angular forces in a natural way.

Harrison [15] has attempted to provide a minimal tight-binding theory with only four parameters (in addition to four dimensionless universal constants) that can describe qualitatively a wide range of materials and properties, and he has also emphasized the necessity of including the nonorthogonality of the local environment in multicoordinated structures. This important factor has been generally overlooked by those seeking a transferable scheme. In a generalized tight-binding molecular dynamics method of Menon and Subbaswami with a minimal number of transferable parameter, the approach has been developed for bulk and cluster simulations of systems containing Si, C, B, N, and H [16,17]. The main advantage of this method is that it can be used to find an energy-minimized structure of a nanoscale system without any symmetry constraints. It turns out that sometimes symmetry-unconstrained dynamic energy minimization of a system helps in finding the global energetic minimum of the system, which is not easily conceptualized on the symmetry consideration alone. Later we show that same Hamiltonian has also been used for electronic transport simulations as well.

The parallelization of the TBMD code involves parallelization of the direct diagonalization (of the electronic Hamiltonian matrix) part as well as the parallelization of the MD part. The parallelization of a sparse symmetric matrix giving many eigen values and eigen vectors is a complex bottleneck in the simulation of large intermediate range system and needs development of new algorithms for this specific purpose.

2.2.2 Atomistic Structure and Molecular Dynamics Simulations

For larger size CNT systems (hundreds of thousands of atoms), the classical atomistic or MD simulations, which refer most commonly to the situation wherein the motion of atoms or molecules is treated in approximate finite difference equations of Newtonian mechanics, are used. Except when dealing with very light atoms and very low temperatures, the use of classical MD methods is well justified. In MD simulations the dynamic evolution of the system is governed by Newton's classical equation of motion, $d^2\mathbf{R}_I/dt^2 = \mathbf{F}_I = -dV/d\mathbf{R}_I$, which is derived from the classical Hamiltonian of the system, $H = \Sigma \mathbf{P}_I^2/2M_I + V(\{\mathbf{R}_I\})$. The atomic forces are derived as analytic derivatives of the interaction energy functions, $\mathbf{F}_I(\{\mathbf{R}_I\}) = -dV/d\mathbf{R}_I$, and are used to construct Newton's classical equations of motion, which are second-order ordinary differential equations. In its global structure a general MD code typically implements an algorithm to find a numerical solution of a set of coupled first-order ordinary differential equations given by the Hamiltonian formulation of Newton's second law [18]. The equations of motion are numerically integrated forward in finite time steps using a predictor-corrector method.

The interatomic forces are described with explicit or implicit many-body force-field functions, using the embedded atom method (EAM)- or modified embedded atom method (MEAM)-type functions for metals and semiconductors and Stillinger-Weber (S-W) and/or Tersoff-Brenner (T-B)-type potentials for semiconductors [19]. The T-B type potentials [20,21] are parameterized for and specially suited for carbon-based systems, such as CNTs, and have been used in a wide variety of scenarios with results in good agreement with experimental observations. One needs to be careful, however, when true chemical changes (involving electronic rearrangements) with large atomic displacements are expected to occur during the dynamics.

In its global structure, a general MD method typically implements an algorithm to find a numerical solution of a set of coupled first-order ordinary differential equations, given by the Hamiltonian formulation of Newton's Second Law. The equations of motion are numerically integrated forward in finite time steps. For reactive force-field functions, chemical bonds can form and break during the course of a simulation. Therefore, as compared with some other molecular dynamics methods, the neighbor list describing the environment of each atom includes only a few atoms and needs to be updated more frequently. The computational cost of the many-body bonded interactions is relatively high compared with the cost of similar methods with nonreactive interactions with simpler functional form. Efficient parallel codes have been developed to handle solutions of millions of coupled first-order differential equations with strategies depending on the nature of interatomic force-field functions [8,22].

The static and dynamic structural simulations of nanoscale systems are only an essential part of a range of simulations needed to carry the simulated materials and devices to the level of applications. The remaining part is obviously the mechanical, chemical, electronic transport, and thermal characterization that also come through intelligent or judicious use of the above described techniques, and the dynamic response properties such as transport under external fields.

2.2.3   Electronic and Thermal Transport Simulations

The low-bias electronic transport can be calculated as a generalization of the transmission amplitude $T(E)$ of an incident electron with energy E in simple one-dimensional potential barrier problem. The Landauer expression is generally used to obtain quantum conductance from the transmission function $T(E)$ as a function of the injected electron energy. The transimission function is obtained using the Green's function formalism that has been described in detail recently [23]. In the nanotube devices under current consideration, a connection to the outside world is made through metallic leads. A realistic treatment of a nanotube interaction with metal electrodes must involve a judicious construction of the Green's function and is an involved process. To maintain a consistency in the simulations, it is proper to use the tight-binding (TB) formulation for both the Hamiltonian and the Green's function. The nonorthogonal TB Hamiltonian mentioned above consists of N by N matrices, where $N = N(at) \times N(orb)$, $N(at)$ is the number of atoms in the embedding subspace, and $N(orb)$ is the number of orbitals on each atom. Contrary to earlier theoretical works on quantum transport, which use $N(orb) = 1$ (only one pi-electron orbital per atom) [24], for accuracy recent calculations have used $N(orb) = 9$ that includes $1s$, $3p$, and $5d$ orbitals for C and Ni (a representative material for metal leads) interface and $N(orb) = 4$ for C atoms [25,26].

For consistency, it is better to use the same TB Hamiltonian to perform full symmetry unconstrained molecular dynamics relaxations for the SWNT systems, which is used for conductivity calculations, because dynamic relaxation is found to have given results significantly different from the cases where dynamic relaxation was not allowed. An efficient transfer matrix formalism for obtaining the quantum conductivity of SWNTs, which makes explicit use of the nonorthogonality of the basis functions within the above-mentioned nonorthogonal TB scheme [27]. This new formalism allows the symmetry unconstrained structural relaxation and even dynamics in a consistent manner with the same Hamiltonian as used in the quantum conductivity calculations. Additionally, for both the structural relaxation and the quantum conductivity simulations the periodic boundary condition or infinite size approximations are not needed. The method is therefore especially suitable for finite-size CNT systems with doping, defects, and chemical functionalization because the Hamiltonian has been parametrized for a variety of interacting atomic species. Current vs. voltage characteristics of single-wall CNTs have been obtained in the presence of topological defects as well as chemical adsorbates [27].

The thermal transport simulations in CNTs are in the early stages of development. Most of the attempts, so far, have been through direct methods for thermal conductivity simulations. The underlying approach to the simulation of thermal transport is through MD simulations with the assumption that at room temperature the electronic contribution to the overall thermal conductivity is small and the thermal transport occurs mainly through phonon driven mechanisms. In direct simulation methods [28,29], appropriate heat baths [30] are simulated at the two ends of CNTs resulting in a flux of thermal energy from the hot region of the tube to the cold. The ratio of heat flux and the temperature gradient in equilibrium condition is sufficient to compute the thermal conductivity at the simulation temperature. The analysis of the phonon-driven heat flux is also aided by the computations of the phonon spectrum and vibrational amplitudes, which are succinctly described recently for a variety of SWNTs using the MD approach [31]. In bulk semiconductors and nanowires, the response function–based approach has been applied with some success, and a comparison between bulk thermal conductivity computed with direct- and response function–based approaches has been discussed [32]. The response function–based methods so far have not been applied to CNTs, because these form quasi one-dimensional structures and applying direct simulation methods is not too difficult.

2.3 Structure and Symmetry

The structure and symmetry of CNTs has been described in detail in Chapter 1. In this section a summary of salient features is provided, which will be used later for understanding some of the characterization and application scenarios that form the focus of this chapter. A SWNT is best described as a rolled-up tubular shell of graphene sheet (Figure 2.1a) [4–7]. The body of the tubular shell is mainly made of hexagonal rings (in a sheet) of carbon atoms, whereas the ends are capped by dome-shaped half-fullerene molecules. The natural curvature in the sidewalls is due to the rolling of the sheet into the tubular structure, where as the curvature in the end caps is due to the presence of topological (pentagonal ring) defects in the otherwise hexagonal structure of the underlying lattice. The role of the pentagonal ring defect is to give a positive (convex) curvature to the surface, which helps in closing of the tube at the two ends and also make the end caps chemically more reactive compared with the cylindrical walls of the CNTs. Later, however, we show that the topological defect-free curved structure of the cylindrical walls also has chemical reactivity that is a function of the natural curvature or strain-induced curvature of CNTs. A MWNT, similarly, is a rolled-up stack of graphene sheets into concentric SWNTs, with the ends again either capped by half-fullerenes or kept open.

A nomenclature (n,m) used to identify each SWNT, in the literature, refers to integer indices of two graphene unit lattice vectors corresponding to the chiral vector of a nanotube [4–7]. Chiral vectors determine the directions along which the graphene sheets are rolled to form tubular shell structures and perpendicular to the tube axis vectors as explained in Reference 4. The nanotubes of type (n,n), as shown in Figure 2.1b, are commonly called armchair nanotubes because of the _/_/ shape perpendicular to the tube axis and have a symmetry along the axis with a short unit cell (0.25 nm) that can be repeated to make the entire section of a long nanotube. Other nanotubes of type $(n, 0)$ are known as zigzag nanotubes (Figure 2.1c) because of the/\/\/shape perpendicular to the axis and also have a short unit cell (0.43 nm) along the axis. All the remaining nanotubes are known as chiral or helical nanotubes that have longer unit cell sizes along the tube axis. Details of the symmetry properties of the nanotubes of different chiralities are explained in detail in References 4 and 5.

Detailed relation between the representation and the unit lattice vectors of a graphene sheet with the basic structural units of CNTs has been discussed in Chapter 1. It turns out that the symmetry of rolling up a graphene sheet into tubular structure determines a unique electronic behavior of CNTs such that some behave like metals and some behave like small band-gap semiconductors. Details of how this is accomplished and the conceptualization of CNT heterojunction–based nanoelectronic devices will be

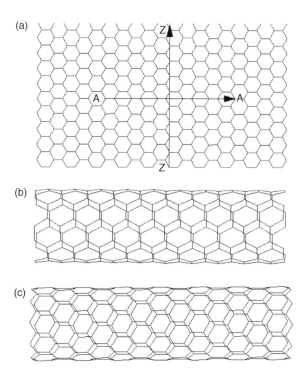

FIGURE 2.1 Making of armchair and zigzag carbon nanotubes with (a) rolling directions of a (5,5) armchair nantube (b) shown with AA and of a (10, 0) zigzag nanotube (c) shown with ZZ . (From D. Srivastava, M. Menon, and K. Cho, Comput. Sci. Eng., 3, 42, 2001. With permission.)

explained later. Here it is noted that this presents a unique opportunity to conceptualize a wide variety of carbon-based electronic switching, modulation, sensing, and actuation devices; however, the same chirality-dependent electronic characteristics also make it practically harder to incorporate SWNTs in any electronics application because, so far, there is not much control over the type of helicity or chirality of CNTs during their growth.

Nevertheless, the SWNTs and MWNTs are interesting nanoscale materials from applications perspective because CNTs (a) have very good elastic-mechanical properties for use as light-weight reinforcing fibers for functional composite materials; (b) can be both metallic or semiconductor leading to the possibility of use in field-effect transistors and sensors and nanotube heterojunctions in electronic switches; (c) are high-aspect ratio objects with good electronic and mechanical characteristics leading to their use in field-emission displays and various types of scanning probe microscope tips for metrological purposes; and (d) are also hollow, tubular molecules with large surface area suitable for packing material for gas and hydrocarbon fuel storage devices, gas or liquid filtration devices, and molecular-scale controlled drug-delivery devices.

In the following sections, the computer modeling– and simulation (computational nanotechnology)–driven advances in nanomechanics, nanoelectronics, sensors and actuators, and gas and energy storage applications of CNTs are discussed. For completeness, the simulations of CNTs in molecular machines and motors are also described very briefly.

2.4   Nanomechanics and Thermal Properties

In this section, we focus on the mechanical properties of individual single- and multiwall CNTs and simulations of CNT-reinforced polymer composite materials. The earliest atomistic simulations of CNT mechanics predicted unusually large Young's modulus (of up to 5 Terra Pascal [TPa] or five times larger than the modulus of diamond) and elastic limits (of up to 20 to 30% strain before failure) [33]. These predictions immediately raised the intriguing possibility of applying the nanotubes as super-strong reinforcing fibers with few orders of magnitude higher strength and stiffness than any other known material. Subsequently, more accurate simulations using TB molecular dynamics methods and *ab-initio* density functional total energy calculations with realistic strain rate, temperature dependence, and CNT sizes have provided more realistic values of 1 TPa as the Young's modulus and 5 to 10% elastic limit of the tensile strain before failure [34]. In this section, computational modeling– and simulation-based investigations of the strength and stiffness modulus, plasticity and yielding behavior, vibrational and thermal transport behavior, reinforcing of polymer composites by CNTs and mesoscopic simulations of CNTs are described.

2.4.1 Modulus of Carbon Nanotubes

The modulus of the CNTs is a measure of the strength and stiffness against small axial stretching and compression strains as well as nonaxial bending and torsion strains. Contributions to the good elastic-mechanical characteristics of CNTs come mainly from the strength of in-plane covalent C–C bonds in graphene sheet and facile out-of-plane deformations of the structure. For large-diameter or small-curvature CNTs, the modulus, strength, and stiffness should therefore be comparable with the in-plane modulus and strength of graphene sheet. In a tubular geometry, however, the elastic strain energies are also affected by the intrinsic curvature of the surface. Using the Tersoff [20] and T-B [21] potentials, Robertson et al. [35] showed that the elastic energy of a single-wall CNT scales as

$$\frac{1}{R^2},$$

where R is the radius of the tube. This is similar to the results deduced from the continuum elastic theory [36]. For axial strains, the Young's modulus of a SWNT is defined as

$$Y = \frac{1}{V}\frac{\partial^2 E}{\partial \varepsilon^2},$$

where E is the strain energy and V is the volume of the nanotube. Initial computational studies [33] using the same T-B potential reported the value of Young's modulus to be as high as 5.5 TPa. This was mainly due to a very small value of CNT wall thickness (~0.06 nm) used in this study [33]. It turns out that this was partly because of the attempts to use continuum theory to describe equivalence between a shell model and the atomistic descriptions of the elastic properties of CNTs. In many later works, this discrepancy [33] has been corrected. The van der Waals radius of C atoms can be used to define the spatial extent of the atoms, and the single-wall thickness comes to be about 0.34 nm. Using an empirical force constant model, Lu [37] found that the Young's modulus of a SWNT is around 970 GPa, which is close to that of an in-plane modulus of a graphene sheet, and is independent of tube diameter and chirality. Portal et al. [38] used a better description for the interatomic forces through a nonorthogonal TB method and found the Young's modulus to be around 1.2 TPa, which is larger than that of in-plane graphene sheet value and is slightly dependent on the tube size for small-diameter (D <1.2 nm) CNTs. The Young's modulus of a variety of carbon and also noncarbon nanotubes as a function of tube diameter in Hernandez et al.'s study [39] are shown in Figure 2.2a. Recently, it has been suggested [39] that by investigating the value $\partial^2 E/\partial\varepsilon^2$, instead of the Young's modulus, the ambiguity of thickness of a CNT wall can be avoided completely. However, the value of 0.34 nm for the wall thickness of a SWNT gives atomistic simulation–based results of Young's modulus that are in broad agreement with experimental results so far. Using a nonorthogonal TB, MD, and DFT methods, the axial compression of SWNTs and boron nitride nanotubes were carried out recently. The Young's modulus of SWNTs was found to be 1.2 TPa and that of boron-nitride nanotubes was 80% of that of equivalent CNTs [40]. These results are in qualitative and quantitative agreement with the DFT results and general experimental observations of

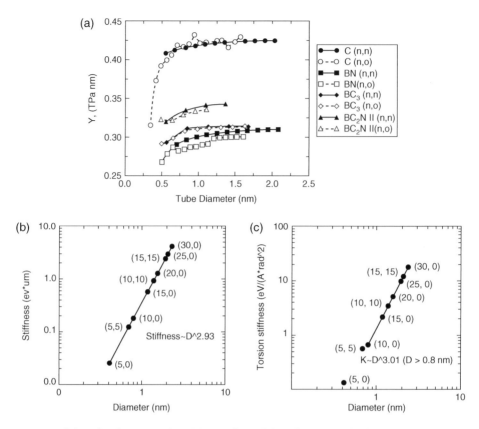

FIGURE 2.2 Modulus of carbon nanotubes, (a) Young's modulus of C, BN, and other heteroatomic nanotubes. (From E. Hernandez et al. Phys. Rev. Lett. 80, 4502, 1998. With permission.) (b) Bending stiffness, and (c) torsion stiffness as defined in the text and described in Reference 34.

Portal et al. known thus far. Another recent work by Harik [41] examines the validity of using continuum elastic theory for nanomechanics of CNTs and suggests that small- and large-diameter CNTs need to be described with different approximations such as solid rod model for small diameter CNTs and the shell model for large diameter nanotubes. This is also consistent with using 0.34 nm for wall thickness, because thin nanotubes with about 0.7-nm diameter would behave as a solid rod and not a cylindrical shell as was assumed earlier.

The bending stiffness of a single-wall CNT is defined as

$$\frac{1}{L}\frac{d^2 E}{dC^2},$$

where E is the total strain energy, L is the length, and C is the curvature of the bent nanotube, which is related with the bending angle θ as

$$C = \frac{\theta}{L}.$$

From the elastic theory of bending of beams, the strain energy of a bent nanotube can be expressed as

$$E = 0.5 YhL \oint t^2 C^2 dl,$$

where Y the Young's modulus of the SWNT, and h is the thickness of the wall [42]. The integral is taken around the circumference of the nanotube, and t is distance of atoms from the central line (or the bent axis) of the tube. From this expression, the bending stiffness K is found to be equal to $Yh(\pi r^3)$ and scales as cubic of the radius of the tube. Results from the molecular dynamics simulations, with Tersoff-Brenner potential in Figure 2.2b, show that stiffness K scales as $R^{2.93}$, which is in good agreement with scaling predicted by the continuum elastic theory. The corresponding bending Young's modulus (Y_B) of SWNT with varied diameters can be calculated from the above equation. For a small-diameter SWNT, Y_B is found to be about 0.9 TPa, which is smaller than the stretching Young's modulus calculated from the TB method or first principle theory. The computed smaller value is also similar to what Robertson et al. [35] showed in their study of the elastic energy of SWNTs, and the qualitative agreement is rather good. Poncharal et al. [43] have experimentally studied the bending Young's modulus of MWNTs (diameter >10 nm) using electrically induced force and found that the bending Young's modulus is in the range of GPa and decreases sharply with the increase in tube diameter. The large angle bending of SWNTs and MWNTs leads to elastic collapse of the structure that has been investigated in molecular dynamics simulations [44].

The torsion stiffness of a CNT is defined as

$$K = \frac{1}{L}\frac{d^2E}{d\theta^2},$$

where E is the total strain energy and θ is the torsion angle. The shear strain is related with torsion angle as

$$\varepsilon = \frac{R\theta}{L},$$

where R is the radius of tube and L is its length. From continuum elastic theory, the total strain energy of a cylinder can be written as

$$E = \frac{1}{2}G^* \iiint \varepsilon^2 dV,$$

where G is shear modulus of the tube. The torsion stiffness thus is related with G as

$$K = \frac{1}{L}\frac{d^2E}{d\theta^2} = G(2\pi h)\frac{R^3}{L^2},$$

where h is the thickness of the wall of the nanotube. The recently computed values of the torsional stiffness of several armchair and zigzag CNTs using T-B potential show the torsional stiffness to be about

$$0.1\frac{eV}{A\ rod^2}$$

for (5,5) and (10,0) CNTs. The dependence of the torsion stiffness on the radius of the tube is found to be as $K \propto R^{3.01}$ (for tube diameter > 0.8 nm) and is shown in Figure 2.2c. This is in good agreement with the prediction of cubic dependence from the continuum elastic theory. The shear modulus of CNTs is found to be around 0.3 TPa and is not strongly dependent on diameters (for $D > 0.8$ nm). This value is smaller than that of about 0.45 TPa in Reference 37 calculated with an empirical force constant model. For small-diameter tubes, such as a (5,5) nanotube, the shear modulus deviates from the continuum elastic theory description.

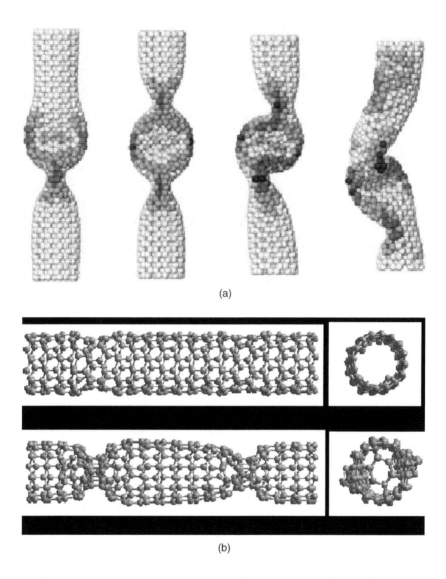

(a)

(b)

FIGURE 2.3 Collapse of CNTs under compressive strain with (a) sideways buckling or fin-like structures under elastic limit. (From B.I. Yakobson, C.J. Brabec, and J. Bernholc, Phys. Rev. Lett. 76, 2511, 1996. With permission.) (b) Local plastic collapse of the structure driven by graphitic to diamond-like bonding transition at the location of the collapse while the CNT remains essentially straight. (From D. Srivastava, M. Menon, and K. Cho, Phys. Rev. Lett. 83, 2973, 1999. With permission.) (c) TEM micrographs of locally collapsed thin CNTs under tensile strain. (From O. Lourie, D.M. Cox, and H.D. Wagner, Phys. Rev. Lett. 81, 1638, (1998). With permission.)

2.4.2 Buckling, Collapse, and Plasticity of Carbon Nanotubes

As the axial strain is increased gradually to more than a few percent, nanotubes have been observed to undergo two kinds of structural changes.

First, under compressive strains, nanotubes exhibit structural instabilities resulting in the sideways buckling and collapse (flattening into finlike structures) of the structure, as shown in Figure 2.3a, but the deformed tubular structure remains with in elastic limit for more than 20% compressive strain [33]. Experiments have shown sideways buckling, collapse, and breaking feature in compressed and tensile stretched MWNTs in polymer composite materials [45,46]. The positions of these deformations can be well predicted with continuum mechanics–based description of CNTs, and a reasonable comparison with

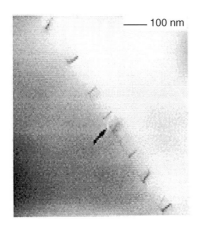 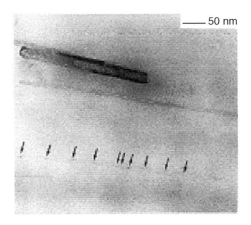

(c)

FIGURE 2.3 (CONTINUED)

MD simulations using T-B potential is obtained [33]. Recently, more extensive MD simulations at higher temperatures [47] and simulations with higher accuracy tight-binding molecular dynamics methods [40] showed that the above results [33] were inherently limited by the shortcomings of continuum mechanics–based approaches. Specifically, in the same experiment on compressed MWNT in a polymer composite [45], another mode of plastic deformation of compressed thin nanotubes was observed, where the CNTs remain essentially straight but the structure locally collapses at large compression. Srivastava et al. [40] used a nonorthogonal tight-binding molecular dynamics method, and found that within Euler buckling length limitation a straight CNT can locally collapse without undergoing either sideways buckling modes or through formation of fin-like structures. The local plastic collapse is due to a graphitic (sp^2) to diamond-like (sp^3) bonding transition at the location of the collapse, which is driven by a compressive pressure as high as 150 GPa at the location of the collapse as shown in Figures 2.3b and 2.3c [40]. Additionally, using the same tight-binding molecular dynamics method for C-B-N hetero-atomic systems [48], it has also been shown that atom-specific chemical effects are very important in describing the correct structural and dynamic behavior. For example, in a BN nanotube under large compressive strain, an anisotropic mechanism of compressive load transfer from one end of the tube to the other has been found [48]. This phenomenon is totally governed by the chemical nature of the B and N atoms in a BN nanotube system and predicted an effect that could be important in designing nano-structured smart materials capable of anisotropic damage and load transfer under large compressive shock conditions [48].

Second, under large tensile strain, Nardelli et al. [49] found that the mechanism of plastic failure is through the formation of Stone-Wales (SW) bond rotation induced defects in the nanotube lattice as shown in Figure 2.4a. The SW defects can be conceptually considered to be formed due to a 90° rotation of a C–C bond in the lattice plane. The initial simulations showed that the formation energy of such defects is decreased with the applied strain and is also dependent on the diameter and chirality of the specific CNT under consideration [49]. At high temperatures, plastic flow of the thus formed SW defects occurs, and that can even change the chirality of the nanotube as shown in Figure 2.4b. Further stretching leads to a plastic flow, the necking, and the eventual breaking of the tube at the location of the defects [50,51]. In the initial simulation-based investigations, the emphasis was on the formation energy of such SW defects through energetics of static structures [49–51].

However, it turns out that the breaking of CNTs under tension, via the formation of SW defects, is inherently a kinetic phenomenon, and a more relevant quantity is the activation energy or barrier to the formation of the SW-type defects in the CNTs under tension. The activation barriers through the static total energy simulations, through the tight-binding or *ab-initio* methods, have been carried out for in-plane rotation of the bond for strained and unstrained CNTs. The static in-plane activation barriers are

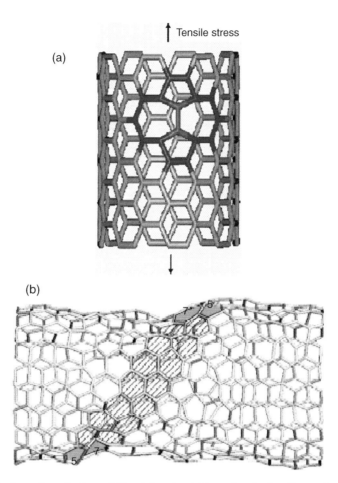

FIGURE 2.4 (a) Configuration of a Stone-Wales defect that initiates the plastic collapse of a CNT under tensile strain [56], and (b) a series of SW defects form and slide apart from each other as CNT is continued to be stretched. (From Nardelli et al. Phys. Rev. Lett., 81, 4656, 1998. With permission.)

found to be about 10 eV in nanotubes with no strain, and the value drops to about 5 to 6 eV for tensile strain between 5 and 10% [49–51]. Using simple transition-state theory, the breaking strain of SWNTs under tensile stretch is predicted to be about 17%, [52], which is smaller than the earlier predictions of about 30% breaking strain simulated by the same group [33] but still much larger than the breaking strain observed in experiments on ropes or bundles of SWNTs and on MWNTs [53–55]. The experimental breaking or yielding strain of a rope or bundle of SWNTs has been found to be about 5 to 6% [53,54], whereas similar measurements on MWNTs show the breaking strain to be about 12% or lower [55]. The reported lower yielding strains in experiments could be partly due to defected tubes in the ropes or bundles or could be due to other factors such as many orders of magnitude difference between the rates at which strain is applied in experiments as compared to what is feasible in MD simulations.

Yielding strains are found to be very sensitive to both the strain rate and the kinetic temperature during the simulation. At the strain rate of 10^{-5}/ps the yielding strain can vary between 15% at low temperature and 5% at high temperature (Figure 2.5a) [46]. Very extensive MD simulations over large variations in the temperature and strain rate changing by three orders of magnitude were carried out and reveal a complex dependence of the yielding strain on the temperature and strain rate [56]. For example, yielding strain of a 6-nm long (10,0) SWNT at several temperatures and strain rates varying between 10^{-3}/ps and 10^{-5}/ps were simulated and are shown in Figure 2.5b. The yielding strain at low temperature and slow strain rates are significantly higher than the yielding strain for high temperature

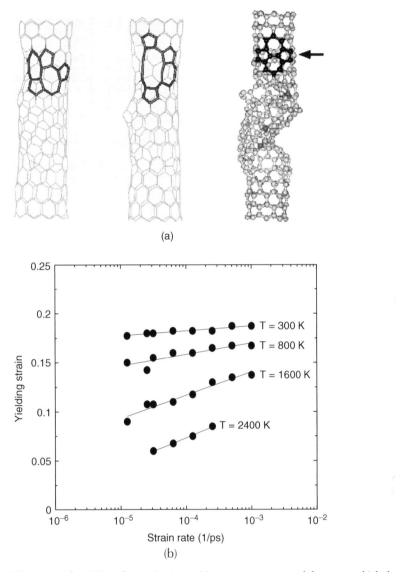

(a)

(b)

FIGURE 2.5 The yielding strain of a CNT under tension is sensitive to temperature and the rate at which the strain is applied, (a) a tensile strained CNT at two different temperatures yield at different strains and with different type of structural defects, (b) yielding strain of a CNT as a function of strain rate varying over 3 orders of magnitude and temperature. [56].

and higher strain rates. Moreover, a linear behavior of the yielding strain as a function of strain rate at a given temperature indicates the tensile yielding to be a kinetic, or activation barrier–driven, process. A strain rate and temperature-dependent model of yielding for SWNTs, within the transition state theory (TST)-based framework, has been derived and gives the yielding strain as

$$\varepsilon_Y = \frac{\overline{E}_v}{VK} + \frac{k_B T}{VK} \ln(\frac{N\dot{\varepsilon}}{n_{site}\dot{\varepsilon}_0})$$

where the yielding strain ε_y depends on the activation volume V, force constant K, the temperature and time-averaged dynamic activation energy $\overline{E}_v$, the temperature T, the number of activation sites n_{site}, the intrinsic strain rate $\dot{\varepsilon}_0$, and the strain rate $\dot{\varepsilon}$. The tensile yielding or breaking of CNT is considered to be through a number N of successive SW defect formations. It turns out that some of the intrinsic parameters such as dynamic activation energy, activation volume, and intrinsic strain rates can be fitted from the MD simulation data as shown in Figure 2.5, whereas the remaining parameters such as the temperature, number of possible activation sites, and strain rates can be chosen to reflect the experimental reality. The room temperature yielding strain of a 1-μm long (10,10) CNT comes out to be about 9% for a realistic and experimentally feasible strain rate of 1% per hour [56]. This compares very well with the yielding strain of 6 to 12% observed in experiments on bundles of SWNTs and MWNTs. Details of the model and the MD simulations are provided in Reference 56, and the advantage of the above model is that for other CNTs one could directly compute the activation energy for yielding defect formation and get the yielding strain from the above model [56]. Within the simulation and experimental error bars, this model is in very good agreement with experimental observations so far. The model has been extended to MWNTs as well, and finds that the MWNTs are stronger than the equivalent SWNTs by a couple of percentages of yielding strain at experimentally feasible strain rate and temperatures as shown in Figure 2.6 [57].

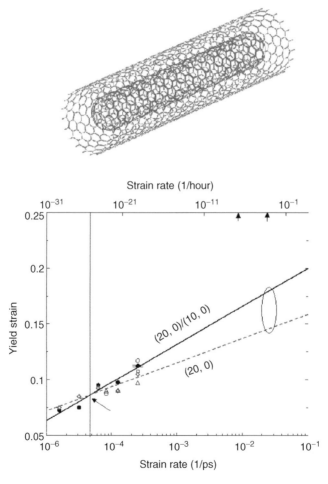

FIGURE 2.6 The yielding strain of a multiwall CNT through a model of a double wall CNT where the contact is made only on the outer most wall (top), and the yielding strain of the double-wall CNT as a function of strain rate (bottom). The experimentally feasible strain rate and yielding strain is indicated with the elliptical area [57].

2.4.3 Simulations of Vibrational and Thermal Properties

The small- and large-amplitude nanomechanics of CNTs described above focuses on the strength and stiffness characterization as well as mechanism and rate of the failure of the nanotubes under external load. Very small-amplitude harmonic displacements of C atoms from their equilibrium position, on the other hand, keep the structural integrity but are mainly responsible for the vibrational and thermal properties of CNTs. In this section, a brief summary on the simulations of vibrational or phonon spectra and thermal conductivity and heat pulse propagation in CNTs is provided.

The simulated phonon and vibrational spectra of any newly discovered material are generally useful for characterization and were simulated first to identify and assign the peaks in resonance Raman experiments on CNTs [58]. The phonon spectrum is obtained by constructing and diagonalizing dynamical matrix from position-dependent interatomic force constants. The eigen values of the dynamic matrix gives the frequency of vibrations, and the corresponding eigen vectors give nature of the corresponding modes of the vibration for atoms within a unit cell. The accuracy of the computed vibrational or phonon spectra at 0 K obviously depends on the accuracy of the interatomic forces used in constructing the dynamical matrix. Many studies on the zero-temperature vibrational or phonon spectra of CNTs using the *ab-initio* density functional [38], tight-binding [58–61], and harmonic spring constants satisfying hooks law [62] have appeared. Agreement between the experimental peaks and the zero-temperature vibrational peaks is generally good because only the configurations at or very close to equilibrium structures are investigated.

Thermal transport through CNTs, on the other hand, is sensitive to the choice of good atomic interaction potential, including the anharmonic part because the atomic displacements far from equilibrium positions are also sampled. The room temperature (300 K) phonon spectra or density of states and vibrational amplitudes have been computed recently [31]. The spectra were simulated through Fourier transform of temperature-dependent velocity autocorrelation functions computed from MD trajectories using T-B potential for C–C interactions [20,21]. Good agreement with zero-temperature phonon spectra, computed with higher accuracy *ab-initio* DFT and TB methods, [38,58–61], was obtained. Additionally, the line width of the computed spectra provides temperature-dependent life and correlation time of the phonon excitations involved. The spectra were used to assign the low-frequency Raman modes of CNT bundles as well [31].

The simulations of heat transport through CNTs under equilibrium or shock pulse conditions are rather few and relatively recent. The thermal conductivity of SWNTs have been computed using the direct MD simulation methods [63–65]. The typical room temperature thermal conductivity of SWNTs of 1- to 3-nm diameters is found to be around 2500 W/mK as shown in Figure 2.7. The thermal conductivity shows a peaking behavior as a function of temperature. This is because as the temperature is raised from very low values, more and more phonons are excited and contribute toward the heat flow in the system. However, at higher temperatures, phonon-phonon scattering starts to dominate and streamlined heat flow in the CNTs decreases, causing a peak in thermal conductivity in the intermediate temperature range. For all the simulated CNTs (of 1 to 2 nm diameter) as a function of the tube diameter and chirality, it was found that the peak position is around room temperature and sensitive to the radius of the CNTs and not the chirality or the helicity [65]. This means that the thermal transport in SWNTs is mostly through excitations of low-frequency radial phonon mode and the coupling of the radial mode with the axial or longitudinal phonons in the low-frequency region [65,66]. The convergence of the thermal conductivity as a function of the length of the CNT, however, is still an issue in all the direct MD simulation results reported so far. A significant variation in the room temperature thermal conductivity of (10,10) CNT has been reported partly because some of the studies have used very small CNT lengths in the simulations [63]. Preliminary investigations of the convergence with respect to the temperature gradient (dT/dX) across the CNT length have shown an inverse power-law dependence of the thermal conductivity on the thermal gradient. Rigorous, long-time, and length-scale simulations are required to investigate the convergence behavior as a function of CNT length. It is noted, however, that only the absolute value

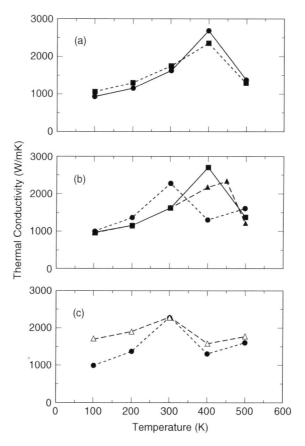

FIGURE 2.7 The direct MD-simulated thermal conductivity of CNTs as a function of strain rate with (a) comparison between a (10,10) nanotube and graphene sheet; (b) a comparison between (5,5), (10,10), and (15,15) nanotubes; and (c) a comparison between a (5,5) and (10,0) nanotubes [65].

of thermal conductivity changes and not the qualitative behavior of the relative peak heights and position as a function of temperature.

Another type of heat transport is feasible and investigated when CNTs are subjected to intense heat pulses [67]. A heat pulse propagating in a crystalline material at low temperatures is expected to generate propagation of a second sound wave that is not attenuated by dissipative scattering processes. The MD simulation technique is ideally suited for the investigation of such phenomenon because the temperature, shape, and duration of heat pulses can be simulated in a controlled way in a CNT. The heat-pulse duration was taken to be 1 ps, during which the temperature was ramped up and down according to a desired shape. The first sound, or the leading waves, move at higher speeds, but the shape of the wave changes and intensity also decays. The leading wave is followed by a second sound wave that maintains its shape and intensity for the entire duration of the simulation. The propagation of both the first and the second sound waves was investigated. In zigzag CNTs, the leading waves are found to move at the sound velocity of longitudinal acoustic (LA) phonons, where, as the in armchair CNTs, the leading wave moves with a sound velocity of transverse acoustic phonon. The main conclusion is that the leading stress waves under heat-pulse conditions travel slower in armchair CNTs compared with those in zigzag CNTs, which is consistent with the earlier thermal conductivity simulations [65] that showed a higher thermal conductivity for a zigzag CNT compared with that of an equivalent armchair CNT of the similar diameter.

2.4.4   Carbon Nanotube Reinforced Polymer Composites

One of the main reasons for studying the mechanical and thermal properties of individual SWNTs or MWNTs is to explore the possibility of using CNTs for light-weight, very strong, multifunctional composite materials (discussed later in Chapter 11). The structural strength or thermal characteristics of such composite materials depend on the transfer characteristics of such properties from the fiber to the matrix and the coupling between the two. In some cases, the coupling is through chemical interfacial bonds, which can be covalent or noncovalent whereas in other cases the coupling could be purely physical through nonbonded van der Waals (VDW) interactions. Additionally, the aspect ratio of fiber, which is defined as L/D (L is the length of the fiber and D is the diameter), is also an important parameter for the efficiency of mechanical load transfers, because the larger surface area of the fiber is better for larger load transfer. The limiting value of the aspect ratio is found related with the interfacial shear stress τ as $L/D > \sigma_{max}/2\tau$, where σ_{max} is the maximum strength of the fiber. Recent experiments on MWNTs or SWNT ropes [53–55] have reported the strength of the nanotubes to be in the range of 50 GPa. With a typical value of 50 MPa for the interfacial shear stress between the nanotube and the polymer matrix, the limiting value of the aspect ratio is 500:1. Therefore, for an optimum load transfer with a MWNT of 10 nm diameter, the nanotube should be at least 5 µm long, which is the range of the length of nanotubes typically investigated in experiments on nanotube-reinforced composites.

Earlier theoretical studies of the mechanical properties of the composites with macroscopic fibers are usually based on continuum media theory. The Young's modulus of a composite is expected to be within a lower bound of

$$\frac{1}{Y_{comp}} = \frac{V_{fiber}}{Y_{fiber}} + \frac{V_{matrix}}{Y_{matrix}}$$

and an upper bound of

$$Y_{comp} = V_{fiber}Y_{fiber} + V_{matrix}Y_{matrix},$$

where V_{fiber} and V_{matrix} is the volume ratio of the fibers and the matrix, respectively. The upper bound obeys the linear mixing rule, which is followed when the fibers are continuous and the bonding between the fibers and the matrix is perfect. Therefore, an upper limit can be reached if the nanotubes are long enough and the bonding with the matrix is perfect. On the other hand, short nanotubes with Poisson ratio of about 0.1 to 0.2 are much harder material compared with the polymer molecules with a Poisson ratio of about 0.44. Therefore, for an axial tensile or compressive strain, there is a resistance to the applied strain perpendicularly to the axis of the tube. For short nanotubes, this can provide a mechanism of load transfer even with weak nonbonded interactions. Detailed MD and structural simulations have been performed recently to investigate some of the above-discussed behavior.

For example, MD simulations of the structural and mechanical properties of polymer composite with embedded nanotubes have been performed recently (Figure 2.8A) [68]. The coupling at the interface was through nonbonded VDW interactions. Shown in Figure 2.8B is the strain-stress curve for both the composite system and the pure polyethylene matrix system. The Young's modulus of the composite is found to be 1900 MPa, which is about 30% larger than that of the pure polymer matrix system. This enhancement is within the upper and lower bound limits discussed above. A further enhancement of the Young's modulus of the same sample can be achieved by carrying the system through repeated cycles of the loading-unloading of the tensile strain on the composite matrix. In agreement with the experimental observation, this tends to align the polymer molecules with the nanotube fibers, causing a better load transfer between the two. Frankland et al. [69] have studied the load transfer between polymer matrix and SWNTs and found that there is no permanent stress transfer for 100-nm long (10,10) CNTs within polyethylene if only van der Waals interaction present. They estimated that the interfacial stress could

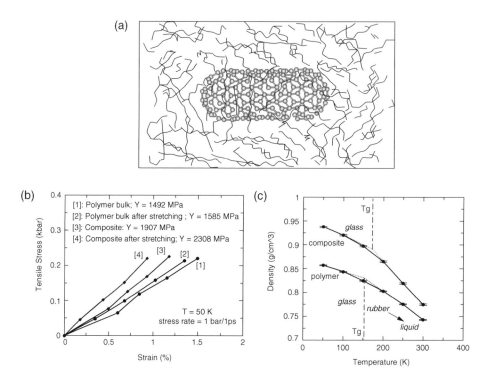

FIGURE 2.8 (a) A typical configuration of CNT in a polyethylene composite, (b) increase in the Young's modulus of the composite over the pure polyethylene value with or without work hardening, (c) structural phase transition of the composite shown through changes in the density as a function of the temperature — the temperature corresponding to a change in the slope refers to glass transition temperature [34,68].

be 70 MPa with chemical bonding between SWNT and polymer matrix, with only 5 MP for the non-bonding case.

Additionally, the density of CNT-polyethylene composite for short (10 monomers) and long (100 monomers) polymer chains has been simulated as function of temperature [68]. For example, the density vs. temperature plot for the long-chain CNT-polyethylene composite is shown in Figure 2.8C. The position of the change in the slope of the linear behavior in the two cases corresponds to the glass transition temperature for pure polymer and the nanotube-polymer composite. The results show that, due to mixing of 8% by volume of SWNT in polyethylene, the glass transition temperature increases by about 20% and, more significantly, the thermal expansion coefficient above glass transition temperature increases by as much as 42%. The enhanced thermal expansion coefficient of the composite is attributed mostly to an equivalent increase in the excluded volume of the embedded CNT as a function of temperature [68]. Because both excited vibrational phonon modes and Brownian motion contribute to the dynamic excluded volume of the embedded CNT, as the temperature is increased their contributions toward excluded volume increase significantly. The cross-linking of polymer with CNT was not allowed in these initial simulations. It is possible that the cross-linking of polymer matrix with embedded CNTs may further reduce the motions of polymer molecules or the CNT, the predicted changes in the glass transition temperature, and the thermal expansion coefficients in that scenario could be different.

2.4.5 Mesoscopic Scale Simulations of CNTs

The simulations of individual CNTs and CNT-reinforced polymer composites described so far are done with MD simulation methods where the advantage is that the structure and dynamics can be described very accurately at the atomistic level. Even electronic or quantum effects can be incorporated by using the interatomic force fields described by more accurate TB or *ab-initio* DFT approaches. The limitations are that, on a regular basis, simulations with millions of atoms are not feasible when long-range or large-size structural, dynamic, and temporal effects are incorporated. Such effects may determine the characteristic behavior in composites involving bundles of CNTs, MWNTs, or dispersion-diffusion of many SWNTs. These issues are important from both the processing and the applications point of view. A bridge to macroscopic size systems is feasible through continuum mechanics–based approaches, but such approaches generally neglect all atomic- and molecular-level interactions that could primarily determine the much-promised novel application capabilities of these nanostructured materials.

In the intermediate "mesoscopic" (~ a few hundred nanometers) length scale regime, therefore, CNTs and composites are generally "too large" to be simulated with quantum or atomistic approaches on a regular basis and are "too small" for the direct applicability or even accuracy of the methods based on continuum mechanics. It turns out that currently there are no mesoscopic structural and dynamic methods especially suited for CNTs or composites based on CNTs. There have been some attempts to find an equivalence between the large scale "structural mechanics" that is based on beams and struts and the atomistic scale "molecular mechanics" based on atoms and bonds. A preliminary hybrid approach, "molecular structural mechanics," has been developed. So far it has been applied only for static structural simulations, and it is not clear how much computational efficacy is achieved when individual atoms and molecules are replaced by equivalent but individual struts and beams in the alternative description [70].

Alterntively, a shape-based mesoscopic simulation approach has been developed recently for both structure and dynamics of CNTs or CNT-based composite systems. In this method a shape-based mesoscopic model of SWNTs is developed [71]. The cylindrical CNT is broken into smaller cylindrical segments, which interact with each other via a mesoscopic force-field (MFF) function. The form of the MFF function allows for all feasible stress-strain distribution in dynamically distorted or strained CNTs. In the spirit of full-scale MD simulations for atomistic dynamics of the same system, both low- and high-frequency vibrational modes of the CNT are incorporated. For example, in Figure 2.9, a comparison between the mesoscale MFF dynamics and full atomistic scale MD simulation of the edge of a bent 198-nm SWNT is shown. The CNT was bent in both the simulations in a similar way, and the velocity of one of the ends as a function of time shows an overall good agreement in not only the low-frequency phonon modes but also the high-frequency phonon modes as well. The differences come in for the very high-frequency atomic vibrations, which are, of course, not represented in the coarse-grained mesoscopic dynamics method. Once fully developed, in the near future, such approaches will allow the micron size simulations of the CNTs and CNT-based composite materials or mechanical systems.

2.5 Chemical Functionalization, Physisorption, and Diffusion in Carbon Nanotubes

Many of the CNT's proposed applications in composite materials, molecular electronics, sensors, etc., require an understanding and control of the chemistry or chemical reactivity of CNTs. The efficiency of the chemical bonding of the end caps or the sidewalls of CNTs will determine where the chemically sensitive reactions occur and how these change the mechanical or electrical properties. For example, the functionalization or chemisorption on the sidewalls of CNTs can be used to increase bonding or linking of nanotubes in a nanotube-matrix composite material and can also be used for chemical or gas sensor applications (see Chapter 9). The operations of CNT electronic device applications may also be sensitive to the presence of chemical species in the surrounding environment, and for medical applications the CNTs could be made biocompatible by chemisorbing or physically coating with biological molecules. In this section, topics such as chemisorption and curvature-dependent localized reactivity, chemical functionalization through radical bombardment, hydrogen physisorption for gas storage, and molecular diffusion/filtration through carbon nanotubes for gas separation and drug delivery applications are discussed.

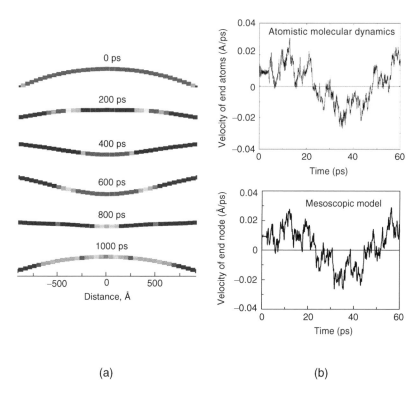

FIGURE 2.9 A mescopic model of (a) the bending vibrations of a 198-nm long (10,10) CNT, (b) detailed comparison of the vibration amplitude as a function of time as computed from the mesoscopic MFF dynamics approach and full-scale MD simulations using TB potentials [71].

2.5.1   Chemisorption and Curvature-Dependent Reactivity of CNTs

Recent computational modeling and simulation efforts to investigate chemical reactivity of CNTs are broadly classified into three categories. First are the direct and accurate calculations of the energetics and structures for the atomic or molecular adsorption on the CNTs of different diameter and helicity by the *ab-initio* quantum chemical, semiempirical or the TB or reactive atomistic simulation approaches. The chemical binding energies are computed for different chemical species such as F, H, N, and O, etc., as a function of tube diameter, reaction site, and surface coverage. The thrust of these approaches has been to test or predict (a) if any given chemisorbing species will react with the CNT and (b), if the reaction occurs, information on the stabilization energy and structure of the final end product. Detailed studies using DFT methods have been carried out to investigate the thermochemistry of F chemisorption on SWNTs. The thermodynamic stability of F on SWNTs increases with decreasing CNT diameter; on the other hand, mean bond dissociation energy of the C-F bond increases for thinner CNTs [72]. The F chemisorption on the inside of a CNT was also investigated, and it was found that the F atoms do not bond covalently to the concave surfaces. A number of simulation based studies for H chemisorption on CNTs have been carried out recently. Froudakis et al. [73] use mixed quantum mechanics/molecular mechanics approaches to show that hydrogen atoms bind on the exterior of the CNT wall and do not enter the interior of the CNT. The binding of H atoms in a zigzag pattern is found to be the most favored high-coverage H chemisorption on CNTs. This causes the length of the CNT to extend by as much as 15% along the tube axis. Bauschlicher and So [74] also have investigated the 50 and 100% coverage of atomic H chemisorption on CNTs of different diameter and chirality using the semiempirical AM1 and the ONIOM approaches. Whereas the 100% coverage outside the CNT is found to be unfavorable, the 50% coverages (both inside and outside the CNT) are found to be a favorable bonding situation. The H atoms arranged linearly along the tube axis, or spiraled around the tube, were found to be the most

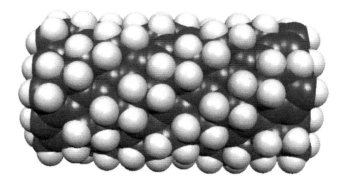

FIGURE 2.10 The 50% H (light grey atoms) chemisorbed on a CNT (dark grey atoms) in a spiral pattern as computed from quantum chemistry methods. (From C.W. Bauschilicher, Jr. and C.R. So, Nano Lett., 2, 337, 2002. With permission.)

favorable patterns for the high (50%)-coverage H chemisorption [74]. The nature of the relative stability of the patterns found in this work is certainly different from the earlier work of Froudakis et al. [73]; however, the common finding is that at high coverage, H atoms prefer to chemisorb in patterns as opposed to being chemisorbed randomly [73,74]. The binding of an H atom on a site neighboring to an already-chemisorbed site is thus expected to be favorable compared with the binding on an isolated site, because of the partial bonding induced surface relaxation near the already chemisorbed site. For example, a stable pattern for H chemisorption with 50% surface coverage is shown in Figure 2.10. This is an important prediction if the 50% atomic H coverage on the surface of the CNT could be used to support additional subsequent physisorbed layers of molecular hydrogen for hydrogen storage applications. Similar quantum chemistry approaches have been used to investigate chemisorption of atomic N and O on a CNT [75]. For both N and O chemisorption, in their ground electronic state, it has been found that chemisorption or addition to a CNT requires an electronic surface crossing with a barrier. The addition of electronically excited states does not require any barrier, and both N and O bond in bridged structures [75].

The second category involves the modeling and simulation-based efforts focused toward (a) understanding and (b) possibly controlling the site-specific local chemical reactivity of CNTs for device-type applications or for cross-linking purposes in CNT-reinforced composite materials. The flat surfaces of graphene layers in bulk graphite are generally inert to chemisorption. The chemical reactivity in CNTs, on the other hand, is governed by the local atomic structure and surface curvature. Pristine CNTs are preferentially reactive at end caps because of the effect of surface curvature on the nature of (a) spatial overlapping of atomic p orbitals of neighboring C atoms leading to conjugation and (b) a shift in the hybridization of the atoms from graphitic (sp^2) to something intermediate between graphitic (sp^2) and diamond like (sp^3). The graphitic bonding nature is generally inert whereas reactivity increases in proportion to the diamond-like nature of the local bonding. The first attempts to qualitatively relate the chemical reactivity with local surface curvature were done by Haddon [76] in the context of the reactivity and functionalization of fullerene, for which the structure is similar to that of an end cap of a CNT. According to this model, the reactivity of a fullerene is proportional to the local pyramidal angle (explained below) of the carbon atom on which the chemisorbing species or functional group can react and form a stable bond [76].

Based on the above concept of curvature-dependent local chemical reactivity of fullerenes, it has been proposed [77] that if the reactivity is driven by the local curvature or strain of the surface C atom, it may be possible to control the reactivity by mechanically straining the fullerene or CNT. The mechanical strain or kink-driven chemistry of a bent or flattened CNT at the location of the bending or flattening site was thus called "kinky chemistry" of CNTs [77]. Using reactive MD and TB total energy calculations, it was shown that the binding energy of an incoming H atom increases by as much as 1 to 1.5 eV, and reaction probability increases by a factor of 5 at the location of the kinked or strained site. A proof of principle experiment on the etching of MWNTs laid out on a ridge feature showed that the CNT is bent

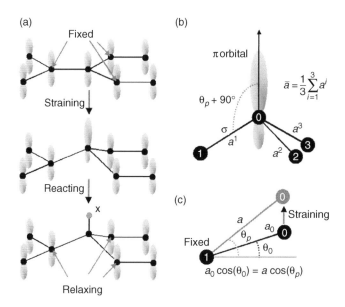

FIGURE 2.11 (a) Decomposition of the total chemisorption energy into straining a C atom, reaction on the strained site, and relaxation of the full configuration; (b and c) definition of the pyramidal angle in terms of the changes in the bond angles and bond lengths. (From S. Park, D. Srivastava, and K. Cho, Nano Lett., 3, 1273, (2003). With permission.)

similar to the shape of the ridge on the underlying surface and etched only at the most strained site when exposed to the nitric acid vapor. This makes it possible to etch the CNT locally at the mechanically strained spot because the cohesive energy of the CNT is also reduced at the location of the kink [77]. Using more accurate DFT methods to compute the binding energy as a function of local deformation, the same concept has been later reported as "tunable" chemistry of CNTs [78].

More recently, a general relationship has been developed [79] between the binding energy of a chemisorbing species and the local surface curvature represented by the local pyramidal angle (shown in Figure 2.11a) at the chemisorption site. The total reaction energy of a chemisorbing species is broken into three components: (a) the strain energy of the reacting C atom on the CNT as it goes from a graphitic bonding in the unreacted state to a diamond-like bonding in the postreaction or product state, (b) the binding energy of the chemisorbing species on the strained C atom of the first step, and (c) the relaxation of the product state after reaction has occurred. The separation of the reactivity into three contributing terms is the basis of the quantification of each as a function of the pyramidal angle. The separation, however, is for conceptual purpose only, and three processes could occur simultaneously instead of occurring in the order described above. All these three terms have been expressed in terms of pyramidal angle as shown in Figure 2.11, and a generalized form of the total reaction energy in terms of pyramidal angle has been derived [79]. The total reaction energy is written as:

$$E_{total} = E_{strain}\left(\theta_p^{min}\right) + E_{C-X}\left(\theta_p^{min}\right) + E_{relax}$$

where θ_p^{min} is the value of the pyramidal angle for which the sum of the strain and binding energies is a minimum, and the strain and the binding energies in terms of pyramidal angle θ_p are

$$E_{strain} = \frac{2}{3}k_b\left(\theta_p - \theta_0\right)^2 + \frac{3}{2}k_s\bar{a}^2\left(\frac{\cos\theta_0}{\cos\theta_p} - 1\right)^2$$

$$E_{C-X}\left(\theta_p\right) = E_{graphite} - \alpha\theta_0^2 + \sqrt{2}\tan\left(\theta_p\right)E_{sx} + \sqrt{1 - 2\tan^2\left(\theta_p\right)}E_{px}$$

For a reaction with species X, this approach has only two parameters, E_{sx} and E_{px}, which are evaluated by fitting the numerical reaction energy of the species X on a graphene sheet to the above expression for $E_{C-X}\left(\theta_p\right)$. The reference parameter, $E_{graphite}$, is the same for all reaction types and needs to be evaluated only once. The total reaction energy expressions of low coverage reaction of a variety of CNTs with H, OH, and F groups were formulated and give the reaction energies shown in Figure 2.12, which are within ~ 0.1 to 0.3 eV errors compared with the fully relaxed *ab-initio* simulation results for the same reactions [79]. The model was also applied to the chemical reactivity of mechanically strained CNTs as described above. The chemical reactivity of a bent CNT was computed as a function of the bending angle and changes in the local pyramidal angle at the bending site. The computed reactivity by the above-described generalized model at the bent site was within a few percentages of the reactivity computed from the reactive atomic MD methods described above. It is possible that this model could be extended in the future to include other reacting species and to chemisorption at high coverages where the interactions between the reacting species at the neighboring sites need to be incorporated as well.

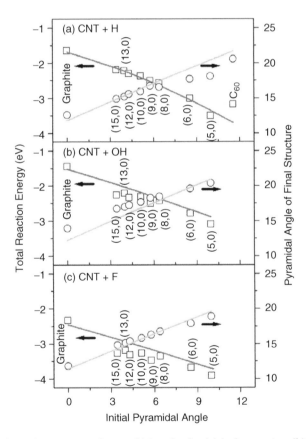

FIGURE 2.12 The total reaction energy and pyramidal angles for (a) hydrogenation, (b) hydroxylation, and (c) fluorination of a variety of CNTs computed with the generalized reactivity model (straight lines) and the fully relaxed *ab-initio* DFT simulations (squares and open circles). (From S. Park, D. Srivastava, and K. Cho, Nano Lett., 3, 1273, 2003. With permission.)

Lastly, another way to functionalize CNTs in a matrix material, such as in polymer-CNT composites, has been investigated through energetic particle beam bombardment of the composite materials. This has been investigated through MD simulations of CH_3 radical collisions of bundles of CNTs and also of polymer matrix [80]. In both cases, probabilities of reactive trajectories for incident energies of 10 to 80 eV were computed. Analysis of reactive trajectories showed chemical adhesion of radicals or heavy fragments from the radicals such as CH_2, CH, or C can occur at all incident energies. The functionalization of the wall was found to be more probable for the incident energies of 10 and 45 eV. At higher energies defects in CNTs are formed and also are cured during the radical collision followed by the simulated annealing of the samples [80].

2.5.2　Hydrogen Physisorption and Gas Storage

The physisorption on CNTs and other carbon-based materials has been investigated vigorously for many years because these are low-density and high surface–area materials, which could be uniquely suitable for gas storage or gas filteration applications. SWNTs are especially attractive for gas storage applications because, conceptually, both the inner and outer surfaces of an open-ended CNT could be accessible for gas storage applications. Dillon et al. [81] reported for the first time excellent molecular hydrogen storage capacity of SWNT samples. The storage capacity was estimated to be as high as between 5 and 10 mass % of the total sample. This started a series of vigorous experimental and modeling and simulation-based investigations to (a) understand the mechanism of hydrogen storage and (b) explore the upper limit of hydrogen storage for environmentally friendly alternative fuel applications supported by many govern-mental agencies. It turned out that the measured hydrogen desorption in the TPD experiments was only 0.01 mass %, and content of SWNTs in the sample was only 0.2 mass %. The errors were attributed to the lack of inclusion of hydrogen storage possibility in the rest of the high-porosity carbon and catalyst materials in the sample. Chambers et al. [82] also reported room temperature and 11.35 MPa pressure hydrogen storage capacity of 11.26 mass % for CNTs, 67.55 mass % for herringbone carbon nanofibers, and 4.52 mass % for graphite. These were exceptionally high storage capacity values but could never be reproduced by any other group since then. Nevertheless, the above experiments launched a significant number of research efforts into understanding physisorption mechanisms and storage of molecular hydrogen in CNTs through both experiments and simulation-based investigations.

From a modeling and simulation point of view, it was a significant challenge because no estimation provided thus far could support such high-percentage storage of hydrogen in CNT samples under ambient conditions. The physisorption and storage of molecular hydrogen on CNTs is not through chemical bonding interactions but through VDW forces. The simulated heat of adsorption of 6.3 kJ/mol in SWNTs [81] compares well with 4 kJ/mol for H_2 on graphite [83], indicating that the VDW interactions available for such simulations are reasonably accurate. Classical molecular dynamics and energy minimization methods [84,85] using these VDW interactions give H_2 adsorption and storage capacity to be two orders of magnitude less than the high values reported above [81] and subsequent work by the same group. It was estimated that the strength of CNT-H_2 nonbonding interaction potential would be conceptually need to increase by a factor of four to achieve 2 wt % adsorption capacity under reported experimental conditions.

Rzepka et al. [85] computed the amount of adsorbed hydrogen for slit-pore and tubular geometries and found the maximum to be about 0.6 mass % at $T = 300$ K and $p = 6$ MPa. This compares well with recent experiments performed under similar conditions [86,87]. It has been estimated that at lower temperature of 77 K the amount of adsorbed hydrogen can be increased to about an order of magnitude higher than the value at room temperature reported above.

2.5.3　Gas, Polymer, and Electrolyte Diffusion

This section briefly summarizes simulations of gas separation; quantum, or molecular sieving; and polymer translocation and diffusion through SWNT pores. There are only a few simulations geared

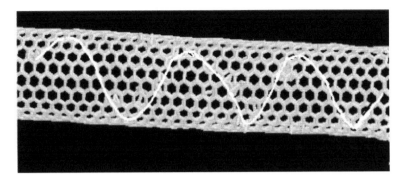

FIGURE 2.13 The average spiral path taken by hydrocarbon gas molecules diffusing through a CNT; the spiral nature of the path depends on the chirality of the CNT. (From Z. Mao and S. Sinnott, Phys. Rev. Lett., 89, 278301, 2002. With permission.)

toward hydrocarbon gas or gas-isotopes separation and polymer mass-spectroscopy or drug-delivery types of applications. The mixtures of light isotopes could be selectively adsorbed by the weak VDW forces in nanopores leading to a novel separation scheme called quantum sieving [88]. Simple theoretical models and detailed path integral simulations show that molecular tritium readily adsorbs in appropriate size nanopores, such as interstitial channels in a closed pack CNT rope or bundle that excludes molecular hydrogen through quantum sieving. Similar simulations were also performed for quantum sieving effects in open-ended CNT nanopore channels, and sieving effect was observed only for CNTs of very small diameters that are not easily made in experiments [88]. The diffusion and separation dynamics of small hydrocarbon gas molecules (molecular sieving) has been simulated using reactive atomistic MD with T-B potential for bonded hydrocarbon interactions and appropriate nonbonded VDW and coulombic interactions [89,90]. The diffusion occurs through the open-ended CNTs, and, depending on the size and shape of the molecules in some cases, as shown in Figure 2.13, diffusive trajectories follow spiral paths within the CNTs to maximize bonding overlap with the C–C bonds within the nanotubes. The specific molecular mixtures included CH_4/C_2H_6, CH_4/n-C_4H_{10}, and CH_4/i-C_4H_{10}. The main results are that for a given binary mixture the amount of separation of molecular species increases as either the CNT diameter decreases or the difference in the relative size of the molecules in the binary mixture increases [89,90].

Long-chain biological molecular species such as DNA through biological nanopores have been recently investigated for DNA sequencing or genomic applications. The shape, size, and functionality of the biological nanopores are not much controllable, and therefore efforts are made to extend toward DNA diffusion and dynamics through solid-state nanopores. In the same spirit, open-ended CNTs offer a natural, structure-free, micron-size nanopore or capillary channel for the nanoscale separation, sequencing, delivery-type applications of long-chain biological or nonbiological molecules such as polymers. A theory of translocation and diffusion of long-chain polymer molecules through SWNTs has been developed using the combined Fokker-Planck equation and direct molecular dynamics simulations approach [91]. The mean transport or translocation time τ of the long-chain polymer molecules through CNT channels, as shown in Figure 2.14, has been derived as:

$$\tau = \frac{b^2}{\left|\frac{V\mu}{k_B T} + \ln\frac{q_2}{q_1}\right|}\left[\frac{N^{v_1+1}}{(v_1+1)D_1} + \frac{N^{v_2+1}}{(v_2+1)D_2}\right] + o\left\{(\gamma_i - 1)N^{v_i}\right\}$$

$$\left(N\left|V\mu/k_B T\right| \gg 1, \quad V\mu\right)$$

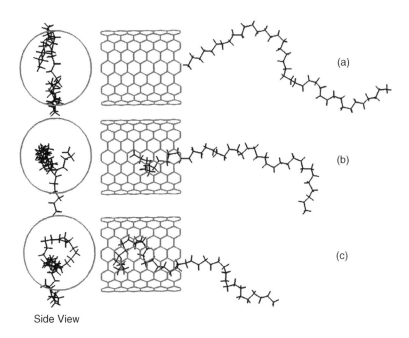

Side View

FIGURE 2.14 Translocation of a 40-monomer long polyethylene chain in a CNT capillary channel. (From C. Wei and D. Srivastava, Phys. Rev. Lett., 91, 235901, 2003. With permission.)

where b is the length of a single unit of the polymer; $\Delta\mu$ is the difference of energy per unit between the two configurations with polymer inside/outside CNT; q_{eff} is the effective coordination number for each unit; γ is a scaling parameter; D_1 and D_2 are diffusion coefficients for a single monomer inside and outside CNT, respectively; and ν_i is an exponential coefficient constant, which has a value between 0 and 2 based on some commonly used models of polymers in the literature. Thus, the mean translocation time of a polymer chain in a CNT channel at a given temperature is found to depend on the chemical potential energy, entropy, and diffusion coefficient. Overall, a power law dependence of the translocation time $\sim N^2$ has been derived and also found through direct MD simulations where N is the number of monomer units in the polymer chain. A fast diffusion of the polymer chains through CNT channels has been observed during the simulations. The fast diffusion in the CNT channels is a few orders of magnitude larger than the diffusion of similar chain molecules through nanopores in silicate-based zeolites [91].

In a realistic scenario, or for biology-based applications, the dynamics and diffusion of long-chain molecules needs to be investigated in the presence of solvents or ionic solutions passing through CNT channels. MD simulations have been used recently to study ionic flow in CNTs as artificial protein channels found in cell membranes. The ionic current under an applied potential results when ions move inside or outside a channel and when a target molecule binds to a binding site in or on the nanopore the ionic current is modulated. The modulated ionic currents under external potentials can be used to generate and send signals on the nanochannels. CNTs have been simulated to reproduce this functionality of ion channels in a much simpler way. The initial studies in this area focused on fundamental transport of water through CNT channels. Even though CNTs are hydrophobic, simulations have shown that water enters CNTs as small as 0.8 nm diameter [92]. Water and molten potassium iodide are shown to form networks of connected crystalline lattice structure for helical ice sheets or solid potassium iodide, respectively, in CNTs of small diameter [92–94]. Recent MD simulations of similar systems, however, show that ion occupancy in a CNT solvated in an electrolyte is very low for a larger-diameter (16,16) CNT. However, if the CNT is modified at the open ends by placing partial charges on the rim atoms and an external field is applied, the ion occupancy inside the CNT is increased significantly [95]. To mimic an ion channel, when functional groups are attached at the ends and an external field is applied, the CNT allows ion transport through the CNT-based nanochannels.

2.6 Nanoelectronics and Sensors

The basic symmetry properties and the metallic/semiconducting behavior of CNTs were discussed briefly earlier. In this section, further discussion is presented on the electronic characteristics of individual CNTs; explanation of the electronic behavior from the symmetry properties; CNT heterojunctions for nano-electronic devices; and CNT-based gas, chemical, and pressure/vibration sensors.

2.6.1 Modeling and Simulations of Electronic Characteristics

Carbon nanotubes exhibit unique electronic properties depending on the diameter as well as helicity or chirality of the underlying graphene structure. The metallic and semiconducting behavior of CNTs was explained by Hamada et al. [96] with a simple tight-binding model and symmetry considerations. The two-dimensional electronic band structure of a graphene sheet reflecting hexagonal symmetry is given by:

$$E(k) = E_f \, (+/-) \, Vp \, (3 + 2 \cos k.a_1 + 2 \cos k.a_2 + 2 \cos k.(a_1 - a_2) 1/2,$$

where $\mathbf{k}$ is any two-dimensional wave vector (for electron momentum) in hexagonal unit cell reflecting the symmetry of the graphene sheet in the Fourier space (called Brilliouin zone), and $(\mathbf{a_1},\mathbf{a_2})$ are the basis vectors of the unit cell in the momentum space. The above construction is based on the knowledge that the Fourier transform of a real-space object with a symmetry of a hexagonal unit cell (of a graphene sheet) is also hexagonal, and that instead of being defined by the real-space coordinate vectors it is defined by the momentum or wave vectors associated with electrons. As explained in Chapter 1, a flat graphene sheet is rolled up to make tubular structure with indices (m,n). For perfect matching of the boundary conditions, the relative positions of the atoms from the two edges reflected by $\mathbf{C} = n\mathbf{a_1} + m\mathbf{a_2}$ must overlap, leading to $\mathbf{C} = 2 \pi \, \mathbf{q}$, where q is an integer. For different values of q in the momentum space, one gets parallel lines with relative separation $\mathbf{C}/2 \pi$. Because $\mathbf{C}$ depends on the indices n and m, the K point reflecting the condition $kK = (a - b)/3$ corresponds to the situation where π and $\pi^\star$ bands of graphene sheet meet defining the Fermi energy (in this picture) of the system. In simplest terms, the Fermi energy is the level up to which electronic states are completely filled (generally referred to as valance band) and above which the states are empty (generally referred to as conduction band). Because electron transport occurs through conduction bands, if there is a gap between the Fermi energy and the lowest edge of the conduction band, the system behaves as a semiconductor; otherwise, it behaves as a metal. If the gap at the Fermi energy is too large, the same material would behave as an insulator for all practical purposes. In the context of the CNTs, at K points defined above, the π and $\pi^\star$ bands meet and tubes behave as metallic. The wave-vector lines corresponding to $m = n$ or m-n as an integer of 3 are metallic. The remaining tubes are semiconductor. This makes all armchair ($m = n$) and 1/3 of the zigzag (n or $m = 0$) CNTs metallic. This model is very simple but works relatively well in explaining the general behavior. Of course, the deviations from the general behavior are due to higher-order effects through mixing of in-plane σ and out-of-plane π orbitals.

Until recently, many of the initial studies of the electronic transport characteristics of CNTs were done with simple π orbital TB model approach [97,98], whereas most of the recent modeling and simulation efforts have started to include more orbitals and overlap between orbitals in the approach. As a general rule, for large-diameter CNTs, the overlap between atomic orbitals can be neglected, and TB models with only π orbitals generally give reasonable description. For small-diameter CNTs or for CNTs with defects or under mechanical strain, it is important to include the overlap between atomic orbitals in either a TB description or even to use a more-accurate DFT approach for electronic band structure calculations [99]. The typical electronic band structures for metallic (armchair) and semiconducting (zigzag) CNTs have been computed by many groups [100]. SWNTs, bundles of SWNTs, or MWNTs have also been explored for quantum conductivity or interconnect-type applications. The intertube or wall-wall interactions in CNT bundles or MWNTs, respectively, are important. The wall-wall interactions within a metallic CNT bundle have been simulated to open a small pseudo-gap [101,102]. Structural deformation of CNTs and chemisorption on CNTs can also significantly change the electronic band structure and transport properties, allowing for their use in mechanical vibration or pressure sensors or chemical sensor applications, respectively. These will be briefly discussed later.

On the experimental side, SWNTs and MWNTs have been laid out on insulating layers with a back-gate applied from underneath in a field-effect transistor (FET) configuration. These devices, circuits, and architectures are described in Chapter 7. Here, the focus is on electronic devices due to heterojunctions of CNTs of metallic and semiconducting nature.

2.6.2   Nanoelectronics: Heterojunctions of Carbon Nanotubes

As explained earlier, in general, SWNTs are rolled-up graphene sheets (a single layer of C atoms arranged in a hexagonal lattice structure) with various chiralities [4–7]. The electronic structure of these tubes can be either metallic or semiconducting, depending on the nature of the rolling direction of the sheet into a cylindrical or tubular structure. The possibilities of connecting nanotubes of different diameters and chiralities (rolling directions) to form two terminal nanotube heterojunctions were proposed [103–106] very early from the theoretical studies.

The simplest way to connect two dissimilar nanotubes is found to be via the introduction of pairs of heptagons and pentagons in an otherwise perfect hexagonal graphene sheet structure [103–106]. The resulting junction still contains three-fold coordination for all carbon atoms, and the heterojunction between a semiconductor and a metallic carbon nanotube can act like a rectifying diode or a switch. An example of such a two-terminal heterojunction between semiconductor and metallic CNTs is shown in Figure 2.15 (from Reference 104). The two-terminal CNT heterojunctions were first proposed through topological consideration alone, which were then tested for stability and electronic properties through electronic TB or *ab-initio* DFT approaches. The straight and bent "knee" type junctions were proposed where the topological defects such as pentagons and heptagons were on the same and opposite side along the circumference, respectively. The junction shown in Figure 2.15 is a bent knee-type junction between (8,0) and (7,1) CNTs. The characteristic angle at which these bent heterojunctions were supposed to have been formed was observed in as-produced samples of SWNTs [107]. The electronic density of states, band structure, and transport simulations have been performed across these two-terminal CNT devices by many groups. Recently, these junctions have been constructed by pushing SWNTs with atomic force microscopy (AFM) tip [108]. An experimental investigation of the electronic transport at the junction showed the predicted rectifying behavior as well [108].

There are two ways to create nanotube heterojunctions with more than two terminals. The first approach involves connecting different nanotubes, like in the above, through molecularly perfect but topological defect-mediated junctions. The second approach consists of laying down crossed nanotubes over each other and simply forming the physically contacted or touching junctions. The differences are in the nature and physical characteristics of the junctions forming the device or interconnect. In the first case, the CNTs are chemically or "weld" connected through bonding networks forming a stable junction that could possibly give rise to a variety of switching, logic, and transistor applications in the category of monomolecular electronic or computing devices [109]. In the second case, the junction is merely through a physical contact and will be amenable to changes in the nature of the contact. The main applications in the second category will be in electromechanical switches for cross-bar memory or sensor applications and experimental progress in some of these concepts and applications has been significant [110]. The structure and quantum electronic conductance of two crossed CNTs (Figure 2.16) have been investigated with DFT-based approaches and constraints to reflect the substrate-CNT interactions or external applied force. The intertube conductance is found to depend globally on the overall structural deformation at the junction. For moderate contact forces, the smaller contact distance leads to greater intertube conductance [111]. Thus, in principle, the contact distances or forces could be externally controlled for applications of individual crossed CNT junctions in the electromechanical switching devices or memory applications.

FIGURE 2.15 A knee-type two-terminal junction between (8,0) and (7,1) carbon nanotubes as a model of CNT-based rectifying devices. (From L. Chico et al. Phys. Rev. Lett. 76, 971, 1996. With permission.)

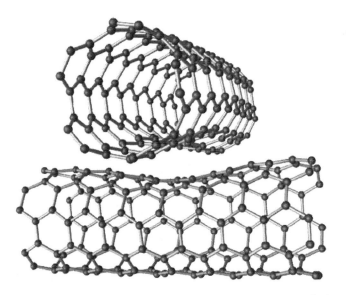

FIGURE 2.16 Deformed structure of crossed (5,5) CNT junction. (From Y-G. Yoon et al. Phys. Rev. Lett., 86, 688, 2001. With permission.)

The proposals and simulations of the prediction and feasibility of 3- or more terminal "weld" connected CNT junctions have been a different challenge altogether. Soon after the discovery of CNTs in 1991 [1,2], the multiterminal CNT junctions in a "Jungle Gym"–type structure [112] or via a nodal fullerene [113] were proposed as topology-based model structures as early as 1992. The 3-terminal branched CNT

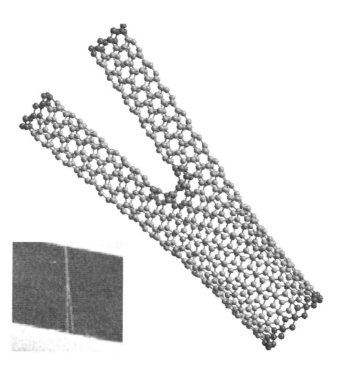

FIGURE 2.17 A SWNT Y-junction simulated for structural stability and current-voltage characteristics for 3-terminal device applications [121–123], and inset shows a similar acute angle Y-junction of MWNTs [117].

T- and Y-junctions shown in Figure 2.17 and Figure 2.18 were the first rigorously proposed structures for 3-terminal molecular electronic devices, such as rectifying, tunnel junctions, or transistor-type applications. Large-scale quantum molecular dynamics simulations with a nonorthogonal TB method showed for the first time that the T- and Y-junctions, if fabricated, (a) will be structurally stable and (b) could form the basis of first all "carbon"-based 3-terminal tunnel junctions for molecular electronics device applications [114,115].

Inspired by the above findings, MWNT Y-junctions were first fabricated in a template-based CVD growth of CNTs in nanoscale anodic cavities that were intrinsically branched. The Y-branching in the template led to a controlled branching of thousands of grown MWNT junctions in the template [116,117]. Since then, many groups using different techniques have produced 3- or more terminal MWNT junctions in different scenarios [116–120] and measured conductance or I-V characteristics under a 2-terminal device scenario [116,117]. In such measurements, the two of the three terminals are placed at the same fixed voltage, and the voltage at the third terminal is varied. The I-V characteristics of Y-junctions, in 2-terminal device configurations, show rectification [117,118]. The I-V characteristics of a variety of SWNT-Y-junctions have been simulated and show similar rectification behavior if the Y-junction is configured as a 2-terminal electronic device, and a zigzag tube forms the stem of the configured Y-junction [121–123]. Leakage current occurs for Y-junctions involving armchair and chiral CNTs, and rectification is found to be an intrinsic characteristic of a smaller subclass of all possible Y-junctions [123]. Additionally, the simulations are also not limited by the current experimental inability to put independent electrical contacts on all the three terminals of the junction. Simulation results in a 3-terminal scenario and show that a gate voltage from a third terminal can be used to modulate the current-voltage behavior through the other two terminals [121,122]. The possibilities of analog logic devices based on the current modulation without any gate isolation have been explored.

More recently, the chemically or weld-connected 3- or 4-terminal junctions of SWNTs also have been fabricated [124]. The experiment shows the feasibility of connecting a SWNT with another SWNT at a variety of angles at the junction by a "nanowelding" technique. The SWNTs are placed over each other,

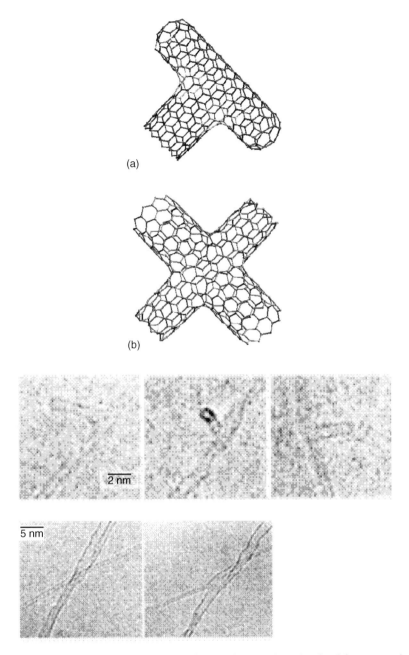

(a)

(b)

2 nm

5 nm

FIGURE 2.18 Examples of single-wall carbon nanotube T-and X-junctions simulated for structural stability and electronic behavior in comparison with experimental single-wall T- and X-junctions fabricated through nanowelding technique [124].

and a simultaneous local e-beam–induced irradiation and annealing in a tunneling electron microscopy (TEM) setup causes the two nanotubes to form a spot weld junction at the contact point. Both T- and Y-junctions of the single CNTs, as shown in Figure 2.18, have been fabricated with this approach. The formation mechanism of such T- junctions via an all graphitic (sp^2)-type low-energy bonding pathway has been investigated recently and gives the activation energy for the formation pathway to be between 5 and 8 eV, which is not too high for an e-beam– or ion beam–induced process [125]. The intrinsic nature of the rectification behavior of the Y junctions has also been investigated [126,127], and it has been noted that the rectification can be determined by a combination of factors such as (a) formation of a quantum dot at the location of the Y-junction, (b) finite length of the stem and branches going out to the metallic leads, (c) the strength of the SWNT and metallic lead interaction, and (d) the asymmetry of the applied bias [127]. Attributing the rectification behavior effect only to the SWNT-metallic lead interactions, as proposed in Reference 126, is neither correct nor complete. The intrinsic structural nature at the junction and asymmetry of the externally applied bias play a significant part in determining the rectification behavior of these devices.

From the architectural view point, a three-dimensional tree or branched network of chemically or weld-connected CNTs [7] has been proposed as a biomimetic architecture for nanoscale computing and sensing systems with functionality similar to that of dendritic neurons in a biological system.

2.6.3   Electrochemical and Electromechanical Sensors

As described earlier, CNTs of different chirality have different electronic properties ranging from metals to semiconductors (1-eV bandgap). Semiconductor SWNTs are promising materials for novel-sensing applications because the surface modifications, due to chemical adsorption or mechanical deformation of the CNTs, can significantly change the electronic band structure and conductance of the nanotubes. This is mainly because in nanoscale covalent materials, such as CNTs and fullerenes, there is a strong coupling between the electronic characteristics and the mechanical deformation of the chemisorbed species. An earlier section on the reactivity of the CNTs featured the strong coupling behavior between the chemical reactivity and the local surface curvature that can be manipulated with mechanical deformation. In this section, a discussion on the simulation-based investigations for chemical and mechanical sensor applications of CNTs is presented.

Recent experiments have shown that the electronic conductivity of semiconducting SWNTs is extremely sensitive to certain gas molecules even in miniscule (parts per million) amounts [128,129]. These have led to applications in chemical and gas sensors with fast response time and ultra-high sensitivity. The sensing mechanism involves detecting conductance change of the CNTs induced by charge transfer from gas molecules adsorbed on CNT surfaces. The initial experiments and simulation-based investigations focused on NO_2, NH_3, and O_2 as adsorbing gases. The changes in the electronic band structures and change transfer, using *ab-initio* DFT-based approaches, due to adsorption of these gases have been computed [130,131]. For example, each NO_2 gas molecule adsorbed on the nanotube surface induces a small amount (about 0.1 e) of electron transfer so that the CNT becomes a p-type doped semiconductor, and conductivity changes significantly [131]. A similar amount of charge transfer for chemisorption of O_2 has also been computed [130] that leads to a change in the conductivity. The O_2 physisorption in this study is found to be of a configurationally restricted nature and is expected to change if a wider search is made using the fully dynamic symmetry unconstrained relaxation techniques. The changes in the computed geometries and the consequent charge transfer are expected to change the conductance behavior of the computed sensor application as well. The applications in the above cases, however, so far, are limited only to the above-mentioned gases because (a) of the intrinsic limitations in the structure of semiconducting CNTs and (b) not many other gases or biochemical species directly adsorb on CNTs.

An *ab-initio* study of H_2O adsorbed on SWNT shows a purely repulsive interaction without any charge transfer so that a SWNT can be fully immersed in water maintaining its intrinsic electronic properties. A route to other chemical or biochemical sensor applications is then proposed through either (a) functionalization of the CNTs (as was discussed earlier) or (b) doping of the CNTs with point defects with different reactivity. It may then be possible to detect many other types of gas- and solution-phase chemical or biochemical species through the interactions with the functionalized or doped moiety on the CNT, respectively [132,133]. For example, using *ab-initio* DFT-based approaches, it is shown that the substitutionally doped boron (B) and nitrogen (N) atoms in the intrinsic CNTs are able to detect CO and H_2O molecules. This is because B or N doping changes the local chemical reactivity around the doped atom, which increases the binding energy of the sensed molecule with the doped species. The

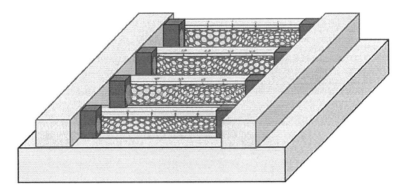

FIGURE 2.19 Schematic of an array of B- and N-doped CNT devices for chemical sensor applications. The chemical, electronic, and structural characteristics for doped CNTs are simulated in [133]. (From S. Peng and K. Cho, Nano Lett., 3, 513, 2003. WIth permission.)

layout of a proposed CNT chemical sensor array with variations in the doping of B and N in CNTs is shown in Figure 2.19. The gas molecules bound to the doped species also cause a charge transfer between the two, which can be then detected by the conductance change. The changes in the reactivity and charge state due to the interaction of the doped species such as B and N on a variety of CNTs with CO and H_2O molecules, respectively, have been computed and tabulated through *ab-initio* DFT based approaches [133].

CNTs are also shown to have a strong electromechanical coupling in recent TB [134] and *ab initio* simulations [135]. As the cross section of (8,0) SWNT is flattened up to 40%, the bandgap of the nanotube decreases from 0.57 eV and disappears at 25% deformation. As the deformation further increases to 40%, the bandgap reopens and reaches 0.45 eV. Similarly, the changes in the electronic band gap as function of axial compression, tensile stretch, and torsional deformations have been investigated in detail with modeling and simulation-based approaches [136]. This strong dependence of SWNT band structure on the mechanical deformation can be applied to develop nanoscale mechanical sensors. Furthermore, mechanical deformation can be used to control the electronic excitation by static electric fields or electromagnetic waves. A recent *ab initio* study of polarons in SWNTs shows that the electron-hole pair creation can induce a tube-length change leading to optical actuation mechanism of nanotube mechanical properties. The strong coupling of electronic, optical, and mechanical properties of SWNTs offers a great opportunity to develop novel nanodevices that can be controlled and actuated by electromagnetic waves as shown by a recent experimental demonstration [137].

2.7 Summary and Outlook

The examples discussed in this chapter provide a narrow window on a very broad field of applications and areas that are opening up and directly fueled by the efforts and advances in the computational nanotechnology–based investigations of CNTs. Some of the covered areas and applications in this chapter include nanomechanics and thermal properties of CNTs and CNTs in polymer composite materials; chemical reactivity or functionalization of pristine or deformed nantubes; gas physisorption, diffusion, and storage devices through CNTs; polymer translocation and diffusion in CNT nanochannels for sequencing or drug–delivery applications; molecular electronic device components through nanotube heterojunctions; and nanosensors for miniscule amounts of gas- and biochemical-detection applications, etc. Equally impressive progress has also been made in applications of CNTs in field-emission devices and as scanning probe microscopic (SPM) tips for metrology purposes, and these developments are covered in later chapters.

In the future, it may be possible to conceptualize nanoscale synthetic machines and motors that could be powered and controlled through external laser, electric, or magnetic fields and operate in a chemical

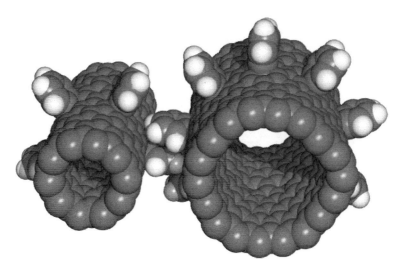

FIGURE 2.20 CNT-based gears simulated at NASA Ames in the beginning of the program; the structural details, operating conditions, and laser-based motor operations are explained in References 138 and 139.

solution phase or inert gas environment. Such a concept of nanogears is shown in Figure 2.20 [138,139]. The design of the gears is based on the analogous chemical reactions that are known to occur in bulk phase chemistry. No fabrication pathway for making such nanoscale gears has been suggested to date. Computational modeling and simulation is useful to investigate the optimal structural, material, and operating conditions. The result of such investigations [139] is that CNT-based nanogears, if made, could be more robust than similar gears and other machines fabricated and operated in the macroworld. This is an example illustrating that the future of CNT-based devices and systems can be shaped by the advances in computational nanotechnology.

Acknowledgments

This work is supported by NASA contract DTT559-99-D-00437 to CSC at Ames Research Center. The author gratefully acknowledges various colleagues, collaborators, students, and postdoctoral students in recent years who have contributed to the development of some of the work described and reviewed in this chapter. In particular the author is thankful to Ajayan, PM (RPI); Anderson, C (University of Minesotta); Andriotis, A (Crete, Greece); Brenner, DW (NC State); Chernozatonskii, LA (Russian Academy of Sciences, Moscow); Cho, KJ (Stanford University); Cummings, A (Washington State University); Globus, A (NASA Ames); Menon, M (University of Kentucky); Noya, EG (University of Santiago, Spain); Osman, M (Washington State University); Park, SJ (Stanford University); Ruoff, R (Northwestern University); Tamma, K (University of Minnesotta); Wei, C (NASA Ames); and Zhigilei, L (University of Virginia) for their continued collaboration and enhancing the author's knowledge of CNTs.

References

1. S. Iijima, Nature, 354, 56 (1991).
2. S. Iijima, Nature, 363, 603 (1993).
3. D.S. Bethune et al., Nature, 363, 605 (1993).
4. R. Saito, G. Dresselhaus, and M.S. Dresselhaus, *Physical Properties of Carbon Nanotubes*, Imperial College Press, London, (1998).
5. M.S. Dresselhaus, G. Dresselhaus, and Ph. Avouris, Eds., *Carbon Nanotubes: Synthesis, Structure, Properties, and Applications*, Springer-Verlag Berlin, Heidelberg, (2001).

6. M. Meyyappan and D. Srivastava, Carbon nanotubes, in *Handbook on Nanoscience, Engineering, and Technology*, CRC Press, Boca Raton, FL, (2003). Chap. 18.

7. D. Srivastava, M. Menon, and K. Cho, Comput. Sci. Eng., 3, 42, (2001).

8. A. Nakano et al., Comput. Sci. Eng., 3, 56, (2001).

9. R.E. Rudd and J.Q. Broughton, Physica Status Solidi B, 217, 251, (2000).

10. M. Payne et al., Rev. Mod. Physics, 68, 1045, (1992).

11. P. Hohenberg and W. Kohn, Phys. Rev., 136, B864, (1964).

12. W. Kohn and L.J. Sham, Phys. Rev., 140, A1133, (1965).

13. J. Hafner, VAMP/VASP Guide, http://cms.mpi.univie.ac.at/vasp (current 5 June 2001), (1999).

14. R. Car and M. Parrinello, Phys. Rev. Lett., 55, 2471, (1985).

15. W. A. Harrison, Electronic Structure and Properties of Solids, W. H. Freeman, San Francisco, (1980).

16. M. Menon and K.R. Subbaswami, Phys. Rev. B, 55, 9231, (1997).

17. M. Menon, J. Chem. Phys., 114, 7731, (2001).

18. M.P. Allen and D.J. Tildsley, *Computer Simulations of Liquids*, Oxford Science Publications, Oxford, (1987).

19. B.J. Garrison and D. Srivastava, Ann. Rev. Physical Chem., 46, 373, (1995).

20. J. Tersoff, Phys. Rev. B, 39, 5566, (1989).

21. D.W. Brenner, O.A. Sherendova, and A. Areshkin, *Quantum Based Analytic Interatomic Forces and Materials Simulations*, Rev. in Comp. Chem., VCH Publishers, New York, 213, (1998).

22. D. Srivastava and S. Barnard, Molecular Dynamics Simulation of Large Scale Carbon Nanotubes on a Shared Memory Archi-tecture, Proc. IEEE Supercomputing **97** (SC '97 cd-rom), (1997).

23. S. Datta, *Electronic Transport in Mesoscopic Systems*, Cambridge University Press, Cambridge, (1995).

24. L. Chico et al., Phys. Rev. B, 54, 2600, (1996).

25. A. Andriotis, M. Menon, and G. E. Froudakis, Phys. Rev. Lett., 85, 3193, (2000).

26. A. Andriotis and M. Menon, J. Chem. Phys., 115, 2737, (2001).

27. A.N. Andriotis, M. Menon, and D. Srivastava, J. Chem. Phys., 117, 2836, (2002).

28. F. Muller-Plathe, J. Chem. Phys., 106, 6082, (1997).

29. M. Osman and D. Srivastava, Nanotechnology, 12, 21, (2001).

30. J.C. Tully et al., Phys. Rev. B, 31, 1184, (1985).

31. V.P. Sokhan, D. Nicholson, and N. Quirke, J. Chem. Phys., 113, 2007, (2000).

32. P.K. Schelling, S.R. Phillpott, and P. Keblinski, Phys. Rev. B, 65, 144306, (2002).

33. B.I. Yakobson, C.J. Brabec, and J. Bernholc, Phys. Rev. Lett., 76, 2511, (1996).

34. D. Srivastava, C. Wei, and K. Cho, Appl. Mech. Rev., 56, 215, (2003).

35. D.H. Robertson, D.W. Brenner, and J.W. Mintmire, Phys. Rev. B, 45, 12592, (1992).

36. G.G. Tibbetts, J. Cryst. Growth, 66, 632, (1994).

37. J.P. Lu, Phys. Rev. Lett., 79, 1297, (1997).

38. S. Portal, Phys. Rev. B, 59, 12678, (1999).

39. E. Hernandez et al., Phys. Rev. Lett., 80, 4502, (1998).

40. D. Srivastava, M. Menon, and K. Cho, Phys. Rev. Lett., 83, 2973, (1999).

41. V.M. Harik, Solid State Commun. 120, 331, (2001).

42. L.D. Landau and E.M. Lifschitz, *Theory of Elasticity*, 3rd ed., Pergamon Press, New York, (1986).

43. P. Poncharal et al., Science, 283, 1513, (1999).

44. S. Iijima et al., J. Chem. Phys., 104, 2089, (1996).

45. O. Lourie, D.M. Cox, and H.D. Wagner, Phys. Rev. Lett., 81,1638, (1998).

46. O. Lourie and H.D. Wagner, Appl. Phys. Lett., 73, 3527, 1998.

47. C.Y. Wei, D. Srivastava, and K. Cho, Comput. Model. Eng. Sci., 3, 255, (2002).

48. D. Srivastava, M. Menon, and K. Cho, Phys. Rev. B, 63, 195413, (2001).

49. M.B. Nardelli, B.I. Yakobson, and J. Bernholc, Phys. Rev. Lett., 81, 4656, (1998).

50. P.H. Zhang, P.E. Lammert, and V.H. Crespi, Phys. Rev. Lett., 81, 5346, (1998).

51. B.I. Yakobson et al., Comput. Mater. Sci., 8, 341, (1997).

52. G.G. Samsonidze and B.I. Yakobson, Phys. Rev. Lett., 88, 065501, (2002).

53. D.A. Walters et al., Appl. Phys. Lett., 74, 3803, (1999).

54. M.F. Yu et al., Phys. Rev. Lett., 84, 5552, (2000).

55. M.F. Yu, Science, 287, 637, (2000).

56. C. Wei, K. Cho, and D. Srivastava, Phys. Rev. B, 67, 115407, (2003).

57. C. Wei, K. Cho, and D. Srivastava, Appl. Phys. Lett., 82, 2512, (2003).

58. A.M. Rao et al., Science, 275, 187, (1997).

59. J. Yu, R.K. Kalia, and P. Vashishta, J. Chem. Phys., 103, 6697, (1995).

60. R. Ernst and K.R. Subbaswami, Phys. Review Lett., 78, 2738, (1997).

61. R. Saito et al., Phys. Rev. B, 59, 2388, (1999).

62. R.A. Jishi et al., Chem. Phys. Lett., 209, 77, (1993).

63. S. Berber, Y.K. Kwon, and D. Tomanek, Phys. Rev. Lett., 84, 4613, (2000).

64. J. Che, T. Cagin, and W.A. Goddard, Nanotechnology, 11, 65 (2001).

65. M.A. Osman and D. Srivastava, Nanotechnology, 12, 21, (2001).

66. E.G., Noya et al., Phys. Rev. B., submitted (2003).

67. M. Osman, A. Cummings, and D. Srivastava, Molecular dynamics simulations of heat pulse propagation in single-wall carbon nanotubes, Proceedings of IEEE-Nano-2003, submitted, (2003).

68. C. Wei, D. Srivastava, and K. Cho, Nano Lett., 2, 647, (2002).

69. S.V.J. Frankland et al., MRS Proceedings, 593, A14.17, (2000).

70. C. Li and T-W. Chou, A computational structural mechanics approach to modeling of nanostructures, H.A. Mang, F.G. Rammerstorfer, and J. Eberhardsteiner, Eds., Proceedings of WCCM V, Austria, (2002).

71. L. Zhigilei, C. Wei, and D. Srivastava, Phys. Rev. B, in preparation, (2003).

72. H.F. Bettinger, K.N. Kudin, and G.E. Scuseria, J. Am. Chem. Soc., 123, 12849, (2001).

73. G.E. Froudakis, Nano Lett., 1, 179, (2001).

74. C.W. Bauschilicher, Jr. and C.R. So, Nano Lett., 2, 337, (2002).

75. S.P. Walsh, Chem. Phys. Lett., 374, 501, (2003).

76. R.C. Haddon, Science, 261, 1545, (1993).

77. D. Srivastava et al., J. Phys. Chem. B, 103, 4430, (1999).

78. O. Gulseran, T. Yildrim, and S. Ciraci, Phys. Rev. Lett., 87, 116802, (2002).

79. S. Park, D. Srivastava, and K. Cho, Nano Lett., 3, 1273, (2003).

80. B. Ni and S. Sinnott, Phys. Rev. B, 61, 16343, (2000).

81. A.C. Dillon et al., Nature, 386, 377, (1997).

82. A. Chambers et al., J. Phys. Chem. B, 102, 4253, (1998).

83. Q. Wang and J.K. Johnson, J. Chem. Phys., 110, 577, (1999).

84. F. Darkim and D. Levesque, J. Chem. Phys., 109, 4981, (1998).

85. M. Rzepka et al., J. Phys. Chem. B, 102, 10894, (1998).

86. Y. Ye et al., Appl. Phys. Lett., 74, 2307, (1999).

87. C. Liu et al., Science, 286, 1127, (1999).

88. Q. Wang et al., Phys. Rev. Lett. 82, 956, (1999).

89. Z. Mao and S. Sinnott, J. Phys. Chem., 104, 4618, (2001).

90. Z. Mao and S. Sinnott, Phys. Rev. Lett., 89, 278301, (2002).

91. C. Wei and D. Srivastava, Phys. Rev. Lett., 91, 235901, (2003).

92. G. Hummer, J.C. Rasaiah, and J.P. Noworyta, Nature, 414, 188, (2001).

93. R.J. Mashl et al., Nano Lett., 3, 589, (2003).

94. W.H. Noon et al., Chem. Phys. Lett., 355, 455, (2002).

95. S. Joseph et al., Nano Lett., 3, 1399, (2003).

96. N. Hamada, S. Sawada, and A. Oshiyama, Phys. Rev. Lett., 68, 1579, (1992).

97. M.P. Anantram and T.R. Govindan, Phys. Rev. B, 58, 4882, (1998).

98. M. Nardelli, Phys. Rev. B, 60, 7828, (1999).

99. X. Blasé et al., Phys. Rev. Lett., 72, 1878, (1994).

100. J. Bernholc et al., Annu. Rev. Mater. Res., 32, 347, (2002).

101. P. Delaney et al., Nature, 391, 466, (1998).

102. Y.K. Kwon and D. Tomanek, Phys. Rev. B, 58, R16001, (1998).

103. Ph. Lambin et al., Chem. Phys. Lett., 245, 85, (1995).

104. L. Chico et al., Phys. Rev. Lett., 76, 971, (1996).

105. J.C. Charlier, T.W. Ebbesen, and Ph. Lambin, Phys. Rev. B, 53, 11108, (1996).

106. R. Saito, G. Dresselhaus, and M.S. Dresselhaus, Phys. Rev. B, 53, 2044, (1996).

107. J. Han et al., Phys. Rev. B, 57, 14983, (1998).

108. Z. Yao et al., Nature, 402, 273, (1999).

109. C. Joachim, J.K. Gimzewski, and A. Aviram, Nature, 408, 541, (2001).

110. T. Rueckes et al., Science, 289, 94, (2000).

111. Y-G. Yoon et al., Phys. Rev. Lett., 86, 688, (2001).

112. L. Chernozatonskii, Phys. Lett. A, 172, 173, (1992).

113. G.E. Scusaria, Chem. Phys. Lett., 195, 534, (1992).

114. M. Menon and D. Srivastava, Phys. Rev. Lett., 79, 4453, (1997).

115. M. Menon and D. Srivastava, J. Mat. Res., 13, 2357, (1998).

116. J. Li, C. Papadopoulos, and J. Xu, Nature, 402, 253, (1999).

117. C. Papadopoulos, Phys. Rev. Lett., 85, 3476, (2000).

118. B.C. Satishkumar et al., Appl. Phys. Lett., 77, 2530, (2000).

119. B. Gan et al., Chem. Phys. Lett., 333, 23, (2001).

120. F.L. Deepak, A. Govindraj, and C.N.R. Rao, Chem. Phys. Lett., 245, 5, (2001).

121. A.N. Andriotis et al., Phys. Rev. Lett., 87, 66802, (2001).

122. A.N. Andriotis et al., Appl. Phys. Lett., 79, 266, (2001).

123. A.N. Andriotis et al., Phys. Rev. B, 65, 165416, (2002).

124. M. Terrones et al., Phys. Rev. Lett., 89, 75505, (2002).

125. M. Menon et al., Phys. Rev. Lett., 91, 145501, (2003).

126. V. Meunier et al., Appl. Phys. Lett., 81, 5234, (2002).

127. A.N. Andriotis, D. Srivastava, and M. Menon, Appl. Phys. Lett., 83, 1674, (2003).

128. J. Kong et al., Science, 287, 622, (2000).

129. P.G. Collins et al., Science, 287, 1801, (2000).

130. S. Jhi, S.G. Louie, and M.L. Cohen, Phys. Rev. Lett., 85, 1710, (2000).

131. S. Peng and K. Cho, Nanotechnology, 11, 57, (2000).

132. P. Qi et al., Nano Lett., 3, 347, (2003).

133. S. Peng and K. Cho, Nano Lett., 3, 513, (2003).

134. L. Yang and J. Han, Phys. Rev. Lett., 85, 154, (2000).

135. S. Peng and K. Cho, J. Appl. Mech., 69, 451, (2002).

136. L. Yang et al., Comp. Mod. In Eng. Sci., 3, 675, (2002).

137. Y. Zhang and S. Iijima, Phys. Rev. Lett., 82, 3472, (1999).

138. J. Han et al., Nanotechnology, 8, 95, (1997).

139. D. Srivastava, Nanotechnology, 8, 186, (1997).

3

Growth of Carbon Nanotubes by Arc Discharge and Laser Ablation

Alexander P. Moravsky

Eugene M. Wexler

Raouf O. Loutfy

MER Corporation

3.1 Introduction

Carbon nanotubes produced from carbon vapor generated by arc discharge or laser ablation of graphite generally have fewer structural defects than those produced by other known techniques. This is due to the higher growth process temperature that ensures perfect annealing of defects in tubular graphene sheets. Multiwalled nanotubes (MWNTs) produced by these high-temperature methods are perfectly straight, in contrast to kinked tubes produced at low temperatures in metal-catalyzed chemical vapor deposition (CVD) processes. While the quality of low-temperature tubes can be improved by prolonged postsynthesis annealing at temperatures above 2000 K, the mechanical and electrical properties of arc-produced MWNTs remain far superior. Very high-quality nanotubes are vital for many applications that do not require multi-ton tube quantities; therefore, high-temperature production methods will not be completely replaced by more productive catalytic methods. On the contrary, the need of industry for the highest quality nanotubes will likely steadily grow, thus making it necessary to scale up the high-temperature production methods. Single-wall nanotubes (SWNTs) are almost lacking in structural defects independent of the synthesis process; therefore, alternative methods already strongly compete with arc and laser techniques and may surpass and substitute for them in future bulk production. The extent of this substitution will depend on other important qualities of the SWNT product, such as uniformity of tube chirality, size distribution, and ease of purification. It remains unclear which production process will eventually better ensure such qualities. Either the arc or the laser method may eventually be preferred. Thus, there are grounds to intensively pursue both further studies and methods of scaling up each process. Knowledge of the nanotube growth mechanism is the basis for success in scaling up efforts and in obtaining desirable nanotube qualities, and therefore it is given much attention in this chapter.

At present, the arc method remains the easiest and cheapest way to obtain significant quantities of SWNTs, but the as-produced nanotubes are less pure than those produced by laser ablation. Hence, many experimental and theoretical studies attempt to improve the SWNT yield in the arc process and to provide a comprehensive understanding of the growth mechanism. Scaling up, optimizing, and controlling the arc process to make it commercially viable is an important challenge.

3.2 Arc Discharge Production of MWNTs

3.2.1 General Technical Features of the Production Process

The carbon arc technique for generating MWNTs appears very simple, but obtaining high yields of tubes is difficult and requires careful control of experimental conditions. In the most common laboratory scale production scheme, the direct current (DC) arc operates in a 1- to 4-mm wide gap between two graphite electrodes 6 to 12 mm in diameter that are vertically or horizontally installed in a water-cooled chamber filled with helium gas at subatmospheric pressure. Helium gas and DC current are more important to maximize yield; the position of the electrode axis does not noticeably affect the MWNT quality or quantity. The 50- to 250-mm long positive electrode (anode) is consumed in the arc, whereas a cylindrically shaped cathode deposit (boule, or slag) is grown on the negative electrode. The linear growth rate of the deposit along the electrode axis is smaller than the rate of anode consumption, so one of the electrodes is advanced automatically into the arc zone with a feed rate (FR) of about 0.5 to 3 mm/min to prevent gap growth. The most stable and high yield arc process requires a constant feed rate and arc current for the duration of synthesis [1]. Under these conditions the arc gap and the voltage drop across the gap remain constant, ensuring constant arc power. In this regime the voltage measured on the electrode that leads outside the chamber grows during the run because the resistance of the anode graphite rod decreases faster than the resistance from the cathode deposit increases. This total voltage change does not alter the properties of the arc plasma and thus the nanotube yield.

The holders for the graphite cathode and anode rods are usually water-cooled to prevent damage that might otherwise occur when the arc is near either the cathode or the anode holder. The cooling affects the properties of the cathode deposit only a few millimeters from its origin; the rest of the deposit up to 10 to 15 cm total length has a consistent cross-sectional structure and composition. During those first few millimeters (up to 1 cm) of the cathode deposit growth, the arc process stabilizes and a steady electrode surface temperature is developed. Subsequently, the cooling of the cathode no longer influences the deposit working surface temperature. Therefore, it is expedient to use a cathode of sufficiently small diameter (i.e., equal to or smaller than the anode rod diameter) that is long enough (i.e., a few centimeters) to more quickly attain a steady temperature at the cathode working surface and in the arc. Furthermore, the cathode deposit is more strongly attached to a smaller diameter cathode than to a larger one. A strong bond is important to prevent the cathode deposit from breaking off of the cathode surface in the course of the production process. This cleavage happens most frequently with long cathode deposits, horizontal electrodes, and vigorous arcs. Accordingly, excessive cooling of the cathode holder is undesirable. The same is true for the anode holder. When the arc zone approaches to within 1 to 2 cm of the cold anode holder, the anode working surface becomes cooler, which alters the plasma properties, thus deteriorating the uniformity of the deposit structure and composition. Low-heat conductive anode holders are therefore equally advisable.

The helium gas atmosphere can be static or dynamic during the process with no difference in product quality observed. The requirements for helium gas purity are much less rigid than for fullerene or SWNT synthesis; 99.5% and 99.999% pure gases produce equivalent MWNTs. The MWNT synthesis is much more tolerant to the presence of oxygen-containing impurities in the buffer gas, because MWNTs are formed in the interelectrode gap, where the gas does not penetrate, whereas fullerenes and SWNTs are formed well outside the gap.

3.2.2 Composition and Structure of Cathode Deposit

The internal structure of the deposit provides the most insight into the MWNT growth process in the arc. Improved understanding of the mechanism of MWNT growth in the arc is essentially based on the comparison of arc characteristics with deposit structure studied initially by scanning electron microscopy (SEM) and transmission electron microscopy (TEM).

Individual MWNTs were first found on the cathode surface by Iijima [2]. Soon after this discovery, Ebbesen and Ajayan found conditions for producing large cylindrical cathode deposits containing gram quantities of MWNTs [3,4]. Such cylindrical deposits consist of a hard gray outer shell, typically about 1.5-mm thick, and a soft fibrous black core. The shape of the deposit cross section closely mimics that of the consumed anode. For example, an anode rod with a square cross section produces a cathode deposit with a similar cross section and a squared core.

TEM analysis identified three distinct structural components within the core material obtained under near-optimal conditions: MWNTs in the amount of approximately 35 wt %, multiwall polyhedral particles (MPPs, ca. 45%), and various kinds of graphitic particles (ca. 25%). Along with tubes and MPPs of approximately isometric shape, various intermediate particle shapes are observed in small quantities, e.g., elongated particles constructed of several fused multilayer cones. Graphitic particles have been mainly found as curved ribbons made of stacked graphene layers. Relative amounts of these components, as well as nanotube and MPPs size, as assessed by TEM observations, strongly depend on the arc parameters as illustrated in Table 3.1. Core material produced at low helium pressure contains a small amount of tubes, and thus the main component of specimen 1 was curved graphite. The tendencies for the tube yield to increase and for the graphitic particles to decrease with an increase in He pressure and a decrease in arc current are clearly revealed by the first three examples in Table 3.1. This is further confirmed by studies performed at higher-than-atmospheric He pressures. Thus, in specimen 6, the main components were nanotubes and MPPs. The range of nanotube and MPP diameters observed at higher pressures is much wider than at low pressures. However, it should be noted that the most probable diameter of MWNTs does not change much with conditions and is about 14 nm. Lower feed rates and thus wider gaps also lead to broader size distributions.

No sole externally controlled parameter may be identified as the crucial factor for deposit component distribution. Comparable yields of MWNTs can be obtained with narrow and wide gaps. Composition and structure of a deposit is dependent on carbon vapor properties inside the gap, such as the pressure and temperature distribution, the presence and mobility of carbon particulates, and the dynamics of graphite vaporization and carbon vapor condensation. These are determined by a multitude of externally controlled factors, including He pressure, arc power, graphite anode qualities, and the arc experiment geometry.

TEM analyses enable selection of nanotube-rich specimens. These are, for example, represented by the entries 5 to 9 in Table 3.1. Three-dimensional (3D) organization of the deposit components cannot be revealed by TEM because the sample preparation completely destroys the deposit structure. Only secondary aggregation of components such as occasional side-by-side bundles of nanotubes is exposed by TEM. SEM is a much more informative technique for the examination of the internal deposit structure, the localization of the components, and exploration of the nanotube orientation. SEM studies have revealed that the soft black core of a deposit obtained under conditions favorable for high yield production of nanotubes is composed of locally parallel columns about 60 μm in diameter with separation between column axes of about 75 μm [1,3–10]. The column diameter and interaxis separation vary with the

TABLE 3.1   Parameters of the MWNT Arc Synthesis with 6-mm Diameter Anode Rods

N	He, Torr	I, A	Feed Rate, mm/min	Gap Width, mm	Voltage Across the Gap, V	Deposit Diameter, mm	Deposit Growth Rate, mg/min	MWNT Amount, %	MWNT Diameter, nm	MPP Amount, %	MPP Diameter, nm	Graphitic Particles, %	Fullerene Yield in Soot, %
1	100	80	4	3.9	18.0	7.8	115	5	8–20	25	—	70	17.9
2	200	65	4	2.2	19.0	6.5	170	10	8–22	40	20–85	50	18.0
3	700	55	4	0.5	20.8	7.4	260	20	8–23	45	20–90	35	5.5
4	500	65	8	0.4	21.3	8.0	350	20	5–29	30	19–93	50	5.8
5	500	65	2.5	3.5	22.3	7.6	90	30	5–46	60	9–93	10	14.3
6	1500	65	1.4	4.2	31.1	5.5	20	35	4–60	40	8–120	25	20.1
7	2000	65	1.4	3.3	32.1	5.5	22	30	4–60	40	8–120	30	20.3
8[a]	500	160	0.65	0.8	22.4	11.6	160	30	5–50	45	10–90	25	12.6
9[b]	500	132	1.0	1.7	22.2	10.9	195	30	5–50	40	15–80	30	10.1

Note: N.A. Kiselev et al. Carbon 37, 1093 (1999). With Permission.

[a] Anode rod diameter 10.6 mm.

[b] Rod 9.8 mm.

temperature of the cathode working surface and properties of arc plasma within ca. 50% of their value. A column axis is normal to the growth surface of the cathode. The surface can be concave or convex, but it is almost flat when the black core is most uniform and abundant. In the latter case, the column axes are parallel to the deposit axis all through the core, and the length of individual columns can reach a few centimeters. This occurs at sufficiently high helium pressures and low arc current densities, which are preferable for a higher yield of nanotubes. The columns terminate at the growth surface with nearly hemispherical caps arranged in a flat pattern of hexagonal symmetry. Much care should be taken while cleaving the core material and preparing samples for SEM observations to preserve the intrinsic nanotube alignment, if any, created in the course of the synthesis. Any sliding touch to the sample surface with an instrument causes such alignment, as nanotube attachment is very weak in all parts of the core.

SEM studies of the samples prepared in a way that does not disturb the intrinsic structure of the core components have produced the following data and concepts [1]. The hemispherical caps of the columns are densely covered with nanotubes that lay along the surface or at angles exceeding 45° from the column axis. Some tubes protrude from the upper layer into the free space, thus leading to the possibility of using a separated column as a small electron field emitter cathode. When embedded in the core, the columns serve as leads for electron current to the cathode surface and column tops as a principal place for emitting electrons into plasma [5]. The highest concentration of nanotubes is found in low-density intercolumnar spaces (denoted as zone 2, with the column itself as zone 1 in [5]) and on the side surface of a column. SEM examination of a cross-sectioned, separate column reveals an area of enhanced emission of secondary electrons along its circumference, thus supporting this observation. This circumferential area is composed of interlaced nanotubes forming an outer braid to the column. This braid is generally several microns thick, with no significant inclusions of MPPs or graphitic particles. The tubes in the braid are oriented preferentially normal to the tube axis, with statistically reliable observations showing that more than 50% of nanotubes are inclined at angles exceeding 45° relative to the deposit axis. The same is true of nanotubes in zone 2. The inner part of a column inside the braid consists of a disordered mixture of graphite-like particles, MPPs, and a small amount of nanotubes.

Such spatial distribution of the MWNTs in the columnar core suggests a method of tube purification [1]. Indeed, the braid can be stripped from the column surface by gentle mechanical shaking of the columnar material on a dense grid, with nanotube enriched powder, which also contains the intercolumn material, coming through the sieve. The braid and zone 2 materials can also be separated by soft ultrasonic treatment of the columnar core in a liquid, followed by decanting of the nanotube-enriched suspension. Both methods are simple and enrich MWNT contents by about 1.5 to 2 times. Thus, purified material is known as "SELECT" grade MWNTs, which is produced by the MER Corporation.

An indirect indication of the direction of nanotube orientation in the core can be derived from deposit magnetization anisotropy measurements, which reveal that the value of longitudinal diamagnetic susceptibility of the core material is about 10% higher than the value of the transverse susceptibility [6]. The experimental [11] and theoretical [12,13] consideration of the tube magnetization have shown that the transverse component of a tube susceptibility is due to the orbital diamagnetism of two-dimensional (2D) conduction carriers and is much larger than the longitudinal component, which is close to the atomic susceptibility value. Accordingly, the tubes are preferentially oriented normal to deposit axis, with $<\sin^2\alpha> = 0.76$, where α is the angle between the deposit axis and that of an individual nanotube [11]. Thus, the tubes are inclined to the deposit axis by about 60° on the average.

Another anisotropic contribution to the magnetization of the core material exists [14]. In this work it has been found that the magnetization curves $M(H)$ of the massive samples of the core material exhibit a pronounced hysteresis loop upon cycling the magnetic field H. The value of residual magnetization at zero field (M_r) depends on the sample orientation in the magnetometer cell and is four times higher when columns in the unperturbed sample are parallel to the field, compared with perpendicular orientation. The magnetic moment M_r remains in the sample upon turning off the field and approaches zero only after about 3 days at room temperature. At liquid helium temperature no decrease in M_r was observed for 20 hours. These observations correspond to the induction of persistent currents in the core material and thus trapping of magnetic flux in microscopic closed circuits that are likely to be formed from several nanotubes having good electrical contact with each other. The conductance of arc-produced MWNTs was shown to be very high [15], approaching the theoretical limit $2e^2/h = (12.9 \ k\Omega)^{-1}$ for ballistic transport of electrons along the nanotube, and is independent of tube diameter and length. The MWNTs can carry a current density above 10^7 A/cm^2 without the release of heat [15]. Thus, persistent currents in closed circuits composed of nanotubes seem quite probable. The observed anisotropy of M_r can be explained if the planes of these circuits are preferentially oriented normal to the deposit axis. Such orientation is quite consistent with the preferential orientation of individual nanotubes in the core that was described previously.

The long columnar structure of the core is formed only under steady arc plasma operation, which can be conveniently recognized by monitoring the arc current with an oscilloscope [6,16]. Instability of the plasma causes various disturbances of the columnar structure, including cross-sectional partitioning of the core by fused graphite layers. These partitions appear as an extension of the fused graphite layers that constitute the outer gray shell of a deposit [9]. A single layer of a partition is commonly about 10 μm thick, depending on the characteristic time of plasma instability. The single layer is formed when a sudden temperature drop and decrease of carbon vapor density occur at the core growth surface because of plasma fluctuations, such as total or partial expulsion of the plasma body from the gap. Such expulsion is fairly common for elevated current densities, high gas pressures, and narrow gaps. Adjacent single layers of fused graphite are spaced by ca. 15 μm, and this interlayer space contains MWNTs, MPPs, and graphitic particles that are sometimes arranged in short (ca. 15 μm) columnar structures. Therefore, nanotubes are found even in the deposits that appear to be entirely composed of densely stacked layers of fused graphite and can be obtained under unstable plasma conditions. Such deposits can be easily cleaved normal to the axis, and a single layer bearing a short columnar structure can be separated. Such specimens seem attractive for manufacturing electron field emission cathodes.

Uneven distribution of temperature and carbon vapor density and composition over the core area causes deterioration of columnar structure quality even with stable arc plasma. For example, much thicker columns can be formed in the central area of the core than on its periphery [6]. Higher temperature in the central area can also lead to sintering of nanotubes and other components together and enhanced graphitization of components. These processes result in hardening and gray coloration of the core material, inseparable columns, and eventually appearance of a strong fused graphite-like cylinder in the center of the core. The cylinder may be as small as 1 mm in diameter or may occupy nearly all the core area, and they extend for the whole deposit length with stable plasma. Further increase of the central area temperature that occurs at low helium pressures (25 to 100 torr) and narrow gaps leads to formation of an empty continuous central hole of 0.3 to 2 mm in diameter in the hard gray cylinder along the whole deposit length. Temperature is so high at the deposit axis that condensation of carbon vapor becomes impossible. In the cross section of such pipelike deposits, the hard gray cylinder is surrounded by a soft black ring containing columnar structures composed of MWNTs, MPPs, and graphitic microparticles. This ring is, in turn, surrounded by the gray outer shell. The MWNTs extracted from such deposits are of very high quality, because the structural defects are more effectively eliminated at higher synthesis process temperature [17].

Radial temperature distribution in the arc gap is not always monotonously descending from the center to the deposit circumference [18]. Under certain arc conditions, deposits can contain, along the whole deposit length, two concentric black cross-sectional rings separated by a gray ring. Clearly, axially symmetric standing waves of plasma temperature and density are established in this case. Their existence is also reflected in the complex shape of anode and cathode surfaces. It may be surmised that standing waves and the columnar structure itself bear some relationship to the so-called dusty plasma systems.

The core structure described above pertains to the stable arc process. Transition into an unstable plasma regime may result from a slight change of the externally controlled process parameters, eventually leading to disappearance of the columnar structure and generally lower nanotube yields. It should be noted, however, that the mere existence of a columnar structure does not guarantee high tube yield. The columnar structure coincides with high yields in most cases but can also exist in cores with low nanotube yields [1].

Under optimal conditions the deposit outer shell consists of curved stacks of graphite layers. Local orientation of the curved graphite layer is nearly normal to the deposit axis on the inner side of the shell, whereas close to the outer side it becomes nearly parallel to the axis. Formation of a layer occurs at 1700 to 2000°C by direct deposition of carbon from the arc plasma on the exposed surface of the previous layer [9]. The low carbon vapor density in the gap above the shell does not allow formation of MWNTs in this area [9].

3.2.3   Growth Mechanism

From the mode of nanotube and other component packing in the core described above, some refinement of the model [5] offered to describe the growth of the deposit in a direct current arc seems necessary [1]. In this model the tips of parallel nanotubes in zone 1 (column top) act as field emitters of electrons into the plasma. This electron injection produces a high degree of carbon ionization resulting in concentration of carbon ion current flow above the columns. This flow provides the principal feedstock for column growth. The helium buffer gas is drawn in by the carbon ion flux to the top of the columns, is swept to the side, and then returns back to the plasma over zone 2, which is the intermediate area between adjacent columns. The refinement of this model is as follows and is illustrated by Figure 3.1. In the absence of a large amount of vertically oriented nanotubes, the emission of electrons from zone 1 is likely to be mainly thermionic in nature and occurs from a variety of constituents of the columns. The dominating abundance of horizontally packed nanotubes over the column top surface probably indicates that that the electric field effect at the open end of a nanotube is not a governing factor for its growth. Those nanotubes that were originally located near the very hot top of a hemisphere could then undergo evaporation, yielding neutral carbon particles. An estimate has shown [5] that the majority of carbon precipitated on columns must evaporate back as neutrals. Evaporation enriches the central part of a column in MPPs and graphitic particles to the extent eventually observed with the SEM, as these non-tube components are thermodynamically more stable than nanotubes [19]. Another reason for the enrichment of the central part of a column with graphitic particles is that they probably carry large positive charges while moving toward the cathode [20] and are strongly attracted by the negatively charged volume over the column top. Helium transports re-evaporated neutral carbon molecules to zone 2, where they serve as a principal feedstock for the growth of new nanotubes [5].

Vaporization of carbon from the anode was the subject of many detailed studies [20]. Under conditions of arc performance useful for fullerene and nanotube production, the most probable scenario for this process is illustrated in Figure 3.1. Carbon is vaporized from the porous surface layer of a polycrystalline graphite anode mainly in the form of graphite crystallites exceeding 3 nm in linear size. Small crystallites serve as the main positive charge carrier in the carbon plasma, as their work function is much lower than the ionization potential of small carbon molecules, which are the minor carbon ingredient. Crystallites undergo vaporization on the way to the cathode, generating small carbon clusters (mainly C_3) and diminishing in size. It takes about 1 ms for carbon species to cross a few millimeters wide arc gap, and during this time the smaller crystallites are fully vaporized, whereas larger crystallites reach the cathode surface and thus are present in the deposit. Structural peculiarities of the graphitic particles found in the deposit may be brought about by delamination of original crystallites and other thermal transformation processes. On the average, the deposit material is enriched in isotope ^{12}C relative to the graphite anode [21]. Graphitic particles thus formed cannot contribute to the shift in isotope distribution, leaving MWNTs and MPPs solely responsible for the shift observed. This strongly suggests that the MWNTs and MPPs have been built from the low-molecular carbon species, mainly from C_3, C_2, and C_1 carbon clusters.

At present it is well substantiated that fullerenes and MWNTs formed from the carbon vapor use these small clusters C_n ($n = 1 - 3$) only as an initial feedstock. The presence of "useless" graphite crystallites in the gas phase seems to be the main limiting factor for high fullerene and MWNT yields under the given conditions. Under experimental conditions that convert more crystallites into small clusters and reduce crystallite production at graphite vaporization, larger fullerene and MWNT yields are obtained. Optimization of fullerene and MWNT production processes for high yields should involve, as a necessary

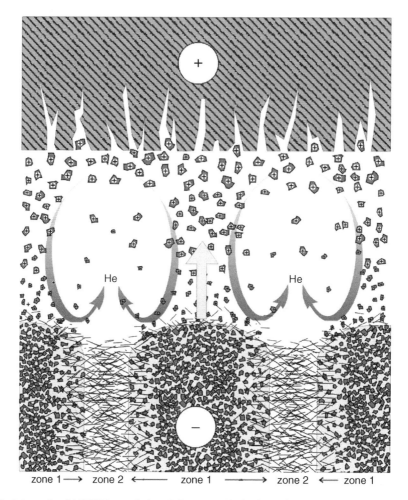

zone 1 ⟶   zone 2 ⟵   zone 1   ⟶ zone 2 ⟵ zone 1

FIGURE 3.1 Schematic of MWNT growth in a DC arc. Positively charged graphitic splinters spalled from the anode partially vaporize on the way to the tops of the columns (zone 1). Intense carbon flux above the columns is due to enhanced thermionic emission from zone 1, which produces a high degree of carbon vapor ionization. Carbon density above the columns is further increased by circular convection of helium. Nanotubes grow in dense gas near the cathode surface from low-molecular carbon species and precipitate in zone 1 and zone 2 with their axes oriented preferentially parallel to the surface. This orientation is held constant in as-produced deposit core in both zone 2 and the nanotube braid of the column [1].

step, the establishment of such experimental conditions. The highest fullerene yield obtained from arc processes is 20 to 25% [22], whereas the yield from laser ablation is 30 to 40% [19]. Better "atomization" of graphite with the laser ablation process contributes strongly to this difference. Generally, the more graphitic particles in the core material, the fewer MWNTs are observed (Table 3.1). Graphitic particles are also found in the fullerene containing soot deposited on the chamber surface. The yield of fullerenes in the soot also tends to anticorrelate with graphitic particle content (Table 3.1), although this dependence is not as pronounced because the formation of fullerenes is mainly governed by other kinetic factors [22].

In the vicinity of the cathode surface, a relatively high carbon-vapor density is built up because the carbon hitting the ca. 3200°C [8] hot surface cannot be totally accommodated in the solid, and thus low-molecular-weight carbon clusters are easily re-evaporated [5,9,23]. The MWNTs are self-assembled from small carbon clusters in this gas layer and then migrate to the cathode surface with the flux of charged and neutral carbon particles and deposit on it preferentially parallel to the surface, an orientation that ensures stronger adhesion of tubes to core components. An adequately high carbon-vapor density in the

FIGURE 3.2 Industrial arc-discharge reactor for producing fullerenes and carbon nanotubes. Features vaporization of up to 1 kg of anode per hour [26].

layer is critical for tube self-assembly to occur [9,24]. Below a certain threshold value of vapor density, the MWNTs are not able to self-assemble, and the only mode of carbon vapor deposition then is formation of two-dimensional graphene structures on vapor-exposed surfaces, which generally takes place in the deposit shell region. This vapor density threshold is very narrow, as indicated by the commonly observed very sharp transition between the core and the shell structures and by independent orientation of the core columns and the shell layers. This transition occurs at around 2000°C, which is the temperature of the inner surface of the shell [9]. The necessity of sufficiently high carbon vapor pressure for MWNT growth was established also in experiments using laser evaporation of graphite. They have shown that the yield of MWNTs increases with better concentration of the laser beam, and that synthesis is possible only with the reaction chamber temperature elevated to at least 1200°C [24]. In the arc under conditions of low anode evaporation rate, the yield of MWNTs also falls drastically [9]. Finding MWNTs embedded in the fullerene containing soot deposited on chamber walls further emphasizes that they are self-assembled in the gas phase. Although usually very rare, such occasions become more common with very wide gaps when the narrow gas layer adjacent to the cathode is more efficiently swept into the chamber volume. Installing an additional negatively biased electrode near the gap also facilitates extraction of MWNTs from the gap [25].

The nanotube-rich columnar deposit formation is a steady-state realization of a nonlinear multi-parameter kinetic process. Therefore, the nanotube growth optimization strategy should address the following factors. All the parameters of the arc process are interdependent, and an attempt to adjust only one of them will lead to inevitable variation of others. Different sets of externally controllable parameters can give similar tube content. It is not productive to optimize tube growth under conditions of unstable arc operation, and it is of prime importance to first identify the stable regions among the set of possible experimental parameters. The conditions listed in Table 3.1 illustrate only a few, though important, stable solutions to the problem of optimal nanotube production [1]. However, variations in materials and equipment may necessitate adjustments of the given conditions to obtain reproducible results.

3.2.4 Scaled-Up MWNT Production Process

A further increase in MWNT production rate was expected by employing thicker graphite rods in larger, more powerful arc reactors. This expectation was realized with the construction of a large-scale arc reactor with a 350 L chamber (Figure 3.2) capable of vaporizing up to 75-mm diameter graphite rods [26]. The

FIGURE 3.3 Cathode deposits produced with industrial apparatus from 65- and 25-mm diameter graphite anodes (two rear deposits) and with laboratory scale arc discharge installation from 8-mm diameter anode (front deposit) during ~ 1 hour runs [26].

cylindrical cathode deposits obtained with this scaled-up process (Figure 3.3) contain soft core within the outer hard shell. Under optimized conditions the core material has the columnar structure. The column diameter and separation are about the same as in small-diameter deposits. The column length in large deposits is commonly much lower (ca. 3 to 5 mm) because of lower plasma stability in the arc gap. Optimal conditions for MWNT production occur in the same range of helium pressure and arc gap widths but at substantially lower current densities compared with laboratory-scale anodes. More efficient thermal insulation of the inner part of the electrodes by the outer shell layer is responsible for lower energy requirements for vaporizing the anode and maintaining sufficiently high temperature and carbon vapor density over the core region. This insulation also results in a more concave cathode growth surface and more convex anode erosion surface than in laboratory scale electrodes. With such gap geometry, a steeper radial fall of temperature and carbon vapor density takes place, resulting in a less-even radial distribution of deposit core components compared with thin electrodes. Nevertheless, the average amount of MWNTs in the core is similar to that in thin cathode deposits, although under the current best conditions it is still 20 to 30% lower. Further optimization of the large-scale process is likely to reduce or eliminate this difference, as no principal limitations have been revealed. At present, the kilogram per run productivity of the large-scale process compensates economically for the somewhat lower MWNT content of the product. The concomitantly reduced cost of arc-derived product stimulates its use in applications that require MWNTs of the highest structural quality. Arc-produced MWNTs are perfectly straight and have much fewer defects than those obtained by other production techniques and thus have higher mechanical strength and thermal and electrical conductivity. The Young's modulus of MWNTs is calculated to be ca. 1000 to 1500 GPa and has been measured at 1000 to 3700 GPa. The measured tensile strength is 11 to 63 GPa for the MWNT outer layer. The thermal conductivity along the length of the tube is ca. 1500 W/m K, and the resistivity is ca. 0.1 $\mu\Omega$ cm. The market niche for arc-produced MWNTs will persist until a method of producing better quality MWNTs appears.

A further increase in vaporized rod diameter and in arc power is technically feasible but does not appear to be economically justified [26]. The large-scale reactor described above has been demonstrated to be close to the optimum size and power for MWNT synthesis in the arc and will probably remain the largest reactor in the world for years to come. In the arc process, the next logical step for increasing MWNT output will be automation of large-scale reactors, which will render the production process nearly continuous.

3.3   MWNT Production by Laser Ablation of Graphite

During fullerene production experiments using a laser vaporization apparatus with an ablated graphite sample positioned in an oven, it was found that closed-ended MWNTs were produced in the gas phase through homogeneous carbon-vapor condensation in a hot argon atmosphere [24]. The laser-produced MWNTs are relatively short (ca. 300 nm), although the number of layers, ranging between 4 and 24, and the inner diameter, varying between 1.5 and 3.5 nm, are similar to those of arc-produced MWNTs. A prevalence of MWNTs with an even number of layers was observed, but no SWNTs were detected. Fullerenes, MPPs, and amorphous carbon are coproduced. The yield and quality of MWNTs declined at oven temperatures below 1200°C, and at 200°C no nanotubes were observed. A similar trend in fullerene yield was explained by the need for sufficiently high temperature to rapidly anneal the imperfect fullerene structures into a closed form through the incorporation and rearrangement of pentagons [27]. It was therefore puzzling how an open nanotube would grow long rather than close off into a short spheroidal structure at 1200°C and in the absence of extrinsic stabilizing factors such as the high electric fields proposed earlier [2,19]. It was suggested [24] that stabilization of the open-end interaction between adjacent tube layers is provided by carbon "adatom" bridges that saturate some of the dangling bonds [28]. Such cohesive lip-lip interactions may explain the tendency of MWNT layers to close in pairs, as well as occasionally observed [29] "turnaround" growth in arc-produced MWNTs. To keep the growing MWNT end open, it is essential to have lip-lip interaction between at least two inner layers. The next layers may grow with some delay, for example, by island nucleation [29] and anneal on the underlying nanotube template. Nanotube thickening by additional layer growth terminates when carbon-vapor density falls below a threshold value corresponding to the saturated vapor pressure of small carbon clusters adsorbed on the tube surface. This threshold value decreases with increasing tube radius and thus may impose a limit for MWNT diameter [24].

The fact that MWNT growth in homogeneous carbon-vapor condensation has been observed implies that electrical field stabilization of open ends [2,19] and the contribution of the flow of directed carbon species to the growth of nanotubes [23] are not necessary for MWNT growth.

3.4   Arc Discharge Production of SWNTs

SWNTs are produced in an arc process utilizing covaporization of graphite and metal in a composite anode [30,31] commonly made by drilling an axial hole in the graphite rod and densely packing it with a mixture of metal and graphite powders. Various pure elements and mixtures have been used to fill the rod, including Fe, Co, Ni, Cr, Mn, Cu, Pd, Pt, Ag, W, Ti, Hf, La, Ce, Pr, Nd, Tb, Dy, Ho, Er, Y, Lu, Gd, Li, B, Si, S, Se, Zn, Sn, Te, Bi, Cd, Ge, Sb, Pb, Al, In, Fe/Co, Fe/Ni, Fe/Co/Ni, Co/Ni, Co/Pt, Co/Cu, Co/Bi, Co/Pb, Co/Ru, Co/Y, Ni/Y, Ni/La, Ni/Lu, Ni/B, Ni/Mg, Ni/Cu, Ni/Ti [32–46], but at present only Ni/Y and Co/Ni catalysts are commonly used in SWNT production.

These preferred systems share many common features. Both perform better in helium atmosphere than in argon, at about 0.6 bar pressure. The process is most efficient with a stable arc discharge and a constant anode erosion rate, corresponding to ca. 2 A/mm^2 current density and ca. 3-mm gap width. This is better achieved by maintaining a constant arc current and anode feed rate and thus constant arc gap width.

A cylindrical deposit grows at the surface of the cathode. The weight of the deposit constitutes about one half of the weight of the anode consumed in the process. The deposit consists of a hard gray shell and a soft core. The core has poorly developed columnar structure and contains MWNTs, MPPs, and graphitic particles. In the Co/Ni system, commonly operating with ca. 2 at% metals in the anode, the core is diamagnetic and does not contain any metal components. High temperature prevents precipitation of metal vapor on the growth surface, and all metal escapes the gap. However, the outermost layer of the gray shell contains trace amounts of metals, introduced by gas convection onto the exposed side surface. In many other systems, especially with high metal loading of the anode, the deposit may contain metal-filled MWNTs and MPPs or bare spherical metallic particles [47]. In the Ni/Y system the deposit contains large amounts of yttrium [48], whereas metal nanoparticles found in the chamber are enriched in nickel [46,48,49].

The mixed carbon and metal vapor, which has escaped the gap, then condenses into the product, which moves to the reactor surfaces and deposits on them. The product is divided into three distinct structural types, depending on the deposition area. A spongy soft belt called *collaret* is formed around the cylindrical deposit and represents about 20% of the product weight. Relatively strong clothlike soot on the chamber walls represents another 70%, and the remaining 10% of the product is a weblike structure suspended in the chamber volume between cathode and walls. All three types contain varying amount of SWNTs, fullerenes, amorphous carbon, empty and metal-filled MPPs, *naked* metal particles, and graphitic nano- and micro-particles. Thermogravimetric analysis (TGA) and near-infrared (NIR) spectroscopy [50] are used to accurately determine the metal and SWNT content in the arc material, whereas electron microscopy (TEM and SEM) and Raman spectroscopy (RS) generally provide only rough assessments of the concentration of product components. These methods appear to be the most useful for analyzing arc-product composition and structure. The collaret contains more SWNTs and metal particles than other components of the product. This difference is about 20% for Co/Ni and 50% for Ni/Y systems. The collaret is formed from an electrostatically deflected flow of soot that is propagating away from the gap. SWNT growth continues in the material that is deflected and captured on the deposit surface for a longer time, leading to enrichment of the collaret in SWNTs. The SWNT yield averaged among all three structural types of the product is about the same in these systems, roughly ca. $20 \pm 5\%$ of the total product weight. Fullerenes C_{60} and C_{70} contribute ca. 5 to 10 wt% to the total product weight in the Co/Ni system and ca. 1% in the Ni/Y system. The average mass fraction of the metal catalyst is ca. 20% of the total product weight in the Co/Ni system and 35% in the Ni/Y system. Amorphous carbon constitutes ca. 50 wt% of the total product weight and graphitic particles ca. 5 wt% in both cases.

SWNTs are generally organized in bundles consisting of a few dozen tubes, tightly compounded in a honeycomb lattice with an average separation between tube axes of ca. 1.7 nm. Rare isolated tubes can be observed. The bundles are covered with an amorphous carbon layer ca. 2 to 5 nm thick, which contains embedded fullerenes. The majority of tubes have diameters in the range of 1.2 to 1.5 nm and lengths reaching up to 5 μm in the Ni/Y system and 20 μm in the Co/Ni system. The tips of individual tubes are closed, in the rare cases when they can be observed. Some bundles terminate with a spherical metal particle of 10 to 30 nm in diameter, which exceeds the bundle diameter. Such termination is characteristic of Co/Ni [51] (Figure 3.4), Ni [70], Ni/La [51], and Ni/Y [49] arc systems, as well as of a nickel-catalyst–based laser vaporization system [52].

SWNT diameter depends on the temperature of the catalytic site at which growth occurs. This temperature is regulated by many factors, including heating of the reaction zone with an externally controlled heat source. The constant temperature thus set by external heating (environment temperature) provides a minimum temperature for the reaction site. It was established that the mean diameter of SWNTs increases with the environment temperature [53,54]. Other factors affecting reaction-site temperature cause the same kind of dependence, namely, production of thicker SWNTs is favored at higher temperatures [49,55–59]. The environment temperature also affects the yield of SWNTs. Higher SWNT production rate at elevated environment temperatures is generally observed both with arc and laser techniques [53,54,60].

3.4.1 Metal Catalyst Particles

The efficiency of SWNT synthesis in a carbon/metal arc is determined primarily by the choice of metal catalyst for the process. It was found [33,34] that catalysts composed of two metals produce much higher yields of SWNTs than did individual metals. The ratio of metals in the original catalyst mixture also produces a drastic effect on SWNT yield. In the Co/Ni catalyst arc system the highest yield is obtained with Co:Ni = 3:1 ratio in the original mixture (Figure 3.5). For the Ni/Y catalyst the optimum molar ratio of the original mixture was determined to be Ni:Y = 4.2:1 [46]. The predominant diameter of metal particles in the arc SWNT product is 10 to 30 nm, and the metal ratio in these particles differs from that

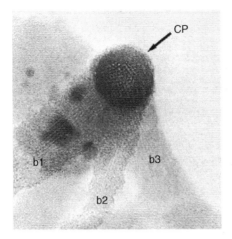
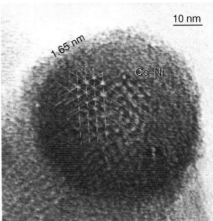

FIGURE 3.4 TEM image of 3Co/Ni catalytic metal alloy particle (CP) with three SWNT bundles (b1, b2, b3) grown from it (left) and magnified view of the same particle showing a bundle root with about 20 active sites arranged in a triangular pattern on the surface from which SWNTs of the bundle b1 originate (right). The 1.65-nm separation of active sites reveals dense grouping of nanotube roots [51].

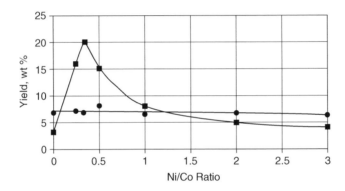

FIGURE 3.5 Dependence of the yields of SWNTs (squares) and fullerenes (circles) on the metal composition of the catalyst with the total metal content held constant at Ni + Co = 2.3 at.% in the 8-mm diameter anode vaporized in the arc-discharge process under optimized conditions [61].

in the original metal mixture. The energy dispersive x-ray fluorescence (EDX) analysis has shown that the metal composition of different particles varies within 30% from the 3:1 ratio of cobalt to nickel in the original mixture [61]. The same-size metal catalyst particles in the Ni/Y system are essentially depleted of yttrium [46,49], regardless of the Ni:Y composition of the original catalyst mixture [49]. This implies that there are additional functions for yttrium in the arc synthesis, besides forming catalytic particles of the optimum composition. The presence of yttrium may facilitate an adjustment of the optimal process temperature through the highly exothermic reaction of yttrium carbide formation. This reaction brings about higher catalytic particle temperature, which results in a generally observed 0.1- to 0.2-nm shift to larger average tube diameter, compared to that produced with yttrium-free catalysts under otherwise similar conditions. The 3Co/Ni mixture produces only long SWNT bundles with 1.2- to 1.3-nm diameter nanotubes. The Ni/Y system, in addition to long bundles, under certain conditions can generate [49] short bundles (~ 50 nm in length) composed of thicker SWNTs, with an average nanotube diameter of 1.8 nm, radiating from metal catalyst particles (sea urchin structures). Long bundles of narrow tubes generally grow from particles containing less yttrium (below 15%) than metal particles in sea urchin structures (over 11% yttrium) [49]. These facts may imply that sea urchin structure tubes grow from hotter metal particles and that the presence of yttrium in the metal particle retards the start of tube growth until a higher carbon supersaturation is attained than for a pure nickel particle. As a result, sea urchin tube growth starts simultaneously over the whole metal particle surface. The sea urchin metal particle cools down very quickly to temperatures at which dissolution of carbon becomes too slow to sustain steady tube growth. In addition, the bundle roots covering the whole metal surface prevent the outside carbon feedstock from coming in contact with the metal particle surface. When the carbon inside the particle is exhausted, growth is terminated, resulting in short bundles. The amount of carbon in these bundles is commensurate with that dissolved in the metal particle at the start of growth.

3.4.2   Dynamics of SWNT Growth in the Arc Process

Carbon vapor flowing from a sufficiently narrow arc gap can be idealized as a turbulent jet of cylindrical symmetry [22,62] and described in the framework of a semiempirical theory [63] of heat and mass transfer in a free turbulent jet. These transfer phenomena control the dynamics of carbon vapor mixing with helium gas and the resulting cooling. In the framework of the turbulent jet model, an analytical relationship between the essential parameters of the arc process is valid [22]. These parameters include the volume rate of carbon vapor flow from the gap V_{soot}, the carbon vapor temperature T_o and velocity U_o at the cylindrical boundary of the gap, the helium pressure in the reactor P, the gap width h_o and electrode diameter $2r_o$, and the characteristic time for turbulent mixing and cooling of carbon vapor τ_{mix}. The value of τ_{mix} is connected to other parameters by Equation (3.1):

$$\tau_{mix} = r_o^{1.5}/U_o h_o^{0.5} = 2\pi r_o^{2.5} h_o^{0.5} P/V_{soot} R T_o. \qquad (3.1)$$

The rate of carbon vapor cooling in a turbulent jet outside the gap is directly proportional to the linear rate of carbon vapor propagation U_o and thus to the volume rate of vapor generation V_{soot} and to the inverse value of the chamber pressure P. Temperature T_o can be either determined experimentally or evaluated under the assumption that the equilibrium carbon/metal vapor pressure in the gap is equal to the ambient gas pressure in the chamber. The volume of carbon/metal vapor produced should allow for vapor composition. It should be noted that the true value of the cooling rate in a real arc system may differ from that calculated with Equation (3.1), because the model considers only turbulent cooling and because of uncertainties in the values of the parameters involved. In this respect, it is noteworthy that laminar diffusion of cold environment helium into the hot zone, which also includes the gap area, can substantially contribute to the rate of reagent cooling [22]. This contribution becomes stronger with wide gaps [22]. However, the laminar diffusion cooling rate has similar pressure and temperature dependence to that of the turbulent case, so in the first approximation this should change only the proportion between the characteristic cooling time and the other process parameters in Equation (3.1), without disturbing the underlying dependence. Assessments show that τ_{mix} value, as determined by Equation (3.1) functional dependence, can be treated as having an accuracy better than a factor of 3 in the analysis of arc synthesis performance in the whole range of arc process conditions.

The parametric study of SWNT yield in the 3Co/Ni arc system was performed in a wide range of helium pressures, arc currents, and FRs, and the data were evaluated in relation to the turbulent jet model [61]. The gap width and rate of vapor generation were measured in each run, and vapor temperature inside the gap was assessed as previously to calculate the τ_{mix} value.

The dependence of the SWNT yield on the τ_{mix} value is presented on Figure 3.6. All SWNT yield data points are grouped around a common curve peaking at ca. 6 ms. This means that τ_{mix} is the sole parameter determining the yield of SWNTs. The largest SWNT yields occurring near $\tau_{mix} = 6$ ms correspond to fairly different sets of helium pressures, arc currents, and gap widths. This implies that no unique set of externally controlled parameters exists that would produce the highest SWNT yield for a given system. The parameter to optimize the process for the highest SWNT yield is τ_{mix}. SWNT production in the arc process is totally controlled by the cooling rate of carbon vapor escaped from the gap, and formation of nanotubes occurs at a distance of a few centimeters from the arc gap. The cooling time $\tau_{mix} = 6$ ms, which

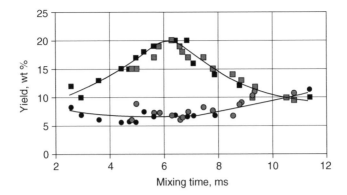

FIGURE 3.6 Dependence of the yields of SWNTs (squares) and fullerenes (circles) on the characteristic time of turbulent mixing of the hot carbon vapor with helium gas in the fan jet flow moving from the interelectrode gap. Data points correspond to a wide variation of externally controlled arc discharge process parameters (helium pressure, arc current, and electrode feed rate) [61].

is optimal for SWNT production, gives characteristic time of the SWNT growth process. Assuming 20 μm as the average SWNT length, the linear growth rate of SWNTs is ca. 3 mm/s. This value is many orders of magnitude higher than the linear growth rate of MWNTs formed on metal catalyst particles at lower temperatures, which implies that diffusion of carbon through the metal particle is the rate limiting step in catalytic nanotube growth.

A similar dependence of fullerene yield on τ_{mix} was observed for fullerene synthesis with a metal-free graphite anode [22]. The optimum value of τ_{mix} for high fullerene yield was determined to be ca. 0.2 ms, at which point the fullerene yield peaked at ca. 23%. Therefore, the characteristic time of fullerene formation in the arc process is much shorter than that of SWNT formation. The carbon component of the mixed carbon/metal vapor totally condenses into fullerenes and soot particles much before the growth of SWNTs is initiated. Fullerenes and soot particles are the only carbon feedstock available for SWNT growth. No low-molecular carbon species — including such components of hot carbon vapor as C_3 and C_2 molecules or other C_n clusters in the form of chains, rings etc. — are available in the SWNT growth zone. Soot particles are used preferentially to fullerenes C_{60} and C_{70} for the carbon feedstock. This conclusion is supported by the following data. Figure 3.6 shows the yield of fullerenes C_{60} and C_{70} in relation to τ_{mix}. The fullerene yield in the total collected soot product is nearly constant under widely varied conditions in the 3Co/Ni system and equal to ca. 7 wt%. In the control runs on fullerene, production performed under identical conditions, but with a metal-free anode the yield of fullerenes was within 1 wt% standard deviation, the same as in the corresponding 3Co/Ni runs. The SWNT production in the 3Co/Ni arc system is independent of the presence of fullerenes. The molar ratio C_{60}/C_{70} for fullerenes produced in carbon arc in helium is perfectly constant and equal to 5.0, independent of arc conditions [64]. The same constancy and value of the C_{60}/C_{70} ratio was determined for SWNT runs presented in Figure 3.6. This provides further evidence that fullerene and SWNT formation processes are independent. Amorphous carbon is ca. 150 kJ/mol less stable than fullerenes [65], thus providing much higher thermodynamic driving force to the process of SWNT growth. Therefore, the observed preference for soot particles over fullerenes for the SWNT precursor in the arc synthesis has a strong thermodynamic basis.

3.4.3 Carbon "Dissolution-Precipitation" Model

The carbon "dissolution-precipitation" (DP) kinetic model developed a long time ago for the growth of carbon filaments and MWNTs on bulk metal nanoparticles [66–69] was subsequently adapted with some modifications for SWNT growth from similarly massive catalytic nanoparticles in the arc discharge and laser ablation processes [49, 51,70–74]. The model includes three consecutive steps: dissolution of carbon coming into contact with the bare metal surface, transfer of dissolved carbon to another location on the particle surface, and precipitation of carbon as nanotubes at this location. Dissolution of amorphous carbon into metals of the iron group is a highly exothermic process. It keeps the particle hot enough to produce SWNTs in the tube formation zone a few centimeters away from the arc, where otherwise the particle temperature would be too low. The round shape of particles found in the product attests to their molten state during the process. Particle melting is facilitated by the incorporation of carbon into metal. The temperature of the carbon-containing, molten catalytic particle in the SWNT synthesis zone is assumed to be near 1300°C, the eutectic temperature of the metal-carbon mixture [49,51,75,76]. Carbon solubility in the particle at SWNT synthesis temperature can be estimated as 2 to 3 wt%, using the binary M-C phase diagrams of cobalt and nickel. Small supersaturated catalytic particles may contain up to four times as much carbon [75,77].

Carbon diffusion through the catalytic metal particle has been shown to be the rate-limiting stage of the overall DP process in the case of vapor-grown carbon fiber (VGCF) and MWNT synthesis by catalytic pyrolysis of hydrocarbons [66–69]. In the framework of the DP model, the rate of SWNT growth can be estimated for arc synthesis as well. The carbon diffusion flux through the catalytic nickel particle at 1300°C can be evaluated as $Q = D \cdot \mathrm{grad}\, C \sim 2 \cdot 10^{-6}$ cm^2/s $\cdot 1 \cdot 10^5$ g/cm$^4 \sim 0.2$ g/cm^{2}s. The carbon diffusivity of $2 \cdot 10^{-6}$ cm^2/s was obtained from the temperature dependence $D = 0.1 \exp (-140\ \mathrm{kJ} \cdot \mathrm{mol}^{-1}/RT)$ for nickel [78]. The carbon concentration gradient $1 \cdot 10^5$ g/cm^4 was estimated using the assumption that carbon concentration at the dissolution spot exceeds that at the precipitation spot by ~ 0.2 g/cm^3, which is about the value of carbon solubility, and the distance between these spots is 20 nm. Division of carbon flux value 0.2 g/cm^2 s by the gravimetric density of the SWNT bundle (~ 1 g/cm^3) yields 0.2 cm/s for the value of the SWNT linear growth rate. This value (0.2 cm/s) practically coincides with the value for SWNT linear growth rate of 3 mm/s independently assessed in Section 3.4.2. This coincidence implies that carbon diffusion through the catalyst nanoparticle is the limiting stage of the overall SWNT production process.

Given the carbon flux of ~ 0.2 g/cm^2 s through the typical size 20-nm particle, and the characteristic SWNT growth time of 6 ms, it is easy to calculate that the fraction of particles active in SWNT production should be only about 1% of the total particles. This is close to what is observed by TEM in the arc synthesis products.

Precipitation of the dissolved carbon as SWNTs is an endothermic process (by ca. 40 kJ/mol), and the particle surface in the precipitation spot is somewhat colder than in the dissolution spot. The temperature gradient in nanometer-size metal particles is too small to cause the directional diffusion of carbon, and the carbon concentration gradient is the sole reason for diffusion in the catalyst particle [68]. Precipitation starts with nucleation of the SWNT caps.

Carbon atoms, brought by diffusion from inside the liquid particle to the surface, start to arrange into a small carbon network containing a few hexagons and pentagons. When this network grows to the diameter of a nanotube and incorporates six pentagons, it has a more or less hemispherical structure with carbon atoms on the circumference of the hemisphere chemically bonded to the surface atoms, which can be either metal or carbon atoms. Other carbon atoms of the hemisphere form a carbon network peeled away from the surface. This scenario is very close to the construction of a SWNT cap by a computer through the molecular dynamics modeling of the segregation of carbon from hot metal-carbon alloy [49]. The SWNT cap diameter is energetically "locked up" at this point, and further addition of carbon atoms to the hemisphere root result only in formation of hexagons in the wall of the growing SWNT, independent of slight changes of synthesis conditions at the root area.

At a higher temperature, a larger number of hexagons are included in the hemisphere structure together with the six pentagons, because the more intense supply of carbon atoms reduces the probability of pentagon formation. Thicker SWNTs therefore emerge at higher temperatures.

Mutual disposition of the six pentagons on the hemisphere determines the tube chirality and is unlikely to easily change after the cap is formed, for the migration of carbon atoms in the isolated hemisphere and Stone-Wales–type rearrangements necessary for this change are essentially suppressed at 1300°C.

Growth of all SWNTs in a bundle starts simultaneously [49] from an active spot on the particle surface. The local temperature of multiple SWNT nucleation sites belonging to an active spot does not vary much, and therefore one-spot bundles consist of nanotubes that are very close in diameter [46,60,79,80] with a strong tendency to contain tubes of uniform chirality [60,79,80]. Several small neighboring spots can coalesce, giving rise to a combined bundle composed of discernible narrow bundles [46,73]. For a substantial portion of their length, bundles originating from different particles can also combine into a thick bundle while drifting in the gas phase or during ultrasonic treatment of the SWNT product in a liquid. Combined bundles of both types can contain SWNTs of substantially different diameters and chiralities [81–88].

The above-proposed molecular model for nanotube growth from a molten metal/carbon particle explains origin of the preferred chirality in tubes. Insertion of carbon atoms into the M-C bonds at the nanotube root cannot proceed without the relative displacement of metal atoms tangentially along the tube circumference, when the tube is chiral. With achiral tubes this displacement can be smaller, or negligible for armchair tubes with a certain type of M-C coordination bonding. For a larger helical pitch of chiral tubes this displacement is larger. The displacement should generate torque that will cause mutual rotation of the tube and the ensemble of metal atoms bound to the tube root, about the tube axis. The rotation will be hindered because of the high viscosity of molten metal and van der Waals interactions among the tubes in the bundle. At a certain bundle lengths the van der Waals retardation of tube rotation prevails over circular metal ensemble drag friction, and mutual rotation of tubes in the bundle will cease. Activation energy of circular metal ensemble motion in the surrounding liquid can exceed or even far surpass that of achiral (armchair) tube growth, thus slowing down or preventing further growth of chiral tubes. As a result, the armchair tubes grow in preference to zigzag and chiral tubes, especially to those chiral tubes that have large helical pitch. This preference has been observed in many studies and is commonly attributed to higher thermodynamic stability of the armchair tubes, particularly the (10,10) tube. The kinetic mechanism for tube chirality selection outlined above finds support in several observations. The core of a bundle is likely composed of only armchair tubes, whereas the outer shell of a bundle consists of chiral tubes [79–88]. This is expected with the kinetic mechanism, as outer tubes in the bundle have more opportunities for rotation due to weakened van der Waals interactions. Also anticipated is more frequent occurrence of nonzero chirality in tubes that are seen lying apart from bundles [79–88], as these tubes most likely have been individually grown or detached from the outer layer of a bundle during sample preparation. In either case, nonzero chirality is presumed by the kinetic selection mechanism. It also predicts that a higher metal particle temperature would favor formation of chiral tubes and larger helical pitch, whereas lower temperature should facilitate armchair tube dominance in the product. Some support for these expectations can be found in the literature; however, more research is required to verify the predictions.

For steady tube growth conditions, the rates of all three consecutive stages of the DP kinetic scheme (dissolution, diffusion, and precipitation) are equal. This rate balance results in production of long tubes and is possible only when particle temperatures are in an appropriately high range and certain other conditions are met. The rate of carbon dissolution should not exceed the maximum achievable carbon diffusion rate in the particle. The maximum achievable rate of carbon precipitation should be higher than the diffusion rate. Failure to observe either of these relationships may, under some circumstances, result in fast termination of tube growth. The particle carbon concentration will increase when either or both of the relationships are violated, and high supersaturation may be reached. This may lead to a sudden precipitation of a graphitic shell, which will eventually prevent further dissolution of external carbon into the particle. The concomitant temperature drop and exhaustion of dissolved carbon lead to fast cessation of the growth process. This temperature drop is likely to occur with acceleration and irreversibly, which is similar to the temperature rise in a thermal explosion; therefore, the process in the particle can be referred to as "cold explosion." This term also presumes the catalytic particle to be a highly nonequilibrium thermal system, with particle temperature substantially exceeding that of the environment.

3.4.4   Scaled-Up SWNT Production Process

The dynamic model of the carbon arc (Section 3.4.1) implies no principal limitations for obtaining high SWNT yields with much larger-diameter graphite/metal anodes than are commonly used in laboratory scale practice. In essence, it predicts that a high SWNT yield would be retained if the value of characteristic cooling time τ_{mix} is maintained at ca. 6 ms while using a larger rod diameter $2r_o$. According to expression (1) for τ_{mix}, to compensate for the growth of the term $r_o^{2.5}$, the value of the term $h_o^{0.5}P/V_{soot}RT_o$ should be decreased appropriately. By adjusting the gap width h_o, helium pressure P, the rate of vapor generation V_{soot} and arc temperature T_o through deliberate variation of externally controlled parameters (electrode FR, helium pressure, and arc current), this compensation may become possible. These adjustments were implemented with 25-mm diameter anodes of the same 3Co/Ni composition and resulted in a product that contained on the average ~15 wt% of SWNTs, which is nearly the same as the SWNT yield obtained with commonly used 8-mm diameter rods under optimal conditions. Experiments were done with the large-scale arc discharge apparatus shown in Figure 3.2. Clothlike soot with ~12 wt% SWNT yield can be peeled off the walls of the reactor chamber as hand towel–sized sheets (Figure 3.7). The average SWNT yield in the large collaret on the cathode deposit is ~25 wt%. The soot production rate with 25-mm-diameter rods makes ~100 g/h; thus a 20-fold scaling factor for the process is attained compared with 8-mm diameter rods [26]. The process using 25-mm-diameter rods can be deemed semi-continuous, as loading a new rod takes a small fraction of the operation cycle. Moreover, in this sense the process can be rendered virtually continuous by arranging the automatic change of rods and continuous harvest of the soot, technical improvements that are already implemented at some laboratory scale arc installations. Finally, a further increase in rod diameter to 50 to 75 mm appears quite feasible and expedient.

3.5   Arc Discharge Production of DWNTs

Selective synthesis of double-walled carbon nanotubes (DWNTs) was a challenge for a long time. Although minute amounts of DWNTs have been identified by high-resolution TEM (HRTEM) in several catalytic CVD systems used for SWNT synthesis, no DWNTs could be recovered at that point from CVD products. The breakthrough in selective synthesis of DWNTs was made with the arc-discharge technique [89].

The laboratory scale arc discharge DWNT production process uses 8-mm diameter graphite rods with a drilled hole of 3.2-mm diameter and 200-mm length, filled with catalyst [89]. The best performing catalyst is a mixture of Fe, Co, and Ni metal and elemental S powders in a molar ratio 2:1:4:1. The components are fused together at ca. 700°C under argon. The cake obtained is ground to micron-size powder and then mixed with graphite powder and packed into the anode hole. Arc synthesis is carried out in a reaction chamber filled with an argon and hydrogen gas mixture at a pressure around one half of an atmosphere, with arc parameters close to those of the conventional SWNT arc synthesis. Quite similar to the SWNT arc synthesis, two major types of product are collected: the clothlike material from the reactor walls and a fibrous collaret around the cathode deposit. A fibrous network is suspended in the chamber volume between wall cloths and collaret. In addition, an unusually strong and dense film is formed on the side surface of the graphite cathode (Figure 3.8). The DWNT film is only 50 to 100 μm thick, but it takes a substantial effort to tear or rip the film. This effort matches that required for tearing or ripping cosmetic tissue paper of the same thickness. The film is of specific interest, as it contains a higher density of DWNTs (up to 70 wt% of the total carbon in the film) than do collaret (ca. 50 wt%), network (ca. 30 wt%), and wall cloths (ca. 20 wt%). Low-resolution SEM imaging reveals the high density of thin (10 to 15 nm in diameter) partially aligned filaments on the surface of as-produced DWNT film (Figure 3.9).

The relative amounts of the film, collaret, network, and wall cloths (~ 0.05:0.3:0.1:1, respectively), as well as the DWNT content of these product components, depend on the size and shape of graphite cathode, cathode deposit, chamber volume, and surface, as well as the arrangement of electrodes and other chamber elements. Therefore, adjusting the process geometry can cause deliberate variations in

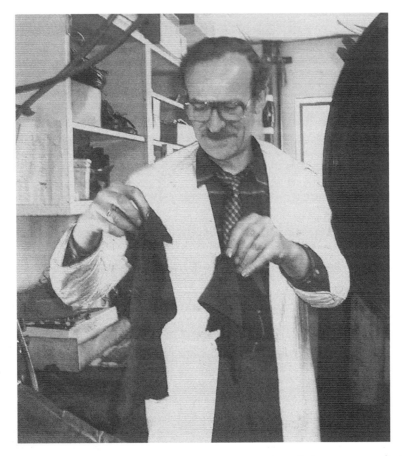

FIGURE 3.7 The SWNT product peeled off the surface of an industrial arc discharge reactor as "wall cloth." The product was obtained with Co_3Ni catalyst and contains ~ 12% of SWNTs by weight.

FIGURE 3.8 A fragment of strong film with metallic shine, which was formed on the side surface of graphite cathode in the arc-discharge synthesis of DWNTs with Fe_2CoNi_4S catalyst under optimal conditions (DWNT film).

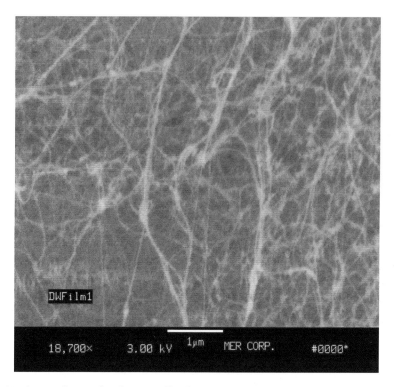

FIGURE 3.9 SEM image of as-produced DWNT film shown in Figure 3.8.

the amount and quality of product components. Solid carbon nanoparticles grown in the flow of gas, ejected from the interelectrode gap, carry uncompensated positive charges, which is the reason for the collaret and cathode film formation. Due to the higher electrophoretic mobility of nanotubes and metal nanoparticle species, their trajectories are more strongly deflected toward the cathode than those of amorphous soot. This is why the collaret and film are enriched in nanotubes and metal, relative to the network and wall cloths. Gas convection in the chamber also contributes to the distribution of solid particles between cathode and noncathode deposition areas. The more directly opposed the convection and the product flows, the better the resulting separation of tubes and metals from amorphous carbon. In particular, when the electrodes are arranged vertically, with the cathode lower, the collaret and film are discernibly enriched in nanotubes and metals. The graphite cathode surface is always colder than the cathode deposit and is partially protected from the carbon flux by the collaret and network, which filter part of the carbon from the gas flow; this causes the differences between the collaret and the film products.

Externally applied magnetic and electrostatic fields also deflect the flow of positively charged particles and can be used to enhance the separation factor. The metal content of product components as measured by thermal gravimetric analysis (TGA) is, from the greatest to the least, film (~70 wt%) > collaret (~65 wt%) > network (~50 wt%) > wall cloths (~45 wt%). Metal is present in the product in the form of 1- to 15-nm diameter nanoparticles mostly encapsulated in carbon graphitic shells. By EDX analysis, the particles are alloys of Fe, Co, Ni, and S in a proportion close to that of the original mixture [89].

Fullerenes are not produced in the DWNT arc discharge synthesis. The amount of C_{60} and C_{70} fullerenes in the product is below 0.01 wt%, by HPLC analysis. Amorphous carbon formed in the gas phase and graphitic nanoparticles originating from spallation of the anode are the nontubular carbon constituents of the product. The tubular constituent of the product is strongly dominated by DWNTs but contains minor amounts of SWNTs, trace amounts of triple-walled nanotubes (TWNT), and occasionally some thin-walled MWNTs, the latter occurring only in the collaret and not found in the wall cloth. Light hydrocarbon gases, including acetylene, ethylene, methane, and ethane, are the gaseous products, which constitute a few percent by weight of the graphite vaporized.

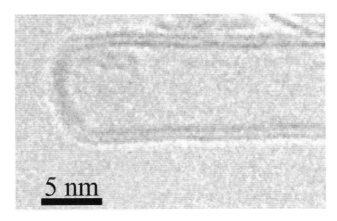

FIGURE 3.10 HRTEM image of the metal-free end of a DWNT in the product obtained with Fe$_2$CoNi$_4$S catalyst.

The HRTEM analysis revealed that most nanotubes in the product are DWNTs, and the ratio of DWNTs to SWNTs is higher than 30 under optimized conditions [89]. Most DWNTs have an outer diameter ranging from 3 to 5 nm. The wall separation distance is 0.39 ± 0.02 nm, which exceeds the 0.35 ± 0.01 nm distance that is usually observed in MWNTs. For a smaller-diameter DWNT, the distance appears to be larger. The SWNT diameter falls in the same 3- to 5-nm range. For a synthesis run, the DWNT diameter varies depending on where in the reactor the product was collected from. In runs performed under different conditions, the average tube diameter also varies. Larger DWNT diameters generally correspond to higher process temperatures.

In large-diameter separate tubes, the circular tube cross section can become energetically unstable, leading to a slow quasi-periodic change in the apparent width of the tube longitudinal section. This distortion can often be observed in HRTEM micrographs where individual tubes are discernable [89]. A DWNT has a rounded terminating cap at one tube end, which is free of catalyst. The cap shape ranges from an almost perfect hemisphere [89] to a flattened hemisphere (Figure 3.10) and consists of two evenly spaced layers, which implies that the tube chirality, whatever it is, is likely similar for the two layers. The chiral angle of the two layers actually differs only slightly, as observed by atomic resolution TEM imaging of a DWNT [90]. Similar but less-definitive information about the chirality of layers in the arc-produced 3- to 5-nm thick DWNTs stems from the analysis of their Raman spectra recorded in the radial breathing mode frequency region 40 to 100 cm^{-1} [91].

In rare cases such as shown in Figure 3.11, the outer layer of a DWNT is abruptly terminated, whereas the inner layer continues as a SWNT. If electrical leads were connected to the SWNT and DWNT portions of this tube, the structure obtained could be, for example, the smallest manufactured capacitor, with a very high-specific capacity. The coaxial cylindrical electrodes in a large diameter (4 to 5 nm) capacitor likely have metallic or pseudo-metallic conductivity, as the energy gap in the electronic structure of that thick semiconducting SWNT becomes smaller than kT. Smaller-diameter DWNT $\rightarrow$ SWNT structures may consist of semiconducting tubes with larger gaps, thus offering the opportunity for manufacturing various nanoelectronic devices, no less interesting than a capacitor that would probably operate even at room temperature. In some DWNTs, long fragments of a third layer were observed. Equipped with electric leads to the SW, DW, and TW portions, such stepped structures could create tremendous opportunities for carbon nanotube–based electronics. Still, selective synthesis of such structures remains a real challenge.

Besides the separate DWNTs, fairly loose bundles of DWNTs composed of a few to a few dozen tubes are observed in TEM images. The observed degree of bundling depends on where in the reactor the sampling was made from and on ultrasonic treatment of the sample in a liquid. The extent of amorphous carbon coverage on single or bundled DWNTs varies, depending on the site sampled, as observed with SEM. Figure 3.12 shows amorphous carbon-coated DWNT bundles in the as-produced collaret. Their outer diameter is larger than that of the filaments in Figure 3.9. For DWNT film filaments, both the

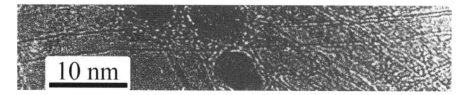

FIGURE 3.11 HRTEM image of the region where transition from SWNT to DWNT structure occurs from left to right in the image.

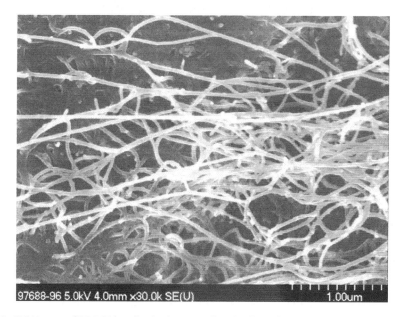

FIGURE 3.12 SEM image of DWNT bundles in the as-produced collaret formed in the arc-discharge process with Fe_2CoNi_4S catalyst.

DWNT bundle and the amorphous carbon layer are somewhat thinner than for collaret filaments. The DWNT bundle length commonly far exceeds 1 µm and can reach several dozen microns. The filaments in the as-produced collaret are multiply branched, as seen in Figure 3.12. Note that many side branches of well below 1 µm in length stem from the long filaments.

The diameter of most metal particles observed on HRTEM images of the arc discharge product closely matches the diameter of DWNTs [89]. This is different from SWNT synthesis in arc discharge, laser ablation, and high pressure CO (HiPco) processes, where the observed average metal particle diameter exceeds that of the tube by an order of magnitude. HRTEM has sometimes revealed equivalent-diameter metal particles attached to the DWNT ends. HRTEM observation of either such metal-capped DWNTs or closed two-layer caps is rare, however, because of the very large tube length. No big metal particles have been observed attached to the ends of DWNT bundles despite a focused search. This implies that DWNTs grow from small metal catalyst particles of matching diameter in the arc discharge synthesis performed in hydrogen-containing gas.

3.5.1   Mechanism of DWNT Arc Synthesis in Hydrogen-Containing Atmosphere

Substantially higher electric power is required to vaporize a graphite anode in argon atmosphere than for equally productive vaporization in helium gas. With the same arc power, a much lower plasma temperature is developed in the argon gas than in helium. Under certain low-power conditions, the carbon arc discharge in argon gas can glow for hours with only slight consumption of the graphite anode. However, when argon gas is diluted with hydrogen, under the same conditions, anode vaporization becomes much faster. This is caused by chemical reactivity of hydrogen gas toward carbon. In the hot plasma region; hydrogen reacts with carbon from the graphite surface and with small carbon clusters C_n producing light hydrocarbons C_nH_m for the primary or intermediate species in the anode vaporization. These low molecular weight hydrocarbon species are thermodynamically more stable at reaction temperature than the corresponding small carbon clusters C_n that are formed as the primary species by anode vaporization in a pure argon atmosphere. The thermodynamic driving force for the vaporization process is higher in an atmosphere containing hydrogen gas, and anode vaporization proceeds much faster.

With the same arc power, the introduction of hydrogen renders the carbon vaporization zone substantially hotter than in pure argon. An approximate temperature gain of 1000 K can be estimated for the hydrocarbon species formation zone, which includes the arc plasma and its close vicinity, from the data [20] on the high temperature equilibrium between graphite and a set of the 23 most important hydrocarbon molecules. Pure carbon vapor condenses at a higher rate than hydrocarbons. For hydrogen-containing systems, the overall arc process of graphite conversion into solid products through vaporization and consequent condensation of vapor flowing from the gap is delayed and moved away from the arc. Hydrogen provides the arc system with a longer arm chemical heat pipe operating in the effluent gas. This leads to a more gently sloping radial temperature profile, which is generally preferable for efficient production of long nanotubes.

Formation of hydrocarbons in the carbon arc in a H_2/Ar atmosphere has been established by MS analysis that identified acetylene, ethylene, methane, and traces of ethane in the gas sampled from the reactor. For a typical initial gas mixture of H_2:Ar = 33.3:66.6 vol.% used in the DWNT arc synthesis, the gas composition in the reaction chamber during the arc process is approximately as follows: H_2 (30.0%), Ar (68.0%), C_2H_2 (1.0%), C_2H_4 (0.35%), CH_4 (0.15%), and C_2H_6 (0.03%). This gas composition establishes itself within 2 minutes of the start of the arc process and remains practically unchanged in the course of the next 1 to 2 hours of the synthesis, as well as after process termination, if the gas chamber is isolated. With dynamic gas atmosphere in the chamber the equivalent gas composition is as quickly established and remains practically unchanged when the gas flow is varied below the standard 0.5 l/min limit. These compositional and kinetic data, supplemented with the rates of carbon erosion from the anode and escape from the gap (ca. 80 and 15 mg/min, correspondingly) and the reaction chamber volume (2.5 l), are consistent with the following mechanism of carbon transfer in the arc synthesis of DWNTs.

Most of the carbon is transferred from the anode through the gas phase to the product-formation site as hydrocarbons. Each hydrogen molecule in the reaction chamber is used many times (about once a minute) in a cyclic process that includes two consecutive steps. In the first, the hydrogen molecule is consumed and a hydrocarbon is formed. In the second, the hydrogen molecule is released by hydrocarbon decomposition, leaving behind solid carbon. In other words, steady-state concentrations of hydrogen and light hydrocarbons in constant mutual ratio are quickly established in the gas phase due to the sufficiently high and equal magnitude of the rate values of hydrocarbon generation and decomposition; in turn, the solid product formation rate has the same magnitude.

Changing essential process parameters alters the stationary gas composition. With a narrower gap and concomitant lower arc power, the acetylene fraction in the light hydrocarbons and the total amount of light hydrocarbons itself both notably decrease. For example, the ratio C_2H_2:C_2H_4:CH_4 ~1:0.3:0.1 with 1.8% total of these hydrocarbons in the gas phase changes to ~1:0.6:0.2 with 0.9% total when the gap width is dropped from ~6 to ~2 mm. This suggests that acetylene is a primary product of carbon hydrogenation in the gap area and is more productive in amorphous carbon and DWNT formation in the cold zone at a distance from the gap, whereas ethylene and methane are inferred to be secondary products.

Acetylene is a more effective vehicle for transporting carbon in the arc process of DWNT synthesis than are ethylene and methane. Acetylene (ethyne) is the lowest homologue of the polyyne $C_{2n}H_2$ family and the most stable. Under certain conditions, polyynes can be selectively generated in the carbon arc and stabilized in bulk amount in the product [91]. Compared with ethylene and methane, the polyynes are thermodynamically more stable and kinetically less prone to conversion into soot at temperatures of over ~1600 K that correspond to the range where light hydrocarbons form and transport carbon in the arc. These facts suggest polyynes as an ideal carbon-transferring agent in DWNT arc synthesis [89]. This expectation is only partially fulfilled by finding substantial amounts of acetylene and disclosing its dominant role in carbon transport. That polyynes have not been similarly detected by MS analysis can be explained by their lower steady-state concentration, because they are more reactive than acetylene in depositing solid carbon products. It should be noted that with the plenum on-line MS sampling scheme the measured stationary hydrocarbon concentration in the reactor volume is far below the concentration that actually occurs in the gas flow ejected from the gap. Local polyyne density in the flow may be orders of magnitude higher than the stationary density in the chamber. Hence, gas sampling directly from the flow region into a MS ionization chamber through a small hole in a separation diaphragm would create a good opportunity to detect polyynes in the DWNT arc synthesis.

The arc system is producing DWNTs in a gas medium containing a substantial amount of hydrocarbons and is, in essence, a CVD system for carbon-nanotube synthesis through catalytic pyrolysis of hydrocarbons over a "floating" metal catalyst [89]. Hydrocarbons provide an efficient means for transporting carbon to the surface of metal catalyst particles to produce nanotubes and to the surface of carbon nanotubes to pyrolytically deposit an amorphous carbon layer. The continuous amorphous carbon layer provides very favorable conditions for long DWNT growth. A metal catalyst particle sorbed by this layer consumes and transforms amorphous carbon into a new tube, preferentially propagating along the parent tube or the bundle. The catalyst particle does not dissolve the parent tube walls because amorphous carbon is much more favorable carbon feedstock. The tube body, anchored in amorphous carbon, elongates and pushes the catalyst particle ahead, thus ensuring continuous contact with the carbon feedstock source when the growth is along the layer. The newly originated tube can grow at so high an angle to the axis of the parent tube that its growth cannot be redirected to be parallel to the axis because of tube rigidity and firm anchoring. When this occurs, the catalytic particle will reach the surface of the amorphous carbon layer and will then propagate only as long as a sufficient supply of hydrocarbon molecules is available. This is more likely with a thick amorphous carbon layer and leads to frequent branching of thick bundles, as shown in Figure 3.12.

So, the DWNT catalyst particle can use both amorphous carbon and hydrocarbon molecules for the feedstock. Amorphous carbon gives rise mainly to bundles, whereas hydrocarbons give rise mainly to discrete tubes. This peculiarity can explain, for example, why the bundles are generally thinner and their branching is more frequent in hydrocarbon-containing CVD systems, including the DWNT arc process, compared with arc and laser systems utilizing a hydrocarbon-free atmosphere. It also explains why the bundles are thinner in the DWNT film than in the collaret and why the collaret filaments are more extensively branched. Indeed, the bundles comprising the film have been growing for a shorter time and therefore have accumulated less amorphous carbon than collaret bundles (Figures 3.9 and 3.12).

DWNT bundles in hydrogen gas form by a different mechanism than SWNT bundles in helium gas arc systems. This mechanism is most likely common for the hydrocarbon feedstock thermal CCVD systems for SWNT and DWNT syntheses, as the tubes are similarly loosely packed in the CCVD bundles, which is distinct from densely packed SWNT bundles grown in inert gas by arc and laser techniques. To verify this conjecture, more TEM studies of the as-produced materials are required, with special care taken in sample preparation because ultrasonic treatment of the sample disturbs original bundling and tube packing. It bends the tubes and bundles, eventually forming "crop circles" and coils of SWNTs and DWNTs and also facilitates metal head detachment from the bundle.

Recognition of selective DWNT synthesis in the arc as a CVD process [89] obviously stimulates the search for technically common catalytic CVD systems capable of selective DWNT production. Several reports on the design of such systems, with the use of supported and floating metal catalysts, have recently appeared [92–100]. Notably, most of these systems use acetylene for the feedstock.

Also due to this recognition, a series of runs with controlled addition of acetylene, ethylene, and methane into the dynamic gas atmosphere of the arc reactor was performed. Slight changes in the product quality occur with the addition of gases in amounts near their steady-state concentration, and the gas phase composition tends to be quickly restored and stabilized after the admixture event. Much higher admixed amounts deteriorate the product quality by causing excessive amorphous carbon production. Still, with moderate addition of gases it is possible to slightly improve product quality and system productivity, even with larger-diameter rods. For example, an $Ar/H_2/CH_4$ system was scaled up to high production rates (~10 mg/min of DWNTs) by utilizing 12.7-mm diameter rods without any loss in DWNT quality.

An extensive parametric study of the process of the DWNT synthesis in the arc has been accomplished. The rate of DWNT production as a function of arc current and power, gas mixture composition and total pressure in the chamber, and gap width between electrodes was determined. The highest rate of DWNT formation achieved is ~4 mg/min for 8-mm diameter rods with 2Fe/Co/4Ni/S catalyst, 82 A arc current, and 350 Torr total pressure of $Ar/H_2 = 2.5$ gas mixture at a constant electrode feed rate 0.7 mm/ min. It was shown that kinetics of DWNT formation is governed by the same hydrodynamic factor as for SWNTs: the rate of cooling the reaction mixture by the buffer gas τ_{mix}. It was established that the maximum DWNT yield is attained at approximately the same τ_{mix} value as in the SWNT synthesis considered in Section 3.4.2.

It was found that 12.7-mm diameter rods under appropriately chosen conditions yield DWNTs of the same quality as standard 8-mm rods but at a substantially higher rate of ~8 mg/min. The conditions were selected in order to keep the value of τ_{mix} constant. Experiments with a large-scale arc apparatus (Figure 3.2) have shown that it is technically feasible to further increase the production rate of similar quality DWNTs by using 25- to 38-mm diameter rods.

Metal catalyst particle performance in the DWNT arc synthesis was found to be in concordance with the DP model regularities described in Section 3.4.3. This enables assessment of the DWNT growth temperature in the arc process. Indeed, the characteristic growth time (τ_{mix} ~ 6 to 8 ms) is equal for DWNTs grown in H_2/Ar gas and SWNTs in helium. In both cases, the growth rate (L/τ_{mix}) is a few millimeters per second because tubes have similar length L ~20 μm. But DWNTs grow from 3- to 5-nm diameter particles and SWNTs from 15 to 20 nm. To allow for equal growth rates, the carbon diffusivity in the metal particle should be four to five times smaller for DWNTs. With 140 kJ·mol^{-1} activation energy of carbon diffusion in nickel, this corresponds to a DWNT particle temperature ~200 K below that of SWNTs. Despite the estimate roughness, this comparison can be of use for understanding and optimizing DWNT production. For example, the comparison implies that DWNT catalytic particles should solidify at lower temperatures than SWNT particles, which conforms to a probable role of sulfur in the DWNT system, specifically, to keep the particle molten for a longer time in the arc process. This role is suggested by a large difference between metal/sulfur and metal/carbon eutectic temperatures for Fe, Co, and Ni metals (989:877:635 and 1153:1309:1318°C, correspondingly). Arc DWNTs can form as well with sulfur-free metal catalyst under similar conditions, though much less efficiently [89], which also supports this conjecture.

The HRTEM observations show that DWNT/SWNT ratio in the arc product appears to increase with process temperature. This trend can be rationalized as follows. To avoid overheating and thus survive catastrophic growth termination by the cold-explosion mechanism, the catalytic particle may need a better heat sink at high temperature. DWNTs provide a better heat sink than do equal-diameter SWNTs. The SWNT catalytic particles, lacking an adequate heat sink and unable to improve it by nucleating a second layer, will be more likely subjects of a cold explosion. Thus the cold explosion in the particle may be the factor determining temperature dependence of selectivity and the selectivity itself of DWNT production in the arc system.

It is interesting to note that, in a similar catalytic arc system, boron nitride double-walled nanotubes (BNDWNTs) can be selectively produced [101]. The DP mechanism for catalytic particle performance can be implied. The process selectivity in DW tubes is very high; practically no SW tubes are observed [101]. Exclusively high selectivity may be due to enhanced effectiveness of the cold-explosion mechanism in choosing between DW and SW boron nitride tubes. This enhancement is indeed expected, as heat conductivity of boron nitride along the layer (~500) is much lower than that of graphite (~1500 W/m K). One boron nitride layer always has insufficient heat conductance, and all BNSWNTs, if nucleated, will soon stop growing by the cold-explosion mechanism.

3.6   SWNT Production by Laser Ablation of Carbon-Metal Target

Evaporative heating of a carbon-metal composite target with laser light pulses, either separate [102–106] or frequently repeated [107–122] and with continuous illumination by laser [123–128] or solar [129–133] light, can bring about SWNT production under the correct conditions. The SWNT yield and properties are reported to be rather sensitive to variations in the process parameters, including light intensity, process temperature and geometry, carrier gas type, pressure, and flow conditions — up to a dozen externally controlled parameters in total. Typical conditions for producing SWNTs in several laser ablation systems are listed in Table 3.2.

A typical 10-ns long and 300-mJ laser pulse vaporizes ~10^{17} carbon and ~10^{15} metal atoms, contained in a ~0.1-μm thick layer over a ~0.2 cm^2 spot, into argon gas preheated by an external furnace to 1200°C. The surface temperature rises during the laser pulse to about 6000 K, with energy continuously deposited into the solid as thermal energy as well as supplying the heat of vaporization for the ablated mass [134]. At the end of the laser pulse, the surface temperature drops rapidly to ca. 4000 K as a large amount of thermal energy is extracted by the additional after-pulse carbon vaporization and then more slowly cools by thermal conduction into the target bulk, reaching ca. 2000 K in about 40 to 50 ns [134]. It was found [60] that application at this moment of a second equivalent laser pulse increases the SWNT yield (Table 3.2, entries 2 to 5). This dual-pulse effect is likely due to additional vaporization of carbon particulates ejected in small amounts from the surface during the first pulse together with initial carbon vapor, composed mainly of C_2 and C_3 species. Ejection of graphite particulates is likely be a consequence of thermally induced stress relaxation, as it is occurring later than the single-pulse termination [102,135]. The graphite particulate amount in the product is generally smaller with laser ablation than with arc discharge. This explains, in part, larger SWNT yields and relative ease of deep postsynthesis purification of the laser product.

During the single-pulse ablation, a flat bubble of ca. 6000 K and 400 bar carbon vapor forms near the target surface and starts to expand with supersonic velocity and compress the background gas [102,134,135]. Gas pressure and temperature in the plume thus formed both quickly decrease until the pressure becomes slightly below that of the compressed background gas [102]. At this moment, less than 300 ns after the pulse, a strong external shock wave is generated and detaches to propagate through the background gas, and an internal shock wave within the plume material starts to propagate in the opposite direction [102]. The internal shock wave brings about backward motion of a portion of the plume toward the target surface. By the time this portion reaches the target (100 to 600 ns after the pulse), the surface temperature is below 1000 K [134], and carbon and metal components of the plume partially precipitate on the cold surface as amorphous carbon and thin metal films or islands that can be seen in SEM images of the target surface [115–119]. Thus depleted of material, the backward-moving portion of the plume reflects from the surface and begins to move out to join the main body of the plume. Simultaneously, the backward motion in the radial direction results in an axial focusing of the plume. In about 200 to 1000 ns this first oscillation of the plume material is completed. Several consecutive oscillations of the same nature, but decaying in strength, can be observed [102]. Following these oscillations, occurring during less than 0.2 ms after the pulse, the plume material is segregated into a "smoke ring" or a bubble, depending on conditions, confined within ~0.1 to 1 cm^3 volume, which detaches from the surface. This volume retains carbon for a time interval between a few milliseconds and seconds and then converts into a shapeless turbulent structure [102].

It is commonly agreed that the SWNT growth time in laser ablation lies in this millisecond-second time scale [102–106,113–121]. There is no spectral feature directly associated with the growth of SWNTs; therefore, more accurate determination of the growth time with time-resolved spectral diagnostics of the

TABLE 3.2 SWNT Synthesis by Ablation of Carbon-Metal Target Containing Co:Ni = 1:1 Catalyst

N	Ref.	Furnace T, °C	Argon Pressure, kPa	Pulse Energy, mJ	Pulse Length, ns	Pulse Repetition, Hz	Laser Spot Diameter, mm	Pulse Fluence, J/cm²	Pulse Intensity, MW/cm²	Average Irradiance, W/cm²	Laser Wavelength, nm	Total Metal in Target, at.%	Production Rate, mg/h	SWNT Yield, Wt% of Product
1	107	1200	66	300	10	10	6.5	0.9	100	9	532	1.2	10	50
2 [a]	60	1200	66	250 300	10 10	10 10	5 7	1.3 0.8	130 80	13 8	532	1.2	n/s	70
3 [a]	108	1200	66	490 550	10 10	30 30	6 6.5	1.7 1.7	170 170	52 50	532 1064	2.0	120	65
4 [a]	108	1100	66	930 930	10 10	30 30	7.1 7.1	2.4 2.4	240 240	70 70	1064 1064	2.0	500	45
5 [a]	54, 48	1050	66	100 100	7 7	20 20	3 3	1.4 1.4	200 200	28 28	532 1064	5.2 1.2 [b]	n/s	65
6	112	1150	105	300	20	20	2	9.5	470	190	1064	1.2	200	60
7	113	1150	65	290	20	15	4.5	1.8	90	27	1079	1.2	100	40
8	104	1200	92	115	6.5	10	1.5	2.0	310	20	532	9.0	25	55
9	104	1200	92	115	6.5	0.5	1.5	2.0	310	1	532	9.0	1	55
10	104	1200	92	115	6.5	0.008	1.5	2.0	310	0.017	532	9.0	0.02	55
11	102	1000	66	140	8	0.016	1.6	7.0	870	0.1	532 [c]	2	n/s	60
12	102	900?	66	140	8	0.016	1.6	7.0	870	0.1	532 [c]	2	n/s	15
13	120	400	80	n/s	8	n/s	n/s	3	380	n/s	532	5 [d]	n/s	5
14	118	1200	80	300	6.5	10	3.5	3.1	470	31	532	1.2	100	50
15	58	25	66	1.0	575	24000	1.0	0.12	0.2	3000	1064	1.2 [e]	n/s	30
16	58	25	66	2.3	300	10000	1.0	0.27	0.9	2800	1064	1.2 [e]	n/s	15
17	58	25	66	3.2	250	6000	1.0	0.37	1.5	2300	1064	1.2 [e]	n/s	4
18	58	25	66	5.3	175	3000	1.0	0.61	3.5	1800	1064	1.2 [e]	n/s	0
19	105	1200	80	20,000	2×10^{7}	s.p.	1	2550	0.13	1×10^{5}	10,600	1.2	1800	60
20	105	25	80	20,000	2×10^{7}	s.p.	1	2550	0.13	1×10^{5}	10,600	1.2	900?	6?
21	125	25	66	∞	∞	∞	1.1	∞	0.025	2×10^{4}	10,600	4	70	30
22	126	25	53	∞	∞	∞	1.1	∞	0.025	2×10^{4}	10,600	4	200	50
23	129	2700	45	∞	∞	∞	5	∞	0.002	2000	Solar	4	40	10
24	61	3500	66 [f]	∞	∞	∞	8	∞	0.003	3000	Arc	2 [g]	5000	20

Note: n/s, nonspecified; s.p., single pulse; ?, denotes rough estimate for a value not specified in the referenced paper; ∞, unidentified.

[a] Dual pulse with 40- to 50-ns gap

[b] Fe:Ni catalyst

[c] Both 532 and 1064 nm simultaneously

[d] 95% C_{60} target

[e] Porous target

[f] Helium gas

[g] Co:Ni = 3:1

laser plume is not possible [102–106,109–111]. Accurate assessment of the SWNT growth time in laser synthesis stems from the comparison of the kinetic data on carbon-metal vapor condensation and further evolution in laser and arc ablation processes. This comparison shows close quantitative similarities in the kinetics of laser and arc processes in the time range of SWNT formation under optimal conditions.

The carbon component of carbon-metal vapor condenses into soot particles and fullerenes during 0.2 ms in the laser process [102,103] and during the same 0.2 ms in the arc process [22,62]. Amount of fullerenes produced in both systems under similar conditions is the same, about 5 to 10% (Figure 3.6, [75,111,113,118,121,135]). Condensation of metal component starts later and is completed in 2.0 to 2.3 ms in laser systems [102,103,111] and in about the same time in the arc process (Figure 3.6). Temporal behavior of laser plume and arc jet flow parameters in the time range of interest, from ~0.2 ms to ~30 ms, is very similar. In this range the ablated material propagates by a few centimeters, decelerating from ~10^3 to ~10 cm/s linear velocity; temperature drops from ~2200 to ~1000°C and carbon particle number density from ~10^{17} to ~10^{16} cm^{-3} both in laser and arc systems [62,102,109–111]. Under optimal conditions, SWNTs of equal length are formed in both systems. This comparison strongly suggests the same growth time for SWNTs produced in laser and arc processes, which is about 10 ms under optimal conditions (Figure 3.6). The bell-shaped dependence of the SWNT yield on the cooling time of the type presented in Figure 3.6 is also implied for laser ablation. That is, dynamics of cooling the laser plume in the 10-ms time scale is the major factor determining the SWNT yield. The comparison eventually leads to the conclusion on the same mechanism of SWNT growth both for laser and arc synthesis, which is described in Section 3.4. A conservative estimate of 3 to 30 ms for the SWNT growth time can be obtained with allowance made for the width of metal catalyst particle diameter distribution and accuracy of cooling-time determination. Accordingly, the linear growth rate of SWNTs is ~2 mm/s (within a factor of 3 accuracy), both for laser and arc processes.

Further support for these conclusions comes from all structural and kinetic results obtained in laser-ablation system studies [102–122]. For example, it was observed that many of the SWNT bundles grow from large rounded metal particles [52,115,117]. No results contradictory to the mechanism described and quantitative estimates of growth parameters made above have been found in the literature. For instance, the SWNT growth time of ~1 s and the linear growth rate of ~0.2 μm/s were estimated from the short length (~100 nm) and low-yield SWNTs (Table 3.2, entry 12) obtained under special conditions. This observation was made when the target was moved with respect to the furnace to shorten the residence time of ablation products in the high-temperature zone [102]. The concomitant slight decrease in the temperature of the gas above the target and of the target itself is likely responsible for this observation. It is known that the SWNT yield sharply decreases, the SWNTs become much shorter, and the diameters of individual SWNTs and their bundles decrease when the ambient temperature set by the furnace falls from 1200 to 900°C, especially near the lower boundary of this temperature range [54,113,114,118,122]. In this respect, the observation of short, thin, and low-yield SWNT bundles [102] is quite conceivable and does not contradict to the mechanism proposed.

The furnace temperature seems to be the most critical externally controlled parameter among those studied with pulsed laser ablation, as it primarily determines the temperature of a few centimeters thick buffer gas layer near the target, in which the plume expands, cools down, and forms SWNTs. No SWNT formation was observed with ablation of graphite-metal target at furnace temperature below 800 or above 1300°C with a maximum SWNT yield lying at about 1200°C (Figure 3.13). As mentioned in Section 3.4.3, the actual temperature of working catalyst particle is higher than environment gas temperature. The temperature excess the particle possesses can be assessed from the shape of the Figure 3.13 dependence. Indeed, the existence of the upper temperature limit for SWNT formation, revealed by the abrupt drop in yield at 1300°C furnace temperature, can be attributed to overheating of metal catalyst particle with subsequent cessation of SWNT growth by the cold-explosion mechanism. In this respect, the width of the bell-shaped dependence on Figure 3.13 near the maximum yield (~200°C) should represent the temperature corridor in which the particle works. Thus the range 1300 to 1500°C can be specified for the actual temperature of the Ni/Co particle working under optimal conditions in laser and arc systems and using amorphous carbon for the feedstock in SWNT formation by DP mechanism. The actual

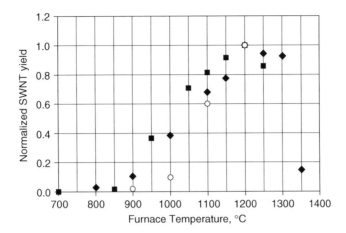

FIGURE 3.13 The influence of the furnace temperature on the relative SWNT yield in the product, obtained under similar laser-pulsing conditions but with different gas-flow velocity near the target. Solid squares [75], 1 cm/s linear velocity; solid diamonds [122], 1.8 cm/s; open circles [116], 4.8 cm/s.

temperature of the particle, not of the gas environment, determines the diameter of the SWNTs. This conjecture was advanced in Section 3.4 based on analysis of SWNT diameter distributions obtained under various conditions, where actual particle temperature or relative value can be assessed. The conjecture finds further confirmation in practically all data on SWNT diameter distributions in laser systems, where such assessments can be made. Thus the actual temperature history of SWNTs can be restored from analyzing their diameter distributions.

It is easy to assess that the flux of carbon in the form of soot particles and fullerenes on metal catalyst is adequate for growing the SWNTs at specified above carbon particle number density 10^{17} to 10^{16} cm^{-3}. Still, both laser and arc ablation produce carbon in the form of spallated graphitic particles and single-walled nano-horn (SWNH) aggregates. These well-graphitized carbon species have free energy of formation so close to that of SWNTs that they cannot be used as feedstock in SWNT growth. The SWNTs grow from amorphous soot particles and do not grow from SWNH aggregates [115]. Metal particles trapped in SWNH aggregates remain inactive toward SWNT formation [115]. Excessive production of graphitic particles and SWNH aggregates in the arc discharge compared with laser-ablation synthesis is an important reason for lower SWNT yields observed.

Thermal conductivity of the buffer gas largely determines the rate of cooling of the plume material. With helium the cooling rate is always too high (τ_{mix} too low, in terms of Figure 3.6 notations) to allow the τ_{mix} to be adjusted to optimal for SWNT production value by varying other laser process parameters. Therefore, argon gas is used in laser experiments. Nitrogen gas is equally efficient [75,121]. Krypton and xenon would probably also provide slow-enough cooling rate to match optimal τ_{mix} value in pulsed laser production of SWNTs. Quite the opposite situation occurs in the arc-discharge production of SWNTs, where high thermal conductivity of helium easily suits the $\tau_{mix} \sim 10$ ms criterion, whereas argon gas cannot ensure sufficiently high cooling rates.

The SWNT yield grows as argon gas pressure increases from ~10 to the optimum value ~ 66 to 80 kPa [113,117]. This behavior corresponds to the adjustment of the optimum cooling rate for the laser plume.

The argon gas flow rate in the laser reactor also influences the SWNT yield [113]. As soon as linear flow rate reaches a sufficiently high critical value, which is equal to ~2 to 3 cm/s [113] for a given experiment geometry, a sharp decrease of SWNT abundance in the soot is observed. This result can be explained by the temperature decrease in a few centimeters long SWNT growth zone caused by high velocity argon.

The SWNT yield depends on metal content in the target [115–118]. Much higher than optimal (~1 to 2 at%) concentration of metals in the target leads to formation of larger metal particles (more than 30 nm in diameter) that are useless in SWNT production under common laser ablation conditions because of too-low carbon diffusion rate through such particles. As a result, the SWNT yield drops down. Thicker bundles are formed by larger particles produced with higher metal content in the target, but they are covered with a thicker amorphous carbon layer, and the resulting SWNT yield is lower [117].

3.6.1 Scaling Up the SWNT Production Using Laser Ablation

Two methods were developed to possibly scale up SWNT production using laser ablation [127,128]. The continuous wave laser-powder method of SWNT synthesis [127] is based on introduction of mixed graphite and Ni/Co powders into 1100°C hot argon gas stream, which is irradiated coaxially by a 2-kW continuous wave CO_2 laser. Because the particles are of micron size, the thermal conductivity losses are significantly decreased and more energy effective ablation achieved. The soot product generation rate is 5 g/h. The SWNT abundance in the soot is 20 to 40% and the tube diameter 1.2 to 1.3 nm.

In the ultrafast pulses from a free electron laser (FEL) method [128], the ~3-μm wavelength light pulses of ~400 fs length and at a pulse repetition rate 75 MHz are generated to vaporize the graphite-metal target. The light beam is focused to give the pulse intensity ~500 GW/cm², which is about 1000 times greater than in Nd:YAG laser systems. In the 1000°C hot zone of quartz reactor, a jet of preheated argon is directed parallel to the surface of rotating graphite target, which contains the Ni/Co or Ni/Y catalyst. Argon gas deflects the ablation plume away from the incident beam direction, clearing the room in front of the target. The produced SWNT soot is collected on a cold finger at a rate 1500 mg/h. If the FEL is upgraded and is working at 100% power, a yield of 45 g/h could be expected [128]. SWNTs of 1- to 0.4-nm diameter and 5- to 20-μm length are produced in 8- to 200-nm thick bundles.

These examples show that a moderate scaling up of laser ablation production is possible.

3.7 Conclusions

The mechanism of the SWNT synthesis in the arc discharge and laser ablation processes appears to be fairly simple. It is governed mainly by the dynamics of hot vapor cooling and carbon diffusion through the bulk metal catalyst particle. It allows deliberate control of the arc and laser processes and is the basis for the successful scaling-up efforts of the arc process. Semiquantitative treatment of the arc and laser processes is possible. It provides a basis for a detailed mathematical modeling of arc and laser ablation reactors, which is necessary in order to design practical processes for maximum productivity, yield, and quality of SWNTs.

Scaling up the arc discharge processes of SWNT, DWNT, and MWNT production is possible and is realized in practice on an industrial scale. A production rate of 100 g/h of the raw product of each kind of these nanotubes is achieved per industrial apparatus. The economically reasonable limit for scaling up is already reached, but the challenge to further increase selectivity and specific productivity of the arc processes remains active.

Synthesis of SWNTs by the arc-discharge method is actually a CVD process that transforms amorphous carbon into a SWNT bundle with an appropriately large metal catalyst particle and proceeds in concordance with the DP kinetic mechanism. Arc synthesis of DWNTs in hydrogen atmosphere is a CVD process that grows one DWNT from an equal-diameter catalytic metal particle, which uses thermally stable light hydrocarbons for the carbon source. The process obeys the regularities of the DP kinetic model. Tube-growth termination in the DP model can be attributed to a cold explosion in the thermally nonequilibrium catalytic particle.

Diffusion of carbon through the bulk metal catalyst particle is the rate-limiting step in all kinetically studied systems that grow long nanotubes. Therefore, in order to reach higher production rates, high-temperature processes should be developed, including CVD processes, ranging up to 1500°C, which appears to be the temperature limit for diffusion-controlled transition metal catalytic systems. For this development, the most difficult problem to resolve is to prevent thermal noncatalytic dehydrogenative condensation of the carbon source. In arc discharge and laser ablation synthesis of SWNTs, this problem is obviated by the use of amorphous carbon for the feedstock. Another important problem is the design of appropriate metal catalysts, which remains a poorly explored area.

A current fundamental and practical challenge is selective synthesis of TWNTs. It does not seem to be an insurmountable problem, as arc DWNT synthesis can produce small but variable amounts of TWNTs, depending on conditions. Study of TWNT occurrence and deliberate variation of arc-system parameters based on the quantitative knowledge of catalyst performance accumulated so far, however scarce, may lead to higher TWNT selectivity and eventually to desired domination of TWNTs in the products.

References

1. N.A. Kiselev et al., Carbon 37, 1093 (1999).
2. S. Iijima, Nature 354, 56 (1991).
3. T.W. Ebbesen and P.M. Ajayan, Nature 358, 220 (1992).
4. T.W. Ebbesen et al., Chem. Phys. Lett. 209, 83 (1993).
5. D.T. Colbert et al., Science 266, 1218 (1994).
6. X.K. Wang et al., Carbon 33, 949 (1995).
7. S. Serapin et al., Carbon 31, 685 (1993).
8. Y. Saito et al., Chem. Phys. Lett. 204, 277 (1993).
9. P.M. Ajayan et al., J. Mater. Res. 12, 244 (1997).
10. G.H. Taylor et al., J. Cryst. Growth 135, 157 (1994).
11. A.S. Kotosonov and S.V. Kuvshinnikov, Phys. Lett. A 240, 377 (1997).
12. H. Ajiki and T. Ando, J. Phys. Soc. Jpn. 62, 2470 (1993).
13. R. Saito, G. Dresselhaus, and M. Dresselhaus, Phys. Rev. B 50, 14698 (1994).
14. V.I. Tsebro, O.E. Omel'yanovskii, and A.P. Moravskii, JETP Lett. 70, 462 (1999).
15. S. Frank et al., Science 280, 1744 (1998).
16. V.P. Bubnov et al., Russ. Chem. Bull. 43, 746 (1994).
17. A.S. Kotosonov, D.V.Shilo, and A.P. Moravskii, Phys. Sol. State 44, 666 (2002).
18. A. Huczko et al., J. Phys. Chem. A 101, 1267 (1997).
19. R.E. Smalley, Mater. Sci. Eng. B 19, 1 (1993).
20. J. Abrahamson, Carbon 12, 111 (1974).
21. J.M. Jones et al., Carbon 34, 231 (1996).
22. A.V. Krestinin and A.P. Moravsky, Chem. Phys. Lett. 286, 479 (1998).
23. E.G. Gamaly and T.W. Ebbesen, Phys. Rev. B 53, 2083 (1995).
24. T. Guo et al., J. Phys. Chem. 99, 10694 (1995).
25. A.A. Setlur et al., ECS Proceedings 98-8, 897 (1998).
26. R.O. Loutfy et al., in: Persp. Full. Nanotech., Ed. E. Osawa, Kluwer, Dordrecht, p. 35 (2002).
27. R.E. Smalley, Acc. Chem. Research 25, 98 (1992).
28. Y-K. Kwon et al., Phys. Rev. Lett. 79, 2065 (1997).
29. P.M. Ajayan, T. Ichihashi, and S. Iijima, Chem. Phys. Lett. 202, 384 (1993).
30. S. Iijima and T. Ichihashi, Nature 363, 603 (1993).
31. D. S. Bethune et al., Nature 363, 605 (1993).
32. P.M. Ajayan et al., Chem. Phys. Lett. 215, 509 (1993).
33. S. Serapin et al., Chem. Phys. Lett. 217, 191 (1994).
34. S. Serapin and D. Zhou, Appl. Phys. Lett. 64, 2087 (1994).
35. C.H. Kiang et al., J. Phys. Chem. 98, 6612 (1994).
36. J.M. Lambert et al., Chem. Phys. Lett. 226, 364 (1994).
37. X. Lin et al., Appl. Phys. Lett. 64, 181 (1994).
38. P.M. Ajayan et al., Phys. Rev. Lett. 72, 1722 (1994).
39. C. Guerret-Piecourt et al., Nature 372, 761 (1994).
40. C.H. Kiang et al., Carbon 33, 903 (1995).

41. Y. Saito, K. Kawabata, and M. Okuda, J. Phys. Chem. 99, 16076 (1995).

42. J.M. Lambert, P.M. Ajayan, and P. Bernier, Synth. Met. 70, 1475 (1995).

43. M. Ata et al., J. Appl. Phys. 34, 4207 (1995).

44. W.K. Maser et al., Synth. Met. 81, 243 (1996).

45. A. Loiseau and H. Pascard, Chem. Phys. Lett. 256, 246 (1996).

46. C. Journet et al., Nature 388, 756 (1997).

47. C. Journet and P. Bernier, Appl. Phys. A67, 1 (1998).

48. M. Yudasaka et al., Chem. Phys. Lett. 312, 155 (1999).

49. J. Gavillet et al., Carbon 40, 1649 (2002).

50. M.E. Itkis et al., Nano Lett. 3, 309 (2003).

51. R.O. Loutfy et al., IWFAC'99, Abstracts, St. Petersburg, 109 (1999).

52. D. Golberg et al., Carbon 38, 2017 (2000).

53. M. Takizawa et al., Chem. Phys. Lett. 302, 146 (1999).

54. S. Bandow et al., Phys. Rev. Lett. 80, 3779 (1998).

55. M. Lami de la Chapelle et al., AIP Conf. Proc. 486: Ed. H. Kuzmany et al., 292 (1999).

56. S. Farhat et al., J. Chem. Phys. 115, 6752 (2001).

57. M. Takizawa et al., Chem. Phys. Lett. 326, 351, (2000).

58. A.C. Dillon et al., Chem. Phys. Lett. 316, 13, (2000).

59. I. Hinkov et al., Proc. 6th ADC/2nd FCT Joint Conf., 816 (2001).

60. A. Thess et al., Science 273, 483 (1996).

61. R.O. Loutfy et al., IWFAC'99, Abstracts, St. Petersburg, 117 (1999).

62. A.V. Krestinin and A.P. Moravsky, Chem. Phys. Reports 18, 3 (1999).

63. G.N. Abramovich, Applied Gas Dynamics, Nauka, Moscow (1969).

64. A.P. Moravsky et al., Fullerene Sci. Tech. 6, 453 (1998).

65. B.I. Yakobson and R.E. Smalley, Am. Scientist 85, 324 (1997).

66. R.T.K. Baker et al., J. Catal. 30, 86 (1973).

67. A. Oberlin, M. Endo, and T. Koyama, J. Cryst. Growth 32, 335 (1976).

68. G.G. Tibbets, J. Cryst. Growth 66, 632 (1984).

69. R.T.K. Baker and J.J. Chludzinski, J. Phys. Chem. 90, 4 (1986).

70. Y. Saito et al., Jpn. J. Appl. Phys. 33, 526 (1994).

71. Y. Saito, Carbon 33, 979 (1995).

72. A. Maiti, C.J. Brabec, and J. Bernholc, Phys. Rev. B Rapid Commun. 55, 6097 (1997).

73. J. Lefebvre, R. Antonov, and A.T. Johnson, Appl. Phys. A 67, 71 (1998).

74. H. Kanzow and A. Ding, Phys. Rev. B 60, 11180 (1999).

75. A. Gorbunov et al., Carbon 40, 113 (2002).

76. H. Kataura et al., Carbon 38, 1691 (2000).

77. O.P. Krivoruchko and V.I. Zaikovskii, Mendeleev Commun. 3, 97 (1998).

78. V. Zait, Diffusion in Metals, Nauka, Moscow, 1958.

79. J.M. Cowley et al., Chem. Phys. Lett. 265, 379 (1997).

80. A. Hassanien et al., Appl. Phys. Lett. 73, 3839 (1998).

81. S. Rols et al., Eur. Phys. J. B 10, 263 (1999).

82. J.W.G. Wildoer et al., Nature 391, 59 (1998).

83. T.W. Odom et al., Nature 391, 62 (1998).

84. D. Golberg et al., Carbon 37, 1858 (2000).

85. G. van Tendeloo et al., Carbon 36, 487 (1998).

86. J. Kurti et al., AIP Conf. CP486 Ext. Abstr., Ed. H. Kuzmany et al., 278 (1999).

87. S. Eisebitt et al., AIP Conf. CP486 Ext. Abstr., Ed. H. Kuzmany et al., 304 (1999).

88. W. Clauss et al., AIP Conf. CP486 Ext. Abstr., Ed. H. Kuzmany et al., 308 (1999).

89. J. Hutchison et al., Carbon 39, 761 (2001).

90. J.M. Zuo et al., Science 300, 1419 (2003).

91. A.P. Moravsky et al., unpublished results.

92. A.P. Moravsky and R.O. Loutfy, US Patent pending (2000).
93. L.J. Ci et al., Chem. Phys. Lett. 359, 63 (2002).
94. W.C. Ren et al., Chem. Phys. Lett. 359, 196 (2002).
95. H. Zhu et al., Carbon 40, 2023 (2002).
96. Z. Zhou et al., Carbon 41, 337 (2003).
97. Z. Zhou et al., Carbon 41, 2607 (2003).
98. J.Q. Wei et al., J. Mater. Chem. 13, 1340 (2003).
99. W.Z. Li et al., Chem. Phys. Lett. 368, 299 (2003).
100. Yi-Feng Shi et al., Carbon 41, 1645 (2003).
101. J. Cumings and A. Zettl, Chem. Phys. Lett. 316, 211 (2000).
102. A.A. Puretzky et al., Appl. Phys. A 70, 153 (2000).
103. A.A. Puretzky et al., Appl. Phys. Lett. 76, 182 (2000).
104. M. Yudasaka et al., Chem. Phys. Lett. 299, 91 (1999).
105. F. Kokai et al., J. Phys. Chem. B 103, 4346 (1999).
106. R. Sen et al., Chem. Phys. Lett. 332, 467 (2000).
107. T. Guo et al., Chem. Phys. Lett. 243, 49 (1995).
108. A.G. Rinzler et al., Appl. Phys A 67, 29 (1998).
109. S. Arepalli and C.D. Scott, Chem. Phys. Lett. 302, 139 (1999).
110. S. Arepalli et al., Appl. Phys. A 69, 1 (1999).
111. C.D. Scott et al., Appl. Phys. A 72, 573 (2001).
112. C. Bower et al., Chem. Phys. Lett. 288, 481 (1998).
113. A.A. Gorbunov et al., Appl. Phys. A 69, 593 (1999).
114. O. Jost et al., Appl. Phys. Lett. 75, 2217 (1999).
115. M. Yudasaka et al., J. Phys. Chem. B 103, 3576 (1999).
116. M. Yudasaka et al., J. Phys. Chem. B 102, 10201 (1998).
117. M. Yudasaka et al., J. Phys. Chem. B 102, 4892 (1998).
118. M. Yudasaka et al., J. Phys. Chem. B 103, 6224 (1999).
119. M. Yudasaka et al., Chem. Phys. Lett. 278, 102 (1997).
120. Y. Zhang and S. Iijima, Appl. Phys. Lett. 75, 3087 (1999).
121. Y. Zhang, H. Gu, and S. Iijima, Appl. Phys. Lett. 73, 3827 (1999).
122. H. Kataura et al., Carbon 38, 1691 (2000).
123. T. Gennett et al., Mat. Res. Soc. Symp. Proc. 633, A2.3.1 (2001).
124. A.C. Dillon et al., Mat. Res. Soc. Symp. Proc. 526, 403 (1998).
125. E. Munoz et al., Carbon 36, 525 (1998).
126. W.K. Mazer et al., Chem. Phys. Lett. 292, 587 (1998).
127. V. Bolshakov et al., Diamond Rel. Mater. 11(3), 6 (2002).
128. P. Eklund et al., Nano Lett. 2(6), 561 (2002).
129. D. Laplaze et al., Carbon 36, 685 (1998).
130. E. Anglaret et al., Carbon 36, 1815 (1998).
131. T. Guillard et al., Eur. Phys. J. AP 5, 251 (1999).
132. L. Alvarez et al., CP 486 Conf., Ed. H. Kuzmany et al., 254 (1999).
133. D. Laplaze et al., Mat. Res. Soc. Symp. Proc. 359, 11 (1995).
134. E.Y. Lo et al., AIP Conf. Proc. 288, Eds. J.C. Miller, D.B. Geohegan, p. 291 (1994).
135. A.A. Puretzky et al., AIP Conf. Proc. 288, Eds. J.C. Miller, D.B. Geohegan, p. 365 (1994).

4

Growth: CVD and PECVD

M. Meyyappan
NASA Ames Research Center

4.1 Introduction

As the applications for carbon nanotubes (CNTs) range from nanoelectronics, sensors, and field emitters to composites, reliable growth techniques capable of yielding high purity material in desirable quantities are critical to realize the potential. Two of the earliest techniques for successful CNT growth are arc synthesis and laser ablation, which were discussed in the previous chapter. Of these two, laser ablation is not amenable for large-scale production, but the arc process is suitable for scale-up to meet the material needs of bulk applications such as composites. The figure of merit for the growth process catering to structural applications is the tons-a-day production ability, in addition to high purity; in contrast, applications such as nanoelectronics, field emission, etc., may require controlled growth on patterned substrates at reasonable rates. This need is satisfied by chemical vapor deposition (CVD) and related techniques, which are categorized according to the energy source. When a conventional heat source such as a resistive or inductive heater, furnace, or IR lamp is used, the technique is called thermal CVD. Note that in the literature the term CVD without any prefix commonly refers to thermal CVD. Plasma-enhanced CVD, or PECVD, refers to the case where a plasma source is used to create a glow discharge.

     Both CVD and PECVD have been extensively used to grow a variety of CNT structures. As discussed in Chapter 1, a single-walled nanotube (SWNT) is a rolled-up tubular shell of graphene sheet, which is made up of benzene-type hexagonal rings of carbon atoms. A multiwalled carbon nanotube (MWNT) is a rolled-up stack of graphene sheets in concentric cylinders. The walls of each layer of the MWNT or the graphite basal planes are parallel to the central axis ($\theta = 0$). In contrast, a stacked-cone arrangement in the central core region (also known as chevron structure, ice cream cone structure, piled cone structure) is also seen in some CVD and most PECVD samples; here the angle between the graphite basal planes and the tube axis is nonzero [1,2]. Nolan et al. [2] suggest that hydrogen satisfies the valences at cone edges in such structures. In contrast, a MWNT has no graphite edges; therefore, there is no need for valence-satisfying species such as hydrogen. Because the stacked-cone structures exhibit only small θ values and are not solid cylinders but are mostly hollow, they can be called multiwalled carbon nanofibers (MWNFs) [1]. Note that the terminologies graphitic carbon fibers (GCFs) and vapor-grown carbon fibers (VGCFs) have long been used to denote solid cylinders. MWNFs are also referred to as carbon nanofibers (CNFs) in the literature, but the use of the terminology multiwall in MWNF is more explicit than CNF in specifying the mostly hollow structure with periodic closures of the central core and in distinguishing from solid cylinders.

The CNT-growth literature has grown rapidly in recent years. A reasonably complete, but not exhaustive, list of references in areas discussed in this chapter is given below.

- Early works on carbon fibers and filaments [3–9]
- Thermal CVD of SWNTs [10–24]
- Thermal CVD of MWNTs [2, 25–64]
- CNT growth on atomic force microscope (AFM) cantilevers [65–67]
- Electric field assistance in thermal CVD [68–70]
- Floating catalyst CVD of SWNTs [71–77] and MWNTs [78–80]
- PECVD of MWNTs and MWNFs [1, 81–115]
  - DC plasma [87–97,115]
  - Plasma-assisted hot filament [81–87]
  - Microwave plasma [98–106]
  - rf capacitive and inductive [1, 20, 107, 109–114]
- Theoretical analysis of CNT growth [116–119]
- Computational models and simulations of CNT growth [120–125]
- Gas phase or plasma diagnostics [1,87,92,110,126]
- Purification [127–139]

This chapter is organized as follows. Section 4.2 presents an overview of the growth apparatus used in thermal CVD and PECVD. Catalyst preparation techniques are outlined in Section 4.3. A floating catalyst approach for large-scale production of nanotubes is also discussed in this section. Growth results from thermal CVD of SWNTs and MWNT and plasma-grown MWNTs and MWNFs are presented in Section 4.4. Purification of the nanotubes is also discussed briefly in this section. Growth mechanism is the subject of Section 4.5, and modeling and diagnostics are discussed in Section 4.6. Finally, a summary of challenges and future directions for investigations is presented in Section 4.7.

4.2   Growth Apparatus

CNT growth equipment at present is primarily a home-made, low-throughput batch reactor because research work is normally done on small pieces (~1 in.) of wafers. This will change as the market for CNT-based products begins to emerge, creating the need for large-scale commercial reactors.

4.2.1   Thermal CVD

The thermal CVD apparatus reported in the literature for CNT growth is very simple [14,18,39]. It consists of a quartz tube (1- to 2-in. diameter) inserted into a tubular furnace capable of maintaining ± 1°C over a 25-cm zone. Thus, it is a hot-wall system at primarily atmospheric pressure CVD and, hence, does not require any pumping systems. The substrate, typically about or smaller than 1 in. now, is placed inside the quartz tube. In thermal CVD, either CO or some hydrocarbon such as methane, ethane, ethylene, acetylene, or other higher hydrocarbon is used without dilution. The feedstock is metered through a mass flow controller. A typical growth run would involve purging the reactor first with argon or some other inert gas until the reactor reaches the desired growth temperature. Then the gas flow is switched to the feedstock for the specified growth period. At the end, the gas flow is switched back to the inert gas while the reactor cools down to 300°C or lower before exposing the nanotubes to air.

Exposure to air at elevated temperatures can cause damage to the CNTs. Typical growth rates range from a few nm/min to 2 to 5 μm/min.

Cold-wall reactors common in semiconductor industry, where the substrate holder is directly heated from below (resistance, inductance, or other heaters), have not been reported in the literature. Because the growth is catalyst-promoted at temperatures of 500 to 1000°C and does not depend on precursor dissociation at these temperatures, either hot- or cold-wall systems could be designed to be effective for CNT growth. Indeed, most batch CVD reactors with well-designed gas inflow (shower-head) and outflow manifolds, wafer holder, and support platform with heating systems underneath should work when growth on large areas becomes desirable. Low-pressure operation is possible if advantages are clearly demonstrated.

For growth on substrates, the catalyst mixture needs to be applied to the substrate (see Section 4.3) before loading it inside the reactor. This is the so-called supported catalyst approach. In contrast, CVD can also be used to grow large quantities of nanotubes using a *floating-catalyst* approach [71–80]. In this case, a nozzle system [72,73] may be used to inject the vaporized catalyst precursor into the flowing CO or hydrocarbon. A second furnace may be required to heat up the catalyst precursor system to its dissociation temperature [71,78]. The floating catalyst approach is amenable to scale-up for large-scale production to meet the *commodity market* demands of nanotubes in a variety of structural applications.

4.2.2 PECVD

The plasma enhancement in CVD first emerged in microelectronics because some processes cannot tolerate the elevated temperatures of the thermal CVD. For example, charring of photoresists on patterned wafers could be a problem at elevated temperatures in integrated circuit manufacturing. So, the most common PECVD processes proceed at substantially lower wafer temperatures (room temperature to 100°C); hence, PECVD has become a key step in semiconductor device fabrication. The low-temperature operation is possible because the precursor dissociation necessary for the deposition of all common semiconductor, metallic, and insulating films is enabled by the high-energy electrons in an otherwise cold plasma [140,141].

In CNT growth by CVD, precursor dissociation in the gas phase is not necessary and, indeed, dissociation at the catalytic particle surface appears to be key for nanotube growth. The growth temperature is necessarily maintained below the pyrolysis temperature of the particular hydrocarbon to prevent excessive production of amorphous carbon. At the commonly used temperature of 900°C for CVD of SWNTs from methane, mass spectrometric analysis [126] and computational fluid dynamics modeling [122] confirm that precursor dissociation in the gas phase is negligible and growth proceeds due to catalytic activity of methane on metal particle surfaces. As a result, thermal CVD temperatures have not been lower than 500°C. That is, there is a minimum temperature necessary to activate the catalyst activity on the particle surface enabling growth. Based on this, arguments have been made [1] that the low-temperature advantage of PECVD may not be of much value if catalytic decomposition of hydrocarbon precursors occurs only at 500°C or more. However, recent PECVD results indicate MWNT/MWNF growth at temperatures as low as 120°C [114,115]. First, it is important to indicate that Reference 114 does not report a wafer temperature. Measuring the wafer temperature *in situ* is difficult and, as such, many works report the temperature under the support (i.e., the heater temperature) using a thermocouple. The wafer, exposed to the plasma heating and ion bombardment, would experience a temperature substantially higher than that of the heater. DC plasmas have very high biases on the wafer and lead to wafer heating by ion bombardment. Indeed, it is not uncommon to see red-hot glowing of the wafer. High-density plasmas with inductive and microwave sources experience considerable plasma heating [142]. Additional issues such as proper thermal contact of the wafer with the support, accuracy of the thermocouple, and temperature of the small piece of wafer (normally used) vs. that of the electrode are all serious issues that need consideration. In any case, the chemistry of the PECVD for the same-source gas is different from that of thermal CVD. Taking methane as an example, References 122 and 126 indicate that a thermal CVD reactor at 900°C has nothing but methane, whereas a PECVD reactor with 20%

methane in H_2 contains a variety of radicals and higher stable hydrocarbons (acetylene, ethylene, ethane, etc., as produced in the plasma), all of which can have lower activation temperature for catalytic reactions at the particle surface. Any reported lower temperature growth suggests that the active radicals and other species produced by the discharge may be responsible for growth. In addition to such possible low-temperature growth, PECVD generally enables more vertically aligned structures than does thermal CVD. Whereas any marginal alignment seen in thermal CVD samples is due to a crowding effect (nanotubes supporting each other by van der Waals attraction), individual, free-standing, and vertically oriented structures are possible with PECVD, as will be seen in Section 4.4.

As in the case of earlier technology waves such as semiconductor materials and diamond deposition, a variety of plasma sources have been used: DC [87–97,143], rf [109–114], hot-filament aided with DC [81–87], microwave [98–106], and inductively coupled plasma reactors [1,20,107]. For a treatment of glow-discharge fundamentals and plasma equipment, the reader is referred to References 140 and 141; here, each type of discharge is described briefly.

A DC plasma reactor consists of a pair of electrodes in a grounded chamber with one electrode grounded and the second connected to a power supply. The negative DC bias applied to the cathode leads to a breakdown of the feed-gas. The resulting glow discharge consists of electrons, positive and negative ions, atoms, and radicals. The electron temperature can be in the range of 1 to 10 eV. The ion temperature is about 0.5 eV, and the neutral species may exhibit 500 to 2500 K depending on the DC bias and pressure. The electrode separation distance (d) is normally determined by Pd = constant where P is the pressure. That is, for a given power or cathode bias, the electrodes have to be pulled apart farther as pressure is decreased in order to sustain the discharge. There is a thick sheath region in front of each of the electrodes, which is characterized by high electric fields and low densities. The bulk region of the plasma, which exhibits the glow, consists of higher densities of all the species and a very weak electric field. The wafer with the catalyst layer may be placed on either the anode or the cathode for CNT deposition. The electrode holding the wafer may need an independent heating source to raise the wafer temperature to the desired growth temperature, though high-cathode bias may lead to substantial wafer heating as well. Instead of a resistive heater underneath the electrode, a tungsten wire suspended in the plasma stream may serve as the heating source. This is the so-called hot-filament system with plasma assistance. These two systems have successfully yielded MWNTs and MWNFs as evidenced from References 81 to 97. However, the bias on the wafer in these DC systems appears to be very high (>300 V). The ions gain substantial energy in the sheath from high electric fields, and the high-energy ion bombardment of the wafer often leads to damage, which is well known from silicon and III-V semiconductors literature. It is for this reason and the poor efficiency of the DC systems that the semiconductor industry abandoned the DC source two decades ago. The use of filaments is almost nonexistent in manufacturing due to contamination problems.

Because the plasma can dissociate the hydrocarbon creating a lot of reactive radicals, pure hydrocarbon feedstock in plasma reactors may lead to substantial amorphous carbon deposition. Therefore, it is desirable to dilute the hydrocarbon with argon [1], hydrogen [1,107], or ammonia [83,93,105]. The pressure in the reactor typically ranges from 1 to 20 Torr with a hydrocarbon fraction of up to 20%. Atmospheric pressure operation of plasma systems is uncommon due to power coupling problems regardless of the nature of the power source. Low-pressure operation in the tens of mTorr has not been common except as noted in Reference 144. So, PECVD reactors are typically operated at 1- to 20-Torr pressure levels for CNT growth. At these pressure levels, inductive coupling to hydrocarbon/H_2 systems appears to be difficult; the coupling appears to have a large capacitive component [1,107]. The capacitive component decreases with an increasing fraction of argon or other inert and diluent gases. On the other hand, conventional rf capacitive discharges — common in semiconductor industry for etching and deposition — have been successfully used for nanotube growth [109–114]. As in the case of DC discharges, the system consists of two parallel plate electrodes with one grounded and the second connected to an rf power supply through a matching network. The most common frequency authorized by the FCC is 13.56 MHz. The oscillating bias at the powered electrode is more favorable than the DC bias in terms of ionization, sustaining a discharge at lower power levels and reducing damage due to excessive cathode

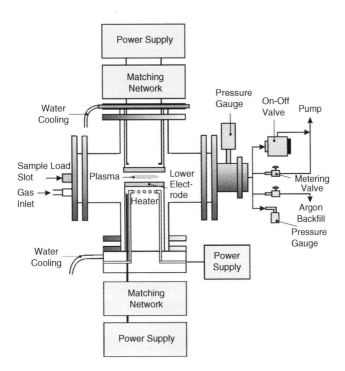

FIGURE 4.1 Schematic of a PECVD setup.

bias. Although the bias at the cathode (both in rf and DC systems) is directly responsible for sustaining the discharge, it is also responsible for accelerating the ions with a lot of kinetic energy toward the wafer, possibly causing the damage. As mentioned before, this is relatively less of a problem in rf than in DC discharges. In any case, it would be ideal if the above functions could be separated; this is precisely what is achieved in inductive and microwave discharges. In these two sources, the power coupling to the plasma is achieved not through the electrode but from an inductive coil or microwave source. The wafer-holding electrode can be independently biased with a DC or rf power supply at the desirable levels, with an option of zero-bias as well.

Microwave sources are very popular at pressures of 1 to 20 Torr and power levels of up to 2 KW and have been widely used for diamond deposition. Following this success, CNT literature also consists of several successful demonstrations of MWNT growth using microwave sources [98–106]. These sources use a 2.45-GHz power supply, which is the frequency allotted by the FCC. The power coupling is achieved through antennas or wave guides. The substrate can be located in the plasma chamber or in a secondary remote chamber to avoid immersing in the plasma. The latter may help reduce the heating from the plasma, if necessary, as well as reduce the effect of ions.

In addition to the plasma source, matching network, and other power-coupling components, the PECVD system consists of mass flow controllers and one or more vacuum pumps. A base pressure of 10^{-5} Torr and operating pressure of 1 Torr or above would require only a roughing or mechanical pump. A desire to eliminate water vapor thoroughly by pumping down to base pressures of 10^{-8} Torr and achieving operating pressures below 100 mTorr would also require a turbo pump. The growth chamber itself is grounded. All plasma reactors are cold-wall systems with the substrate directly heated using some form of heat source from below the substrate holder. A schematic of the PECVD setup is given in Figure 4.1.

For CNT growth, the wafer is first loaded in the reactor and the system is pumped down to 10^{-5} Torr or below to minimize impurities and water vapor. The substrate is then heated to the desired temperature. Then the feedstock is admitted, and the flow rate and chamber pressure can be set to desired levels independent of each other with the aid of a throttle valve. Next, the power from the power source is coupled to the plasma. At the end of the run, the heater, power source, and the gas flow are turned off and the system is purged with argon flow. The wafer is removed after the reactor cools down below 300°C. The reactor can be equipped with a load-lock system to introduce/withdraw the wafer and reduce the downtime associated with reaching the desired vacuum level.

4.3 Catalyst Preparation

Although there have been some studies reporting CNT growth without catalysts in arc discharge processes, it is widely acknowledged that transition metal catalysts are needed for SWNT, MWNT, and MWNF growth by CVD and PECVD. It is also believed that the catalyst on the substrate must be in the form of particles instead of smooth, continuous films. The latter do not appear to yield nanotubes. There have been several studies correlating the catalyst particle size and the diameter of the resulting nanotubes [21,36,43,46,54,92,103,109,143]. The metals used to date as catalysts include Fe, Ni, Co, and Mo. It is possible to apply these onto the substrate from solutions containing them, or they can be directly deposited using some physical techniques. These two approaches are different in terms of needed resources, time, and cost and the nature of the resulting products. Almost every paper cited in this chapter seems to have its own recipe, and a comparative critical evaluation of various catalysts and preparation techniques is not available. A brief overview is provided below on the two routes to supported-catalyst preparation followed by the floating-catalyst approach for large-scale production.

4.3.1 Solution-Based Catalyst Preparation

The literature contains numerous recipes for preparing catalysts from solutions, and one such recipe is given below [41]. First, 0.5 g (0.09 mmol) of Pluronic P-123 triblock copolymer is dissolved in 15 cc of a 2:1 mixture of ethanol and methanol. Next, $SiCl_4$ (0.85 cc, 7.5 mmol) is slowly added using a syringe into the triblock copolymer/alcohol solution and stirred for 30 minutes at room temperature. Stock solutions of $AlCl_3 \cdot 6H_2O$, $CoCl_2 \cdot 6H_2O$, and $Fe(NO_3)_3 \cdot 6H_2O$ are prepared at the concentration of the structure-directing agent (SDA) and inorganic salts. The catalyst solutions are filtered through 0.45-μm polytetrafluoroethylene membranes before applying onto the substrate. The substrate with the catalyst formulation is loaded into a furnace and heated at 700°C for 4 hours in air to render the catalyst active by the decomposition of the inorganic salts and removal of the SDA. Admission of hydrocarbon feedstock into the reactor at this point would initiate nanotube growth.

It is noted that a mixture of transition metal containing compounds along with SDAs is used in the above recipe. It is difficult to come up with optimum concentrations of each constituent in a trial-and-error approach because the number of trials is large. Cassell and coauthors pioneered a combinatorial optimization process for catalyst discovery for the growth of SWNTs [19] and MWNTs [41]. This rapid throughput approach coupled with characterization techniques allows development of catalyst libraries with minimal number of growth experiments.

In general, even with the right formulation known *a priori*, solution-based approaches are time consuming. A typical preparation includes such steps as dissolution, stirring, precipitation, refluxing, separation, cooling, gel formation, reduction, drying/annealing/calcinations, etc. The overall process is cumbersome and time consuming; some recipes even call for overnight annealing. Another problem is the difficulty in confining the catalyst within small patterns.

4.3.2 Physical Techniques for Catalyst Preparation

Physical techniques such as electron gun evaporation [88–90,143], thermal evaporation [92,113], pulsed-laser deposition, ion-beam sputtering [1,18,20,52,64,87,105,107], and magnetron sputtering [83,84, 102,103,109,112] have been successfully used in catalyst preparation. These techniques are quick, easy, and amenable to produce small patterns, in contrast to the solution-based approaches discussed in the previous section. Typically, a thin catalyst film (<20 nm) is applied by these techniques. The eventual particle size and the resultant nanotube diameter seem to correlate to film thickness. Thinner films in general lead to smaller particles and tube diameters [92,143]. Although a small grain size is not guaranteed in as-prepared films, further ensurance steps appear to help break the films into desired particles. For example, in PECVD techniques, an inert gas plasma or hydrogen or ammonia plasma is run first before admitting the feed-gas and initiating growth [83,105]. The plasma ion bombardment will create particles. In thermal CVD, the substrate with the catalyst often first faces a *preparation step* where an inert gas at the growth temperature flows through the reactor for about 10 minutes before admitting the feed gas. This influences the size of the particles. In some cases, particularly with Ni films, a pretreatment with NH_3 has been used where Ni is etched into small particles [83]. Delzeit et al [20,52] have shown that introduction of a metal underlayer (such as Al) can be used instead of any chemical pretreatment steps. Thermodynamics and kinetic studies [2] indicate that alloying a catalyst with a noncatalytic metal increases the number of reactive sites through surface clusters. The metal underlayer may also play the role of barrier layer between an incompatible catalyst metal and substrate, which is the case with Fe and a highly oriented pyrolytic graphite substrate. In addition, an underlayer such as Al allows tuning of the final conductivity of the substrate plus CNTs. In this regard, a variety of metals can be used as underlayer, and the effectiveness of Al, Ir, Ti, Ta, and W as underlayers with each of Co, Ni, and Fe was reported using a combinatorial study in Reference 64.

4.3.3   Floating Catalyst Approach

If a stream of catalyst particles can be injected into the flowing feedstock, it is possible to produce nanotubes in the gas phase. This approach is amenable for scale-up to large-scale production. Sen et al. [78] first reported such a possibility when they used ferrocene or nickolecene as a source of the transition metal and benzene as carbon source. This approach yielded MWNTs, whereas their later work [71] with gas-phase pyrolysis of acetylene using a metallocene yielded SWNTs with diameters around 1 nm. Nikolev et al. [72] used CO disproportionation aided by Fe clusters created from $Fe(CO)_5$. This process, called the HiPCo process, grows SWNTs in the gas phase, and the authors reported nanotubes as small as 0.7 nm in diameter. The transition metal sources vaporize at temperatures much lower than that for the gas-phase pyrolysis of the carbon sources. In using ferrocene or $Fe(CO)_5$, the iron particles condense together and form clusters. An iron cluster of 40 to 50 atoms (~0.7 nm diameter) is about the size of the smallest SWNT. Smaller clusters tend to evaporate and are unstable. Very large clusters are also not ideal for nanotube growth because they favor graphitic overcoating. Declustering or breakup of large clusters also happens in the reactor. It is the competition between various processes (clustering and evaporation) that creates favorable size clusters [120]. Tuning of various parameters such as temperature, flow rates of various gases, injection velocity of the $Fe(CO)_5$, residence time, etc., is done to obtain reasonable quantities of nanotubes. The kinetics of cluster formation and nanotube growth reactions is not well understood, and a preliminary model is presented by Dateo et al. [120]. Availability of reliable kinetics data would lead to reactor and process design simulations, based on traditional chemically reacting flow models, as attempted by Gokcen et al. [121].

4.4   Growth Results

This section discusses a few examples of growth results from thermal CVD and PECVD to introduce various structures possible using these processes. For additional details, the reader is referred to the literature listed in Section 4.1. Also, no attempt is made here to critically evaluate the effectiveness of various *growth recipes* found in the literature. Because carbonaceous and metal impurities accompany the nanotubes, a brief discussion of purification procedures is also provided.

4.4.1   Nanotube Growth

Figure 4.2 shows a typical sample of SWNTs grown by thermal CVD. The feedstock is methane heated to 900°C and catalyzed by a 1-nm layer of iron sputtered on top of a 10-nm Al layer in a silicon substrate.

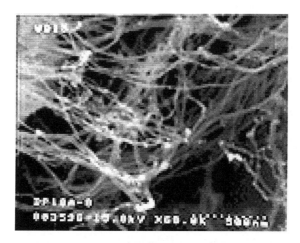

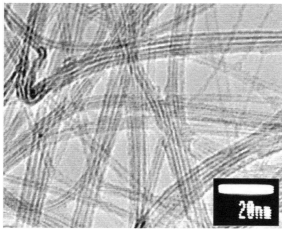

FIGURE 4.2   (Top) SEM image of SWNTs grown by thermal CVD and (bottom) corresponding TEM showing ropes of SWNTs. Image courtesy of Lance Delzeit.

Nanotubes grow like a tangled web, as seen on the scanning electron microscopy (SEM) image. The transmission electron microscopy (TEM) image shows that SWNTs bundle together like ropes. Both TEM and Raman analysis show that the SWNTs in this case are largely 1.3 nm in diameter with distribution ranging from 0.9 to 2.7 nm. The density of growth itself depends on the thickness of the catalyst layer. At growth temperatures, the sputtered catalyst film breaks into small particles of a few nanometers in size, and nanotubes grow out of these particles (see Section 4.5 for growth mechanisms). Regardless of the catalyst preparation technique, particle formation is critical to nanotube growth as mentioned earlier. Indeed, several studies confirm as-prepared catalyst particle distribution using TEM, AFM, or STM and find them well under 10 nm. But what is not clear is the actual particle size or distribution at the growth temperature because there is no *in situ* approach to measure catalyst particle size during growth. Analyzing the size of particles stuck on the base or tip of the nanotubes, it is found that the nanotube diameter approximately correlates to the catalyst particle size.

   Figure 4.3 shows MWNTs grown by thermal CVD at 750°C with ethylene feedstock and using a catalyst preparation recipe discussed in Section 4.3.1. In this solution-based approach, gel-based catalysts form films, and the characteristics depend on the choice of solvents, concentration, the influence of counterions (catalyst ligands), etc. In Figure 4.3, a series of cylinder-shaped towers composed of MWNTs is seen, which are fairly uniform, about 200 μm in diameter and 400-μm tall. When the concentration of the catalyst precursor is increased, a transition occurs from a cylinder to solid towers (not shown here) [41].

FIGURE 4.3   Cylinder-shaped MWNT towers grown by thermal CVD. Image courtesy of Alan Cassell.

Figure 4.4 shows MWNTs formed on a patterned silicon substrate using a mixture of Fe and Ni. Here acetylene diluted with argon is used as feedstock, and the catalyst is pretreated at 750°C for 10 minutes before CVD. A growth time of 5 minutes yields about 25-μm tall patterns shown in Figure 4.4. Although the patterns in Figures 4.3 and 4.4 appear to be vertically well aligned, high-resolution images reveal that the nanotubes indeed grow like *vines* and support each other due to van der Waals forces.

Figure 4.5 shows an SEM image of MWNTs grown using a PECVD approach. Here, the rf-powered electrode (upon which the silicon substrate is held) has a 10-W power. The counter electrode can be grounded (0 W) or connected to an inductive supply (100 W). The feedstock consists of 20-sccm methane diluted with 80 sccm of H_2, and the substrate is heated to 850°C. The catalyst layer consists of 10 nm Fe and 2 nm Mo sputtered on top of a 10-nm Al layer. A thick growth of MWNTs (two to four walls) is observed in Figure 4.5. With such low rf power on the electrode or unbiased electrodes in microwave plasmas, MWNTs can be typically grown. However, most DC discharges with high bias values on the electrode generally yield MWNFs. Figure 4.6 shows MWNFs grown in a DC plasma reactor on templates created by reactive ion etching. Here 1-μm deep trenches in Si_3N_4 have been created by etching, followed by deposition of nickel catalyst at the bottom trenches, and MWNFs have been grown inside these trenches using a DC plasma. In general, PECVD is amenable for growth on patterned substrates. Another example is the patterned nanotube growth (about 2.5-μm tall) shown on this book cover. Here a pattern writing "CARBON NANOTUBES" was generated using electron-beam lithography and lift off with a 7-nm nickel catalyst on a 15-nm indium tin oxide underlayer deposited on n-doped silicon (100) substrate. A 25% acetylene diluted in ammonia is the feedstock, and the growth conditions include 650°C, 5 mbar pressure, and -600 V DC bias on the powered electrode. The MWNFs typically have a bamboo-like inner structure and almost always have the catalyst particles at the top, as seen in Figure 4.7.

4.4.2   Purification

CNTs produced by any of the methods described in this chapter and also Chapter 3 contain impurities, most notably catalyst metal particles and different forms of amorphous carbon. Although it is always a goal to produce CNTs as pure as possible, the research community has developed several postprocessing purification methods [127–139]. This subject is covered here briefly.

Several methods are available for postsynthesis processing for the removal of metal catalysts and other impurities such as amorphous carbon, metal catalyst-containing carbon nanospheres, and polyaromatic carbons including fullerenes. Some common techniques reported in the literature include thermal oxidation in air [127], hydrothermal treatment [128], H_2O-plasma oxidation [129], acid oxidation [130], dispersion separation by microfiltration [131], high-performance liquid chromatrography (HPLC) [132],

FIGURE 4.4 Patterned growth of MWNTs. The bottom image (higher resolution SEM) reveals that nanotubes grow like *vines*. Thermal CVD at 750°C with a mixture of Fe and Ni sputtered onto a silicon substrate. Image courtesy of H.T. Ng.

and a number of combinations of these techniques. In addition to these techniques, Murphy et al. [133] reported the use of a conjugated polymer host system to extract the graphitic particle impurities from CNT soot produced by the arc discharge method. The polymer extraction technique is claimed to be nondestructive because it does not rely on the difference in the oxidation reaction rates for the crystalline graphitic structures vs. the amorphous carbon impurities.

Here purification of SWNTs produced by the HiPCo process (floating catalyst method described in Section 4.3.3. and References 72 and 73) is described first. The major impurity is iron particles (up to 30% by weight). Typical purification procedures [134] consist of an acid treatment to remove the metal, filtration, washing in water, and drying in vacuum. The most effective procedure to date has been reported by Cinke et al. [139] with a final Fe content of ~0.4%, and this procedure is described below.

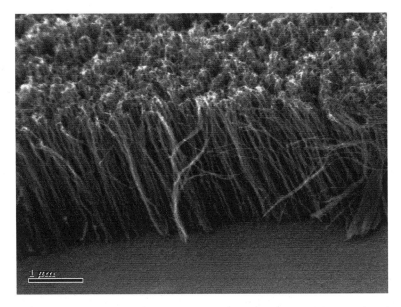

FIGURE 4.5 MWNTs grown by PECVD using CH_4/H_2. The absence of a strong bias on the substrate provides MWNTs with two to four walls. Image courtesy of Lance Delzeit.

FIGURE 4.6 Multiwalled nanofibers grown by a DC plasma. Image courtesy of Alan Cassell.

The procedure in Reference 139 consists of two steps. The first step, designed to debundle the nanotubes, involves suspending raw SWNTs in a mixture of 200 mL DMF (dimethyl formamide) and 100 µl EDA (ethylene diamine), and this solution was stirred for 18 h followed by a 6.5-h sonication. The solution was then centrifuged and the solvent mixture was decanted. The precipitates were centrifuged and decanted twice with methanol as the washing solvent. The entire procedure was repeated once more. The amine and amide groups in these solvents can interact with the π-electrons on the surface of the CNTs. Therefore, this procedure helps loosen the nanotube bundles. In the second step, the DMF/EDA-treated SWNTs were suspended in 250 mL of 37% HCl and sonicated for 15 minutes to get the nanotubes dissolved. The stirred solution was heated to 45°C for 2 h. The solution was then diluted with double-distilled water and cooled to room temperature because the centrifuge tubes cannot tolerate a high concentration of acid. The solution was centrifuged and decanted four times with double distilled water. The SWNTs were dried in air and placed in a quartz boat located at the center of a quartz tube connected to a water bubbler. A stream of wet air was fed into the quartz tube with the tube maintained at 225°C for 18 h, and then the SWNTs were cooled to room temperature. The HCl treatment removes the metals, and the wet-air oxidation removes the amorphous carbon. The combined two-step effort has yielded the largest surface area for SWNTs to date with 1567 m^2/gm.

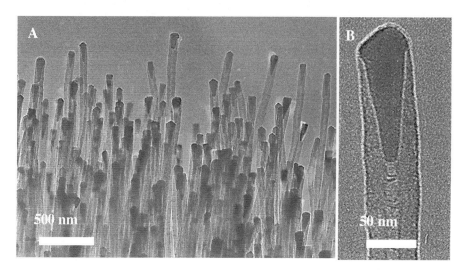

FIGURE 4.7 TEM image showing bamboo structure of the MWNFs and catalyst particles at the top of the fibers. Image courtesy of Alan Cassell.

In general, most procedures described in the literature deal with bulk-produced samples but not arrays. It is desirable to develop procedures to remove metal particles at the top of the PECVD-grown arrays (as in Figures 4.6 and 4.7). Because of the typical end applications of such arrays in electrodes, sensors, etc., polymer extraction and all other purification approaches involving solution dispersion or filtration are not suitable. Moreover, harsh oxidation processes should be avoided because (1) the carbon nanofibers cannot be removed from the substrate and (2) the vertical alignment of the carbon nanofiber array must be maintained. A purification procedure with relatively mild conditions for removing metal catalyst impurities from the tips without damaging the graphitic layers of the vertically aligned arrays has been reported in Reference 138. The MWNFs used in Reference 138 were grown by PECVD and found to contain very little amorphous carbon coating. The plasma contains abundant atomic hydrogen, which etches amorphous carbon but not the more stable MWNFs. TEM images clearly reveal the metal particles at the tip (see Figure 4.7 for example) in need of removal. The first step consists of thermal oxidation in air to convert iron into iron oxide. This is followed by a mild acid treatment in 12% HCl to remove the oxide particles. Final TEM images revealed the absence of particles at the top and no destruction of the graphitic layers while maintaining the vertical orientation of the arrays.

4.5   Growth Mechanisms

Nanotube growth on catalyst particles has similarities to traditional gas-solid interaction processes such as thin film deposition on substrates by CVD and PECVD. The process proceeds according to the following sequence of steps, and one or more of these steps may be rate controlling, which varies from case to case and requires careful experimental analysis:

1. Diffusion of precursor(s) through a thin boundary layer to the substrate
2. Adsorption of reactive species onto the particle surface
3. Surface reactions leading to nanotube formation and gaseous by-products
4. Desorption of gaseous product species from the surface
5. Diffusion of outgassing species through the boundary layer into the bulk stream

In low-temperature plasmas, the bombardment of positive ions on the substrate may provide the energy needed in steps (1) and (3) or aid in the desorption in step (4).

In CNT growth, the above steps may proceed differently in CVD and PECVD. For example, thermal CVD reactors have very little of any other species in the gas phase besides the feedstock hydrocarbon

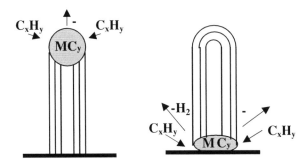

FIGURE 4.8 Base growth (left) and tip growth (right) mechanisms.

itself. In contrast, as discussed earlier, PECVD is characterized by a variety of reactive radicals and atomic hydrogen, along with stable higher hydrocarbons and ions in the plasma. Some of the radicals and higher hydrocarbons may provide the carbon on the particle surface at temperatures lower than in thermal CVD. Whereas carbon dissolution and diffusion into the metal catalyst particle followed by the extrusion of the supersaturated carbon are thought to be the steps in laser ablation and arc discharge, these processes are not deemed essential [119] in CVD and PECVD; instead, hydrocarbons or radicals get rid of their hydrogens, eventually breaking some of their C bonds and assembling on the particle surface to form nanotube structures [119].

In common microelectronics-processing steps such as deposition of silicon, aluminum, silicon nitride, etc., extensive studies have been performed to identify rate-controlling steps and understand surface processes. The diamond literature also features numerous such studies. Unfortunately in CNT growth, there have been no careful experimental investigations to date on this subject. However, earlier studies on carbon filament formation from catalytic pyrolysis in the 1970s discuss the five-step process sequence described above [3–5]. A hydrocarbon such as methane adsorbed on the catalytic particle surface releases carbon upon decomposition, which dissolves and diffuses into the metal particle. When a supersaturated state is reached, carbon precipitates in a crystalline tubular form. At this juncture two different scenarios are possible (Figure 4.8). If the particle adherence to the surface is strong, then carbon precipitates from the top surface of the particle and the filament continues to grow with the particle attached to the substrate. This is called the base growth model. In cases where the particle adherance to the surface is weak, then carbon precipitation occurs at the bottom surface of the particle and the growing filament lifts the particle as it grows. In this case, the top end of the filament contains the catalyst particle, with the resultant scenerio called the tip growth.

Baker and coworkers [3–5] arrived at the above mechanisms for carbon filament growth based on temperature dependent growth rates, activation energy for various steps, and electron microscopy observations. It is commonly believed in the CNT community now that the mechanisms for the filament growth also apply, by extension, to nanotube growth [12]. The most common reason for this belief is the visual observation of catalyst particles on the top or bottom ends of SWNTs/MWNTs/MWNFs, as was the case with filament studies.

In CNT literature the term *vertical alignment* is the most misused. In most cases, any alignment seen is due to crowding effect (i.e., neighboring tubes supporting each other by van der Waals force); in SEM images with large-scale bars, crowded CNTs about 0.5 μm and above in height would look well aligned. But individual CNTs within the ensemble grow almost like vines even though the ensemble looks nicely aligned. This is the case with both thermal and plasma CVD. However, individual, free-standing, vertical carbon nanostructures are also enabled by PECVD [1,83,92,107,143] as seen in Figures 4.6 and 4.7. Invariably, these are MWNFs and, coincidentally, they all follow tip-growth mechanism. In plasma CVD, it is entirely likely that ion bombardment not only creates particles from thin catalyst films but also makes their attachment to the surface weak. Hence, the observed tip-growth mechanism is not surprising. Merkulov et al [90] assert that the presence of the catalyst particle at the tip is essential for the vertical alignment of MWNFs. Combining the effects of electric field and the compressive or tensile stress at the particle interface, they argue that the particle at the tip provides a stable negative feedback mechanism ensuring vertically aligned growth.

4.6 Modeling and Diagnostics

Development of reliable, reproducible, large-scale processes requires an understanding of growth mechanisms, effect of process parameters on growth characteristics, and gas phase and surface kinetics. Such knowledge is possible only with careful diagnostics and modeling studies. Although the CNT literature is full of recipes, growth results, and early application demonstrations, diagnostics and modeling articles are almost nonexistent. A fundamental question in CNT growth relates to the identification of precursor species responsible for growth. In thermal CVD, it is almost unambiguous that the feedstock hydrocarbon reacts at particle surfaces to produce carbon for diffusion into the particles; the temperatures are kept at a level to minimize gas-phase pyrolysis of the hydrocarbons. This has been verified by mass spectrometric analysis [126], which showed negligible decomposition of the methane feedstock. A reacting flow model also confirmed the finding [122]. Detailed numerical simulations of the flow and growth processes have been provided for the first time in References 124 and 125. Large-scale numerical simulations analyzing the HiPCo process for producing SWNTs have been reported [120,121]. The model consists of cluster formation, cluster growth, and breakup for the catalytic particles along with the reactions leading to nanotube formation. This set of reactions is then coupled with conservation equations for mass (of each species), momentum, and energy, and the equations are solved in two dimensions. The results show the temperature profile along the radius and the length of the HiPCo reactor in addition to the information on how well the mixing of the reactants and the catalyst stream is near the nozzle. The predictions compared reasonably with experimental observations.

A simple zero-dimensional model [108,122] was the first to show that in PECVD, in contrast to thermal CVD, the feedstock dissociates extensively and provides the identity and quantities of the numerous radicals, stable species, and ions. A zero-dimensional model is a simple set of mass and energy balances that account for the various sources and sinks for each species as well as the neutral and electron energy. The results are global or reactor-averaged values of the species concentration, electron density, and temperature. The analysis is rather quick and provides a good insight into the plasma chemistry. The results (for electron density and temperature) can be compared with data from a Langmuir probe. The latter is a common diagnostic tool in plasma physics, and References 92 and 110 have effectively used it to characterize the plasma.

It was mentioned earlier that manufacturing reactors avoid using filaments for heating due to contamination problems. Hash et al. [123] modeled a DC PECVD reactor for CNT growth with and without the aid of the filament and concluded that the filament has negligible influence on the system characteristics such as temperature, species distribution, and plasma properties. This modeling result was later confirmed by Cruden et al. [87] who performed a residual gas analysis (RGA) of an acetylene/ammonia plasma with and without the filament. RGA is a relatively simple technique to monitor the downstream products. Another simple diagnostics is optical emission spectroscopy, which was used to show the differences in H-emission intensity between the plasmas that produced MWNTs and MWNFs [1]. As said earlier, there is an extensive knowledge base in the semiconductor processing literature on gas phase, plasma, and surface diagnostics as well as reactor-scale and growth-level modeling. It is important to bring these tools and expertise to the CNT growth field to advance our understanding of mechanisms.

4.7 Challenges and Future Directions

CVD- and PECVD-grown CNT structures have much application potential as described in several chapters of this book. Application development, as expected, is receiving much attention. Successful applications and large-scale production processes depend on understanding of several important issues, which are listed below:

- How can the nanotube diameter be controlled? What is the correlation with the catalyst particle diameter?
- Even if a good correlation is found with catalyst particle size, this size is not retained because heating may result in agglomoration of particles. How can nanotubes of desirable diameter and chirality then be produced?
- For electronics and other device applications, good contact with the metal layer with minimal interface resistance is required. What methods can be used to achieve this and good adhesion to the substrate?
- Is it possible to obtain well-aligned, individually separated, free-standing SWNTs or MWNTs (even few tens to few hundred nanometers tall)? This is possible now only with MWNFs.
- How can the alignment mechanism be utilized?
- How does the catalyst effect issues in terms of the transition metal choice, method of depositing the catalyst, layer thickness, pretreatment (if any), particle creation, effect of particle size on nanotube diameter and growth rate?
- How is/are the determining step(s) in CNT growth rate determined?
- What is the effect of process parameters on tip vs. base growth?
- Is there a preferred substrate heating method (resistive heating, hot filament, IR lamp…)?
- In large-scale gas-phase production techniques, is it possible to do an *in situ* separation of the catalyst particle from the nanotubes?
- In any growth process, can the leftover catalyst metals and amorphous carbon be etched away, *in situ*, selectively?
- What is the role of atomic hydrogen?
- Is amorphous carbon preferentially etched away by atomic hydrogen compared with nanotubes?
- What are the species that are responsible for nanotube growth? This is particularly critical in PECVD because the discharge consists of a variety of chemical derivatives from the feedstock.
- Are the radicals in PECVD deleterious? Do they lead to amorphous carbon contamination due to their high reactivity (high "sticking" coefficient)?
- Is there a specific role for ions in PECVD related to growth? Do they weaken the particle adhesion to the surface? If so, what is the dependence on ion energy? Is the weakened adhesion responsible for more tip-growth observations in PECVD than in CVD?
- Is there a preferred hydrogen-carrying diluent (H_2 vs. NH_3)?
- What is the role of or how effective are other diluents such as argon or nitrogen?
- Is it possible to grow SWNTs by PECVD? How?
- Can disorders be annealed away?
- PECVD grown MWNTs vs. MWNFs: what are the parameters dictating this choice?
- How does electric field influence growth orientation and resulting alignment?
- How low of a growth temperature is possible? This is critical for large-area flat panels on glass substrates or other applications requiring plastic substrates. Is it possible not to compromise on material quality?
- Is a large DC bias appropriate? Does it damage the CNT structure? What is the effect of the substrate bias on growth rate, structure, and alignment?
- Why is the sub–100-mTorr operation common in IC manufacturing not that common in PECVD of CNTs? What is the effect of pressure? What is the effect of other process parameters?
- Is it possible to obtain growth uniformities over large areas common in IC manufacturing?

References

1. L. Delzeit et al., J. Appl. Phys., 91, 6027 (2002).
2. D. Nolan, D.C. Lynch, and A.H. Cutler, J. Phys. Chem. B, 102, 4165 (1998).
3. R.T.K. Baker et al., J. Catal., 26, 51 (1972).
4. R.T.K. Baker, P.S. Harris, R.B. Thomas, and R.J. Waite, J. Catal., 30, 86 (1973).

5. R.T.K. Baker, Carbon, 27, 315 (1989).
6. H.P. Boehm, Carbon, 11, 583 (1973).
7. G.G. Tibbetts, J. Cryst. Growth, 66, 632 (1984).
8. G.G. Tibbetts, M. G. Devour, and E.J. Rodda, Carbon, 25, 367 (1987).
9. G.G. Tibbetts, Appl. Phys. Lett., 42, 666 (1983).
10. H. Dai et al., Chem. Phys. Lett., 260, 471 (1996).
11. J.H. Hafner et al., Chem. Phys. Lett., 296, 195 (1998).
12. J. Kong, H.T. Soh, A. M. Cassell, C. F. Quate, and H. Dai, Nature, 395, 878 (1998).
13. A.M. Cassell et al., J. Am. Chem. Soc., 121, 7975 (1999).
14. A.M. Cassell, J.A. Raymakers, J. King, and H. Dai, J. Phys. Chem. B, 103, 6484 (1999).
15. W. Kim, H. Choi, M. Shim, Y. Li, D. Wang, and H. Dai, Nano Lett., 2, 703 (2002).
16. B. Kitiyanan, W.E. Alvarez, J.H. Harwell, and D.E. Resasco, Chem. Phys. Lett., 317, 497 (2000).
17. M. Su, B. Zheng, and J. Liu, Chem. Phys. Lett., 322, 321 (2000).
18. L. Delzeit, B. Chen, A.M. Cassell, R. Stevens, C. Nguyen, and M. Meyyappan, Chem. Phys. Lett., 348, 368 (2001).
19. B. Chen, G. Parker III, J. Han, M. Meyyappan, and A. Cassell, Chem. Mater., 14, 1891 (2002).
20. L. Delzeit, C.V. Nguyen, R.M. Stevens, J. Han, and M. Meyyappan, Nanotechnology, 13, 280 (2002).
21. Y. Homma et al., Jpn. J. Appl. Phys., 41, 89 (2002).
22. E. Joselevich and C.M. Lieber, Nano Lett., 2, 1137 (2002).
23. B. Zheng et al., Nano Lett., 2, 895 (2002).
24. A. Lan et al., Appl. Phys. Lett., 81, 433 (2002).
25. V. Ivanov et al., Chem. Phys. Lett., 223, 329 (1994).
26. M. Yudasaka et al., Appl. Phys. Lett., 67, 2477 (1995).
27. P.E. Nolan, M.J. Schabel, D.C. Lynch, and A.H. Cutler, Carbon, 33, 79, (1995).
28. J. Jiao, P.E. Nolan, S. Seraphin, A.H. Cutler, and D.C. Lynch, J. Electrochem. Soc., 143, 932 (1996).
29. M. Yudasaka et al., Appl. Phys. Lett., 70, 1817 (1997).
30. P. Chen, H.B. Zhang, G.D. Lin, Q. Hong, and K.R. Tsai, Carbon, 35, 1495 (1997).
31. Z.W. Pan et al., Nature, 394, 631 (1998).
32. S. Fan et al., Science, 283, 512 (1999).
33. J. Li, C. Papadopoulos, J.M. Xu, and M. Moskovits, Appl. Phys. Lett., 75, 367 (1999).
34. C. J. Lee et al., Chem. Phys. Lett., 312, 461 (1999).
35. C.J. Lee, J. H. Park, and J. Park, Chem. Phys. Lett., 323, 560 (2000).
36. C. J. Lee et al., Chem. Phys. Lett., 341, 245 (2001).
37. R.R. Bacsa et al., Chem. Phys. Lett., 323, 566 (2000).
38. K. Hernadi, A. Fonseca, J.B. Nagy, A. Siska, and I. Kiricsi, Appl. Catal. A, 199, 245 (2000).
39. H. Kind et al., Langmuir, 16, 6877 (2000).
40. A.M. Cassell, M. Meyyappan, and J. Han, J. Nanoparticle Res., 2, 387 (2000).
41. A.M. Cassell, S. Verma, L. Delzeit, and M. Meyyappan, Langmuir, 17, 266 (2001).
42. C. Klinke, J.M. Bonard, and K. Kern, Surf. Sci., 492, 195 (2001).
43. J. I. Sohn, C. J. Choi, S. Lee, and T. Y. Seong, Appl. Phys. Lett., 78, 3130 (2001).
44. Y.S. Han, J. Shin, and S.T. Kim, J. Appl. Phys., 90, 5731 (2001).
45. J.M. Ting and C.C. Chang, Appl. Phys. Lett., 80, 324 (2002).
46. G.S. Choi et al., J. Appl. Phys., 91, 3847 (2002).
47. Y. H. Lee et al., J. Appl. Phys., 91, 6044 (2002).
48. C.F. Chen, C.L. Lin, and C.M. Wang, Jpn. J. Appl. Phys., 41, 67 (2002).
49. F. Zheng et al., Nano Lett. 2, 729 (2002).
50. N. Chopra, P.D. Kichambare, R. Andrews, and B.J. Hinds, Nano Lett., 2, 1177 (2002).
51. A. Cassell, L. Delzeit, C. Nguyen, R. Stevens, J. Han, and M. Meyyappan, J. Physique IV, 11, 401 (2001).
52. L. Delzeit et al., J. Phys. Chem. B, 106, 5629 (2002).
53. J.M. Bonard, P. Chauvin, and C. Klinke, Nano Lett., 2, 665 (2002).

54. C. Ducati et al., J. Appl. Phys., 92, 3299 (2002).

55. G.S. Duesberg et al., Nano Lett., 3, 257 (2003).

56. S. Huang and A.W.H. Mau, J. Phys. Chem. B, 107, 8285 (2003).

57. W.B. Choi et al., Adv. Func. Mat., 13, 80 (2003).

58. H.T. Ng et al., Langmuir, 18, 1 (2002).

59. W. Zhang et al., Thin Solid Films, 422, 120 (2002).

60. Y. Yang et al., J. Am. Chem. Soc., 121, 10832 (1999).

61. S. Huang, L. Dai, and A.W.H. Mau, J. Phys. Chem. B, 103, 4223 (1999).

62. S. Huang et al., J. Phys. Chem. B, 104, 2193 (2000).

63. Q. Chen and L. Dai, J. Nanosci. Nanotech., 1, 43 (2001).

64. H.T. Ng et al., J. Phys. Chem. B, 107, 8484 (2003).

65. J.H. Hafner, C.L. Cheung, and C.M. Lieber, J. Am. Chem. Soc., 121, 9750 (1999).

66. C.V. Nguyen et al., Nanotechnology, 12, 363 (2001).

67. C.V. Nguyen, R. Stevens, J. Barber, J. Han, and M. Meyyappan, Appl. Phys. Lett., 81, 901 (2002).

68. Y. Avigal and R. Kalish, Appl. Phys. Lett., 78, 2291 (2001).

69. Y. Zhang et al., Appl. Phys. Lett., 79, 3155 (2001).

70. L. Delzeit, R. Stevens, C. Nguyen, and M. Meyyappan, Int. J. Nanotech., 1, 197 (2002).

71. B.C. Satishkumar, A. Govindraj, R. Sen, and C.N.R. Rao, Chem. Phys. Lett., 293, 47, (1998).

72. P. Nikolaev et al., Chem. Phys. Lett., 313, 91 (1999).

73. M.J. Bronikowski et al., J. Vac. Sci. Technol. A, 19, 1800 (2001).

74. H.M. Cheng et al., Chem Phys. Lett., 289, 602 (1998).

75. H.M. Cheng et al., Appl. Phys. Lett., 72, 3282 (1998).

76. L. Ci et al., Chem. Phys. Lett., 349, 191 (2001).

77. H.W. Zhu et al., Science, 296, 884 (2002).

78. R. Sen, A. Govindaraj, and C. N. R. Rao, Chem. Phys. Lett., 267, 276 (1997).

79. R. Andrews et al., Chem. Phys. Lett., 303, 467 (1999).

80. L. Ci et al., Carbon, 38, 1933 (2000).

81. Y. Chen et al., Chem. Phys. Lett., 272, 178 (1997).

82. Y. Chen, L.P. Guo, D.J. Johnson, and R.H. Prince, J. Cryst. Growth, 193, 342 (1998).

83. Z. F. Ren et al., Science, 282, 1105 (1998).

84. J. Han et al., J. Appl. Phys., 88, 7363 (2000).

85. Y. Hayashi, T. Negishi, and S., Nishino, J. Vac. Sci. Technol. A, 19, 1796 (2001).

86. Z.P. Huang et al., Appl. Phys. A, 74, 387 (2002).

87. B. Cruden, A.M. Cassell, Q. Ye, and M. Meyyappan, J. Appl. Phys. 94, 4070 (2003).

88. V.I. Merkulov et al., Appl. Phys. Lett., 76, 3555 (2000).

89. V.I. Merkulov et al., Appl. Phys. Lett., 79, 1178 (2001).

90. V.I. Merkulov et al., Appl. Phys. Lett., 79, 2970 (2001).

91. K.B.K. Teo et al., Appl. Phys. Lett., 79, 1534 (2001).

92. M. Chhowalla et al., J. Appl. Phys., 90, 5308 (2001).

93. K.B.K. Teo et al., J. Vac. Sci. Technol. B, 20, 116 (2002).

94. M. Tanemura et al., J. Appl. Phys., 90, 1529 (2001).

95. J. Han et al., Thin Solid Films, 409, 120 (2002).

96. J. Han et al., J. Appl. Phys., 91, 483 (2002).

97. Y.Y. Wei et al., Appl. Phys. Lett., 76, 3555 (2001).

98. L.C. Qin, D. Zhou, A.R. Krauss, and D.M. Gruen, Appl. Phys. Lett., 72, 3437 (1998).

99. O. Kuttel et al., Appl. Phys. Lett., 73, 2113 (1998).

100. S. H. Tsai, C. W. Chao, C. L. Lee, and H. C. Shin, Appl. Phys. Lett., 74, 3462 (1999).

101. Q. Zhang et al., J. Phys. Chem. Solids, 61, 1179 (2000).

102. Y. C. Choi et al., J. Vac. Sci. Technol. A, 18, 1864 (2000).

103. Y. C. Choi et al., J. Appl. Phys., 88, 4898 (2000).

104. M. Okai, T. Muneyoshi, T. Yaguchi, and S. Sasaki, Appl. Phys. Lett., 77, 3468 (2000).

105. C. Bower, W. Zhu, S. Jin, and O. Zhou, Appl. Phys. Lett., 77, 830 (2000).

106. H. Cui, O. Zhou, and B.R. Stoner, J. Appl. Phys., 88, 6072 (2000).

107. K. Matthews, B. Cruden, B. Chen, M. Meyyappan, and L. Delzeit, J. Nanosci. Nanotech., 2, 475 (2002).

108. M. Meyyappan, L. Delzeit, A. Cassell, and D. Hash, Plasma Sources Sci. Technol., 12, 205 (2003).

109. G.W. Ho, A.T.S. Wee, J. Lin, and W.C. Tjiu, Thin Solid Films, 388, 73 (2001).

110. H. Ishida et al., Thin Solid Films, 407, 26 (2002).

111. N. Satake et al., Physica B, 323, 290 (2002).

112. Y.H. Wang et al., Appl. Phys. Lett., 79, 680 (2001)

113. L. Valentini et al., J. Appl. Phys., 92, 6188 (2002).

114. B. O. Boskovic et al., Nat. Mater., 1, 165 (2002).

115. S. Hofmann et al., Appl. Phys. Lett., 83, 135 (2003).

116. H. Kanzow and A. Ding, Phys. Rev. B, 60, 11180 (1999).

117. H. Kanzow, C. Lenski, and A. Ding, Phys. Rev. B, 63, 125402 (2001).

118. S.B. Sinnott et al., Chem. Phys. Lett., 315, 25 (1999).

119. V. Vinciguerra et al., Nanotechnology, 14, 655 (2003).

120. C. Dateo, T. Gokcen, and M. Meyyappan, J. Nanosci. Nanotech., 2, 523 (2002).

121. T. Gokcen, C. Dateo, and M. Meyyappan, J. Nanosci. Nanotech., 2, 535 (2002).

122. D. Hash and M. Meyyappan, J. Appl. Phys., 93, 750 (2003).

123. D. Hash, D. Bose, T.R. Govindan, and M. Meyyappan, J. Appl. Phys., 93, 6284 (2003).

124. M. Grujicic, G. Cao, and B. Gersten, Appl. Surf. Sci., 199, 90 (2002).

125. M. Grujicic, G. Cao, and B. Gersten, J. Mat. Sci., 38, 1819 (2003).

126. N.R. Franklin and H. Dai, Adv. Mater., 12, 890 (2002).

127. A. Thess et al., Science, 273, 483 (1996).

128. K. Johji et al., J. Phys. Chem., 101, 1974 (1997).

129. S. Huang and L. Dai, J. Phys. Chem. B, 106, 3543 (2002).

130. L.W. Chiang et al., J. Phys. Chem. B, 105, 1157 (2001).

131. S. Bandow et al., J. Phys. Chem. B 101, 8839 (1997).

132. S. Niyogi et al., J. Am. Chem. Soc., 123, 733 (2001).

133. R. Murphy et al., J. Phys. Chem. B, 106, 3087 (2002).

134. L.W. Chiang et al., J. Phys. Chem. B, 105, 8297 (2001).

135. W. Du et al., Nano Lett., 2, 343 (2002).

136. M. Yudasaka et al., Nano Lett., 1, 487 (2001).

137. C.M. Yang et al., Nano Lett., 2, 385 (2002).

138. C. Nguyen et al., J. Nanosci. Nanotech., 3, 121 (2003).

139. M. Cinke et al., Chem. Phys. Lett., 365, 69 (2002).

140. M.A. Lieberman and A.J. Lichtenberg, *Principles of Plasma Discharges and Materials Processing*, John Wiley and Sons, Inc., New York (1994).

141. *High Density Plasma Sources*, O.A. Popov, Ed., Noyes Publications (1995).

142. D.B. Hash et al., J. Appl. Phys., 90, 2148 (2001).

143. Y.Y. Wei, G. Eres, V.I. Merkulov, and D.H. Lowndes, Appl. Phys. Lett., 78, 1394 (2001).

144. J.B.O. Caughman et al., Appl. Phys. Lett. 83, 1207 (2003).

5

Characterization Techniques in Carbon Nanotube Research

K. McGuire
Clemson University

A. M. Rao
Clemson University

5.1   Introduction

Characterization tools are crucial in the study of emerging materials to evaluate their full potential in applications and to comprehend their basic physical and chemical properties. Typically, a wide range of tools are brought to bear in order to elicit these properties. This chapter presents a variety of techniques that have helped reveal some of the intrinsic properties of carbon nanotubes. The list of techniques that probe these properties is long, and more often these techniques reveal only a portion of the full spectrum of the exciting properties of carbon nanotubes. The following sections describe these techniques; for detailed information on a specific technique, the reader should refer to the original publications in the referenced literature.

5.2   Electron Microscopy

"Seeing is believing" when it comes to proving whether something really exists in nature. This is most certainly the case with carbon nanotubes. How does one really know if such a material exists before tapping into its exciting properties? Using different electron microscopy techniques, one is able to study carbon nanotube structures in great detail and identify their growth mechanism, which in turn helps gain insight into improving the nanotube growth processes or modifying their structure. Scanning electron microscopy (SEM) allows one to image the ropes of single-walled nanotubes (SWNTs) in a sample [1] or to view the highly oriented forest of multiwalled carbon nanotube (MWNT) films grown on quartz or silicon substrates [2]. Both of these examples are seen in Figure 5.1. The typical resolution of a SEM is about 2 to 5 nm. Although the resolution of this technique does not allow for individual SWNTs to be imaged within a SWNT bundle, combined with other techniques it can be used to determine the amount of impurities such as amorphous carbon or carbon-coated catalyst particles, which typically co-exist with SWNT bundles in the sample.

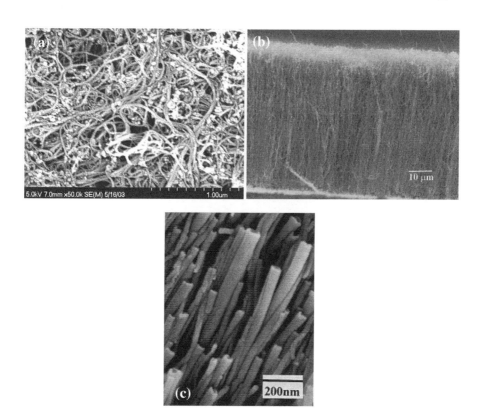

FIGURE 5.1 SEM images of (a) SWNT bundles [1], (b) oriented MWNT film [2] (low resolution), and (c) MWNTs with open ends revealed under higher magnification [2].

Although SEM can be useful in imaging the tubular one-dimensional (1-D) structure of the MWNTs, there is another method of electron microscopy that can be much more helpful in the structural studies of carbon nanotubes. This method is transmission electron microscopy (TEM), and it is a powerful technique that allows one to determine the number of walls in a MWNT [2,3] or image the isolated SWNTs residing inside a SWNT bundle. This allows for careful measurement of the tube diameters as well as investigation of structural defects in carbon nanotubes. Many studies have been done on nanotubes involving TEM [4–15], and Figure 5.2 illustrates the results of some of those studies. TEM has played an important role in investigating new structures such as SWNT peapods (Figure 5.2b) and the effects of nitrogen doping in MWNTs (Figure 5.2d).

Another benefit that comes with using TEM is the use of electron diffraction and electron energy loss spectroscopy (EELS) on the nano scale. Electron diffraction has been used to determine average helicity and the local variations of helicity of the individual SWNT within the ropes of SWNTs [16]. EELS has been particularly beneficial in the area of doped nanotubes [17]. It has allowed researchers to determine the amount of dopant present in the nanotube structure. In conjunction with x-ray energy dispersive spectroscopy, TEM has also enabled the identification of the catalyst composition responsible for nanotube nucleation [18].

It is obvious that electron microscopic (EM) techniques are invaluable for nanotube characterization. There is no doubt that EM will continue to play an important role in future studies pertaining to nanotube research. Recent work has shown that high-energy electron beams in the microscope can also be used to modify the structure of carbon nanotubes. One such example is nano welding of overlapping nanotubes using a high-energy electron beam as described in Figure 5.3 [19] (see color insert following page 146). It is remarkable that irradiation with energetic particles, which is generally believed to create more damage than order, can induce self-assembly processes in graphene cylinders to create well-defined structures.

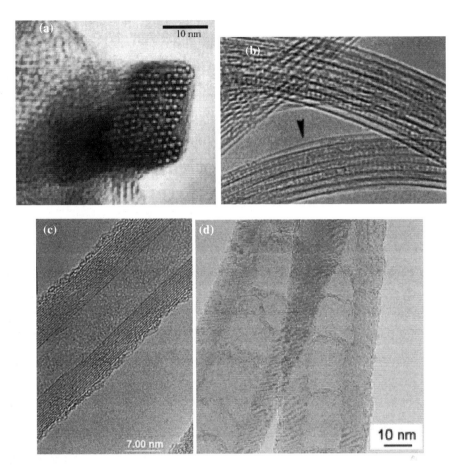

FIGURE 5.2 TEM images of (a) a single bundle containing ~100 SWNTs as it bends through the image plane of the microscope, showing uniform diameter and triangular packing of the nanotubes within the bundle [15]; (b) peapods (C_{60}@SWNT) bundles [8]; (c) a MWNT showing the concentric stacking of several nanotubes with increasing diameters [2]; and (d) bamboo-like morphology of nitrogen-doped MWNTs [5]. (Figure 5.2a reprinted with permission from A. Thess. Science. 293, 483, 1996. ©1996 American Association for the Advancement of Science; Figure 5.2b reprinted from D.E. Luzzi, Carbon, 38, 1751, ©2000, with permission from Elsevier; Figure 5.2d reprinted from C.J. Lee, Chem. Phys. Lett., 359, 115, ©2002, with permission from Elsevier.)

5.3 Atomic Force and Scanning Tunneling Microscopy

The previous section demonstrated the usefulness of EM techniques applied to nanotube research. However, there are two other microscopic methods that are indispensable in nanotube research. These methods are atomic force microscopy (AFM) and scanning tunneling microscopy (STM). Both methods have provided important information of nanotube structure and properties.

AFM has been very useful in imaging isolated SWNTs that have either been grown directly on silicon substrates via a chemical vapor deposition (CVD) process [20–22] (Figure 5.4a) or prepared by processing SWNT bundles (Figure 5.4b) [23]. Although the resolution is not as high as that found in the TEM, diameter estimates as well as the length measurements of the nanotubes within a sample are possible, as can be seen in Figure 5.4.

Interestingly enough, nanotubes have been used as AFM tips [24]. Their 1-D nature allows for them to be used as an improvement over existing AFM tips. Also, nanotubes tend to buckle under mechanical pressure, which will allow a nanotube AFM tip to survive a crash into a sample with little or no damage

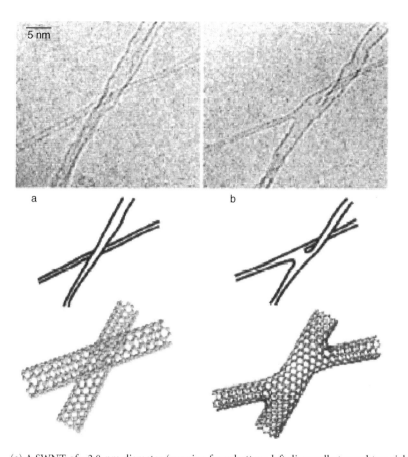

FIGURE 5.3 (a) A SWNT of ~2.0-nm diameter (running from bottom-left diagonally toward top-right) crossing with an individual SWNT of ~0.9-nm diameter; (b) 60 seconds of electron irradiation promotes a molecular connection between the thin and the wide tube, forming an "X" junction. [**See color insert following page 146.**] Schematics show that this junction is twisted out of the plane. Molecular models of each image are provided; heptagonal rings are indicated in red[19]. (Reprinted from M. Terrones et al., Phys. Rev. Lett., 75, 3932, 1999. ©1999 by the American Physical Society.)

to the tip. Thus, not only are nanotubes imaged using AFM, but they also possess qualities that can improve the technique as well. This subject is discussed in detail in Chapter 6.

The last type of imaging technique to be discussed here is STM and spectroscopy, or STM/STS. This technique allows for atomic scale imaging of the nanotube lattice, as well as a direct measurement of the electronic density of states (DOS) of the nanotubes. Several images are given in Figure 5.5 illustrating the usefulness of STM/STS in the field of nanotube research.

STM has proven itself invaluable in studies pertaining to defects within nanotubes, functionalized nanotubes, helicity measurements, and many more [25–32]. Interestingly enough, just as nanotubes serve as efficient AFM tips, they have also been useful as STM/STS tips, which is again due to their 1-D structure and flexible nature [33].

Although imaging techniques certainly can provide useful information about the carbon nanotube structure, other techniques are needed to determine inherent properties of nanotubes. As mentioned previously, some of the imaging methods, such as the STM/STS, can yield information on the fundamental properties as well as structure of carbon nanotubes. However, they all have limitations. For example, isolated SWNTs grown on quartz or silicon substrates cannot be readily studied using high-resolution TEM. Similar limitations can be mentioned for STM/STS. The sample must be dispersed, or grown, on a conducting surface to be examined. However, these limitations are not mentioned to

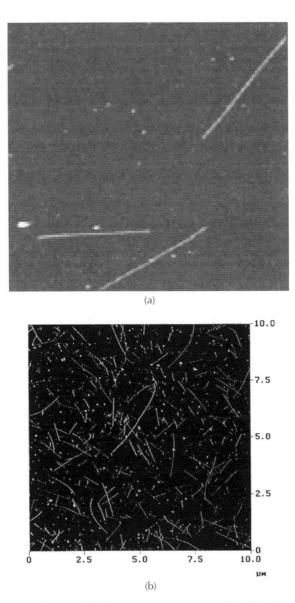

(a)

(b)

FIGURE 5.4 (a) AFM image of isolated SWNTs grown on Si substrates [20]. (b) AFM image of isolated SWNTs prepared from a chemical cutting process of SWNT bundle [23]. (Figure 5.4 reproduced with permission from J. Hafner et al., J. Phys. Chem. B, 105, 743, 2001. ©2001 American Chemical Society; Figure 5.4b reproduced with permission from J. Chen et al., J. Phys. Chem. B, 105, 2525, 2001. ©2001 American Chemical Society.)

downplay the importance of imaging methods. They are listed only to give an idea of which technique is best suited for determining specific information of the nanotube sample. They have played and will continue to play an important role in the research done on all nanomaterials.

5.4 Properties Characterization

Certainly the most exciting aspect of carbon nanotubes is the quantum confinement of the electrons in the circumferential direction leading to its new properties that are distinctly different from planar graphite. These properties have allowed carbon nanotubes to gain an edge over existing materials in future applications. SWNTs are known to be semiconducting or metallic depending on the (n,m) indices

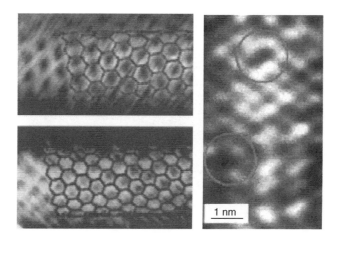

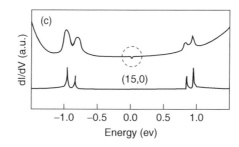

FIGURE 5.5 (a) STM images of two SWNTs with a projection of an ideal SWNT lattice superimposed over the STM image [25]. Such analysis helps identify the chirality of a specific nanotube. (b) STM image showing lattice distortions (circled in the image) in a MWNT due to nitrogen doping [26]. (c) Tunneling conductance for (15,0) zigzag SWNTs (top) compared with the corresponding calculated DOS (bottom). At E_p a new feature is highlighted with a dashed circle demonstrating that the metallic (15,0) nanotube is actually a narrow band gap semiconductor [27]. (Figure 5.5a reprinted from P. Kim et al., Carb on, 38, 1741, ©2000, with permission from Elsevier; Figure 5.5b reprinted from R. Czerw et al., Nano Lett., 1, 457, 2001, with permission; Figure 5.5c reprinted, with permission, from the Annual Review of Physical Chemistry, Volume 53, ©2002 by Annual Reviews. www.annualreviews.org)

that define the nanotube structure [27,34]. Thus, nanotubes are ideal candidates for nanoelectronic applications and devices [35,36], which is the subject of Chapters 7 and 8. Changes in the electronic structure of nanotubes can also occur in the presence of defects in the lattice, intercalation of various compounds, and adsorbed species on the nanotube surface [37–41].

The higher thermal conductivity and enhanced mechanical properties have led to research in the areas of nanotube composites. Examples where nanotubes serve as ideal filler in the composite can be found in the nanotube literature [42–47] and this is discussed in detail in Chapter 11. Thermal, electrical, and vibrational properties divulge important information that can enhance the basic science and applications of carbon nanotubes. Several methods that measure nanotube properties are discussed next. Each of these methods provides varying degrees of information on the aforementioned properties, as will be seen.

5.5 Electrical Conductivity Measurements

Measurements such as the current-voltage (I-V), conductivity, and resistivity are commonly used to identify how a nanotube behaves in a given nanotube-based device. There are already many reports that have used these tools for investigating the behavior of the nanotubes in field-effect transistors (FETs) and composites [48–50]. Figure 5.6a illustrates the I-V characteristics for a single nanotube-based

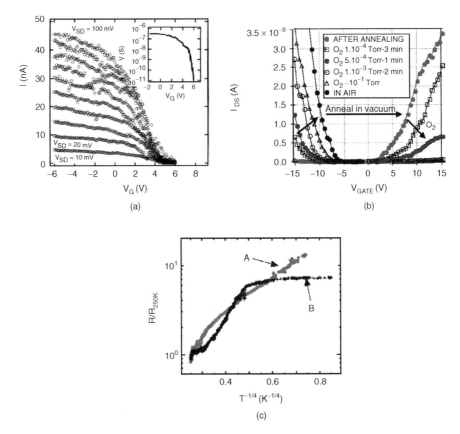

FIGURE 5.6 (a) Room temperature I-V data of a single SWNT FET. The inset shows the variation of the low-bias conductance of the nanotube as a function of the gate voltage [51]. (b) I-V data of a nanotube-based FET as it is operated under ambient conditions (see text). (c) Resistivity data for two different samples (A and B) of C_{60}@SWNT peapods [52]. (Figures 5.6a and 5.6b reprinted from Ph. Avouris et al., Appl. Surf. Sci., 141, 201, ©1999, with permission from Elsevier; Figure 5.6c reprinted from H. Hongo et al., Physica B, 323, 244, ©2002, with permission from Elsevier.)

oxygen-exposed FET in which the current flows through the nanotube for negative gate voltages [51]. Careful studies revealed that freshly prepared FET exposed to ambient conditions exhibit *n*-type behavior, which subsequently begins to exhibit *p*-type characteristics as the device ages in ambient conditions (Figure 5.6b). Upon degassing the air-exposed FET, the *n*-type characteristics are completely recovered. These changes in device characteristics have been attributed to changes in the Schottky barrier of the electrical contacts due to adsorption of oxygen [51]. Returning to Figure 5.6a, the data also indicate that the primary carriers in the nanotube are holes due to the current enhancement at a negative gate bias [51]. Electrical properties of modified nanotube structures, such as C_{60}@SWNTs (peapods) have also been determined (Figure 5.6c). The results indicated that the peapods have a resistivity two to three orders of magnitude higher than SWNTs [52]. This high resistivity is thought to come about due to defects present within the material, possibly induced by the synthesis process itself.

Another interesting structure that has been investigated by means of electrical transport measurements is the carbon nanotube coil. The results of this study found that the conductivity of the nanocoils varied between 100 and 180 S/cm, which was less than that found in regular nanotubes [53].

Obviously, modification of the nanotube structure by any means can lead to significant changes in the electronic structure and properties of these materials. Electrical transport measurements will be necessary in evaluating how these structural changes will affect the feasibility of using carbon nanotubes in future nanomaterial-based devices.

5.6   Thermoelectric Measurements

Thermoelectric measurements have been used for quite some time to identify materials for novel applications, such as refrigeration and waste heat recycling. Now this technique is being used extensively to investigate the properties of carbon. The thermoelectric power is a measure of the electrical potential developed when a temperature gradient (ΔT) is applied across the sample due to a migration of the higher-energy electrons to the cooler side of the sample. As a result, a potential difference (ΔV) is set up across the sample, which can be measured. Thermopower is expressed as

$$S = \frac{\Delta V}{\Delta T}$$

and relates to the electrical potential change induced by the temperature gradient. Because thermopower is related to the energy derivative of electrical conductivity at the Fermi energy, the sign of thermopower directly reflects the type of dominant charge carriers in the sample.

Thermoelectric measurements of nanotubes have been performed on individual SWNTs [54] and on films [55–58], which have been useful in investigating the effect that adsorbed gases can have on the intrinsic properties of SWNTs [55–58]. Figure 5.7 is an excellent example of how adsorbed oxygen on a SWNT sample can lead to a large change in the intrinsic thermopower of the SWNTs. Through the desorption process (solid dots), the thermopower decreases to a negative value, indicating that the natural charge carriers in the nanotubes are electrons. However, as the sample undergoes exposure to oxygen (empty dots), the thermopower begins to increase to a high positive value. This suggests that the adsorbed oxygen brings about a change in the dominant carrier type to holes by removing electrons from the SWNTs. From a detailed Raman study of the O_2 stretching mode frequency, it was concluded that the charge transfer from SWNTs to molecular O_2 is ~0.1e to 0.2e [59]. This information is useful when accessing SWNTs in a particular application in which the charge carrier type may be critical, for example, in nanoelectronic devices. It should be mentioned that the thermopower of SWNT bundles is sensitive to the type of adsorbed gas species and this fact has been exploited in gas detection applications [55,58]; gas sensor applications of carbon nanotubes are covered in Chapter 9.

Thermopower measurements have also been used to investigate the effect of dopants in carbon nanotubes [60,61]. Figure 5.8a shows the results for the thermopower measurements done on nitrogen-doped MWNTs [60]. The nitrogen dopant acts as an electron donor and is expected to make the doped

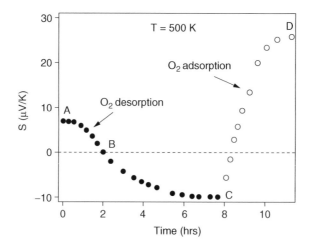

FIGURE 5.7   Effect of oxygen desorption on the thermopower for SWNT bundles [58]. (Reprinted with permission from G.U. Sumanasekara, Phys. Rev. Lett., 85, 1096, 2000. ©2000 by the American Physical Society.)

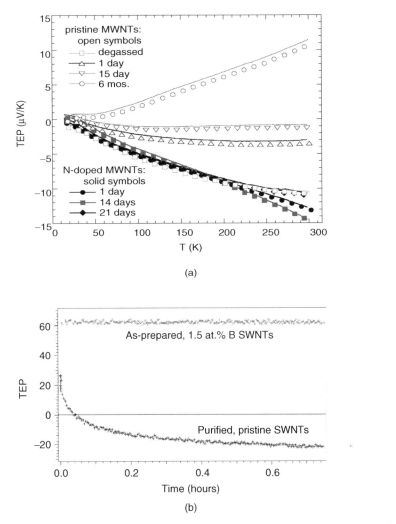

(a)

(b)

FIGURE 5.8 (a) Thermopower for pristine (open symbols) and nitrogen-doped (solid symbols) MWNTs as a function of temperature and exposure time to room air. The resolution or precision of the described techniques is 1 μV/K [60]. (b) TEP data collected from purified, undoped SWNT bundles and as-prepared B-doped SWNT bundles [61].

nanotubes permanently *n*-type. This is clearly illustrated in the figure as the samples (undoped and nitrogen doped) are exposed to air over the course of time. Whereas the thermopower of the undoped MWNTs slowly turns positive (*p*-type), the thermopower of the nitrogen-doped sample stays negative, indicating that the nitrogen-doped nanotubes are permanently *n*-type [60]. Likewise, doping SWNTs with boron makes them *p*-type materials. A comparison of purified, pristine SWNT bundles compared with the boron-doped SWNT bundles is shown in Figure 5.8b. The thermopower of the pristine SWNT bundles turns negative as the sample is degassed of adsorbed oxygen, whereas the thermopower in B-doped SWNT bundles remains strongly positive, implying that the boron-doping caused the sample to become permanently *p*-type.

5.7 Raman Spectroscopy

Over the past 20 years, Raman scattering has proven to be a very useful probe of carbon-based materials and has been used extensively to study the bonding and properties of pristine, metallic, and superconducting

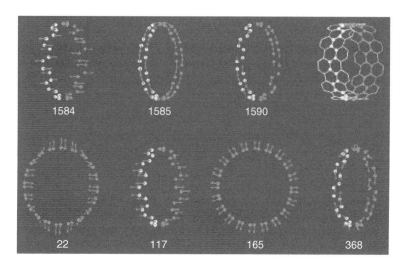

FIGURE 5.9 Raman active eigenmodes along with their frequencies for a (10,10) nanotube. The red arrows show the magnitude and displacement of the carbon atoms in the nanotube lattice, and the unit cell for this nanotube is shown in the upper right-hand corner of the figure [66]. [**See color insert following page 146.**]

phases of graphite intercalation compounds and fullerene-based solids [62–71]. So it seems natural that this spectroscopy may be equally useful to characterize carbon nanotubes. This section provides some of the highlights of recent Raman studies on these fascinating *quantum* wires. Basically, Raman scattering in nanotubes results from the inelastic scattering of light from the nanotubes, leading to an increase or decrease in the energy of the incident light due to an emission or absorption of a phonon present in the nanotube.

The calculated eigenmodes and frequencies for a (10,10) single-wall carbon are depicted in Figure 5.9 [66] (see color insert following page 146). The notation (n,m) defines the atomic coordinates for the 1-D unit cell of the nanotube. For $n = m > 0$, the tube has chiral symmetry. Achiral tubes exists if $m = 0$ or $n = m$. The former and the latter subclasses of achiral tubes are referred to as *zigzag* $(n,0)$ and *armchair* (n,n) tubes. The Raman mode at ~165 cm^{-1} is the calculated radial breathing mode (RBM) in which the displacement of the carbon atoms is purely along the radial direction. The modes in the top row in Figure 5.9 involve purely tangential displacement of the carbon atoms and are referred as the tangential vibrational modes. As an example, the experimentally measured Raman spectrum for a purified SWNT bundle is shown in Figure 5.10. The Raman spectrum was collected in the backscattering geometry using 514 nm Ar laser radiation. Note that sections of the experimental spectrum have been scaled vertically (as indicated) to best expose the rich detail.

Furthermore, the frequency scale has been expanded for the highest frequency region. If the tube would have been *unzipped* and laid out flat, only a single Raman active mode at ~1582 cm^{-1} would have been expected. There are typically two, well-defined peaks seen in the tangential band region for carbon nanotubes. Upon rolling a graphite sheet into a nanotube structure, the tube curvature lifts the degeneracy of the tangential mode, resulting in two peaks that correspond to an axial and circumferential tangential mode. By rolling the graphene sheet into tube, numerous other vibrational modes are made Raman active via the cyclic boundary condition. For comparison with experiment, the calculated Raman spectrum is shown in the bottom panel for $n = 8$ to 11 armchair symmetry tubes. The calculated spectra in Figure 5.10 are for a range of tube diameters consistent with those diameters determined through TEM studies, and the vibrational frequencies are calculated with the same C-C force constants used to fit Raman, neutron, and electron energy loss data for a flat graphene sheet. The small effect of curvature on the force constants was not taken into account [66]. The theoretical Raman intensities were calculated using a bond polarizability model. It may be noticed that some of the experimental Raman bands are narrow and some are broad. This difference in width is attributed to an inhomogeneous line-broadening mechanism, which

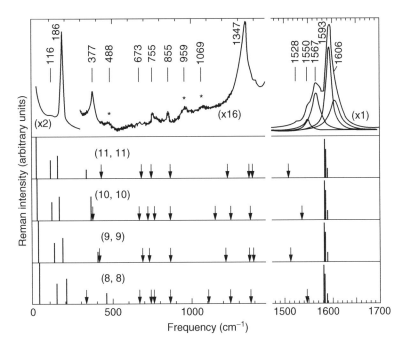

FIGURE 5.10 Experimentally measured Raman spectrum compared with the theoretically determined Raman peak positions and intensities for several armchair nanotubes [66].

stems from the fact that many (but not all) of the mode frequencies exhibit a strong tube-diameter dependence. This can be appreciated by inspecting the calculated frequencies in Figure 5.10. Note that the low-frequency modes exhibit the strongest diameter dependence. Because the frequency of the RBM feature varies as $\omega_{RBM} \propto 1/d_t$, the RBM mode serves as a convenient tool for the determination of the tube diameter, which is one of the parameters needed for the (n,m) characterization.

Besides the RBM and tangential band, the third very distinct and dominant mode that appears in the Raman spectrum of a SWNT bundle is the disorder-induced peak (D-band) around 1300 cm⁻¹. This mode is similar to the disorder mode seen in graphite regarding its position and dispersive nature. As can be seen in Figure 5.10, there are many other smaller modes that come about from rolling a graphite sheet into a seamless tubular structure. However, the modes that are of most importance are the three that have been mentioned and are the main focus, along with their second-order peaks as reported in many studies [72,73].

In the nanotube, the cyclic boundary conditions around the tube waist activate new Raman and IR modes not observable in a flat graphene sheet or in graphite. This *explosion* in Raman activity is the first consequence of 1-D *quantum confinement* in a small-diameter nanotube. A second quantum confinement effect stems from the presence of singularities in the electronic DOS which occur because the electrons must propagate down the tube axis in a true 1-D periodic structure (Figure 5.11a). These singularities are the source of optical resonances in the Raman scattering cross section that occurs when the laser photons drive optical transitions between these singularities. The manifestation of these resonances can be observed in Figure 5.11b.

The resonant Raman scattering in SWNTs is an important issue that deserves some discussion. This can be done briefly with the visual aid of the calculated electronic DOS for SWNTs (Figure 5.11a). Depending on the diameter and chiral vector of the nanotube, the DOS can vary from tube to tube. The sharp, narrow peaks that arise in the SWNT DOS are referred to as van Hove singularities and come about due to the reduced dimensions of the nanotube to 1-D. If the energy of the incident light matches the transition from between two peaks that are mirrored in the DOS plot, then a resonance enhancement of the spectrum occurs. Examples of these types of transitions would be $v_1 c_1$ and $v_2 c_2$ in the (8,8) tube.

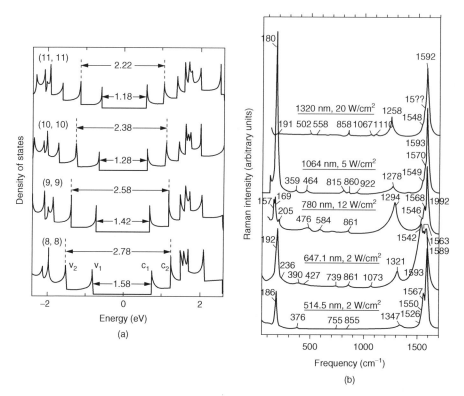

FIGURE 5.11 (a) Calculated electronic DOS for several SWNTs [66]. (b) Raman spectra of SWNT bundles taken with various laser wavelengths [66].

It is also worth noting that these energy transitions between the van Hove singularities are diameter dependent. By finding a resonant enhancement it is possible to measure the energy gaps between these van Hove singularities by varying the laser wavelength, thus giving insight into the electronic structure of the nanotube of a particular diameter. In SWNT bundles, the resonance effect occurs and results in a Raman spectrum for a particular diameter distribution of nanotubes within the sample. Figure 5.11b illustrates the resonant scattering phenomena in nanotubes, which lead to the observation of a different Raman spectrum from the same nanotube bundle as the excitation wavelength is varied [66].

Calculations have also been done to correlate nanotube diameter with the diameter-dependent transition energies of van Hove singularities in SWNTs. The result of these calculations is the Kataura plot [74] (Figure 5.12) in which various $v_i c_i$ energies, now designated as E_{ii}, are plotted as a function of nanotube diameter. Because SWNTs can be either semiconducting or metallic, the E_{ii} notation is changed to E_{ii}^{y}, where $y = S$ or M for semiconducting or metallic SWNTs, respectively. For a given tube-diameter distribution in the sample, this plot is useful in determining whether a given laser excitation wavelength couples to semiconducting or metallic SWNTs.

Because the energies of these sharp features in the 1-D electronic DOS are strongly dependent on the nanotube diameter, a change in the laser frequency brings into resonance a different SWNT with a different diameter that satisfies the new resonance condition. This effect is called *diameter-selective Raman spectroscopy* and led in 1998 to the first use of Raman spectroscopy to distinguish between metallic and semiconducting nanotubes [66]. The concept of diameter-selective spectroscopy follows from Figure 5.12, where each point on the plot of the energy E_{ii} of the van Hove singularity in the DOS vs. tube diameter, d_t, denotes a different (n,m) SWNT. For a given diameter distribution in the sample (e.g., 1.37 ± 0.20 nm, for the sample in Figure 5.13), for E_{laser} below 1.70 and above 2.14 eV, the semiconducting nanotubes are resonant with E_{laser} in agreement with the line shape shown for these traces in Figure 5.13. On the other hand, for E_{laser} in the range 1.83 to 2.14 eV, the G-band shows a very different broad line

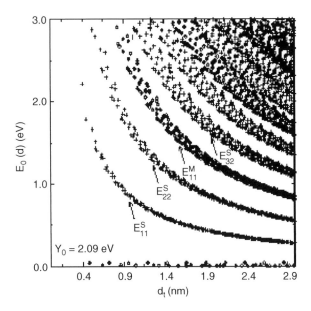

FIGURE 5.12 Kataura plot generated for SWNTs in a diameter range of 0.4 to 3nm [74]. (Reprinted from H. Kataura et al., Synthetic Metals, 103, 2555, ©1999, with permission from Elsevier.)

shape, characteristic of metallic nanotubes [75]. Here the lower-frequency G-band component (designated by G^-) for metallic tubes is described by a Breit-Wigner-Fano lineshape arising from the coupling of the G^- phonons to plasmons through the tube curvature, which allows mixing of out-of-plane phonons with the tangential in-plane phonon modes characteristic of two-dimensional graphite.

As described in Figures 5.11a and 5.12, the electron DOS is sensitive to the chirality of the SWNT. This fact has been cleverly exploited by nanotube researchers to identify nanotube chirality based on a combined Raman/absorption/fluorescence spectroscopy of isolated semiconducting SWNTs. The first study was reported on a SWNT mixture in which each tube was encased in a cylindrical micelle suspended in water [76,77]. As expected, the spectra of the ensemble were composed of overlapping absorption/fluorescence spectra, which obscure the true spectral linewidth (Figures 5.14 and 5.15).

The images in Figure 5.14 show distinct bright spots at different positions, indicating isolated emission sources with different emission energies. Representative fluorescence spectra detected at these bright spots is shown in panel A of Figure 5.16 for the three wavelength regions identified in Figure 5.14. Unlike the fluorescence shown in Figure 5.15, a single fluorescence band is observed for each region and is well described by a fit to a single Lorentzian peak. The corresponding radial breathing Raman-active modes were also single bands (panel B in Figure 5.16) directly reflecting the diameter of the SWNT. Collectively, the data presented in Figure 5.16 allowed unambiguous structural assignment for the SWNTs (Table 5.1).

Theory and experiment have shown that the pristine armchair SWNTs are weakly metallic. Experiments were then designed to investigate whether chemical doping could be used to enhance the electrical conductivity. The dopant atoms or molecules are thought to reside in the channels between the tubes. A SWNT bundle offers at least three different sites where dopant atoms and molecules can reside without substituting the carbon atoms in the nanotube cage with dopant atoms. These sites are schematically shown in Figure 5.17. If the end caps of the tubes in a bundle are opened, the dopant may reside inside a SWNT. These sites are called the endohedral sites. Regardless of the endcaps being opened or closed, the interstitial sites and groove sites are always accessible. In the case of substitutional doping, the dopant atoms substitute for the carbon atoms in the framework of a single SWNT. Beyond a certain doping concentration in the substitutionally doped SWNT bundles, the intensity of the disorder band in the Raman spectrum grows in intensity.

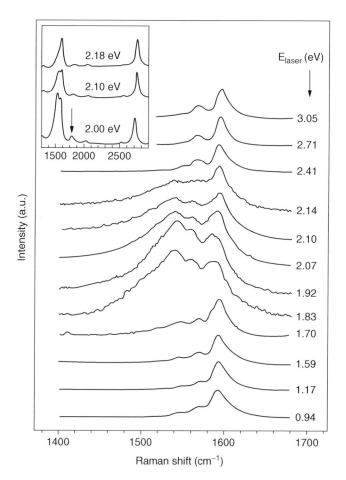

FIGURE 5.13 Raman spectra of the tangential modes of carbon nanotubes with diameters in the range $d_t = 1.37 \pm 0.20$ nm, obtained with several different laser excitation energies E_{laser} [75]. The inset shows low-resolution Raman spectra between 1300 and 2800 cm^{-1} in the range of laser energies 2.00 to 2.18 eV where the metallic nanotubes are dominant, and the arrow points to a low-intensity M-band feature near 1750 cm^{-1}, which has its maximum intensity correlated with the maximum intensity of the metallic G-band lineshape.

In a series of Raman scattering studies and electrical resistivity experiments, the amphoteric nature of the SWNTs was shown for the first time [67,79]. A material is *amphoteric* when it possesses the ability to accept (or donate) electrons from (or to) dopant atoms or molecules. Using Raman scattering, doping-induced shifts in the high-frequency vibrational modes were observed while the electrical conductivity (under identical reaction conditions) increased by over a factor of ten [79]. Donor reagents such as K or Rb were observed to downshift the Raman-active mode frequencies, whereas acceptor reagents (e.g., Br$_2$) were found to upshift these mode frequencies. Similar behavior in the Raman-active modes was observed during the doping of graphite to form graphite intercalation compounds and is associated with an expansion (contraction) in the carbon hexagons.

In the case of Rb and K intercalated SWNTs, the intercalated dopant acts as an electron donor. This extra electron donated by the alkali metal to the SWNT tends to weaken the C-C bonds in the SWNTs, because electrons have been known to soften the C-C bond in all sp^2-bonded carbon materials. The result of a weakened bond is a downshift in tangential band from 1593 cm^{-1} to 1557 cm^{-1} and 1554 cm^{-1} for the Rb and K intercalated SWNTs, respectively (Figure 5.18). In the case of the I$_2$ and Br$_2$ intercalated SWNTs, the opposite behavior is expected as the two halogens act as electron acceptors. This is clearly seen in the Br$_2$ intercalated SWNTs seen as an increase in the tangential-mode frequency to 1617 cm^{-1}.

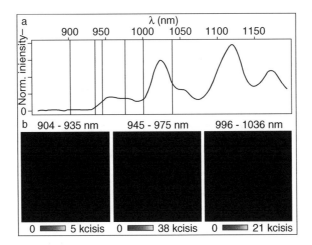

FIGURE 5.14 (a) Fluorescence images and spectra collected from an ensemble of isolated nanotubes [78]. The spectra were excited using the 633-nm laser excitation energy. (b) Confocal fluorescence images of single spatially isolated SWNTs on glass acquired by raster scanning of the sample and simultaneous detection of an optical spectrum at every pixel. Spectroscopic images were obtained by integration of the detected intensities within the spectral windows in (a) [78]. (Reprinted with permission from A. Hartschuh et al., Science, 301, 1354, ©2003 AAAS.)

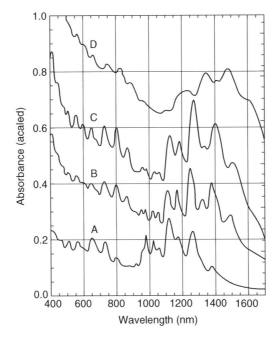

FIGURE 5.15 Absorbance spectra of nanotubes as reported by O'Connell et al. [76]. The spectra are sensitive to the different processing conditions employed in that study. (Reprinted with permission from M.J. O'Connell et al., Science, 297, 593, ©2002 AAAS.)

However, this is not seen clearly in the I_2 vapor intercalated SWNTs. Subsequent research involving a more rigorous intercalation process did observe a clear upshift in the tangential-mode frequency [80]. Other changes that occur in the spectra include the Breit-Wigner-Fano (BWF) lineshape of the Rb- and K-intercalated SWNTs that come about due to interference of Raman scattering between a continuum of excitations and discrete phonons [67]. This line shape is commonly seen in metallic tubes [66,75]. An

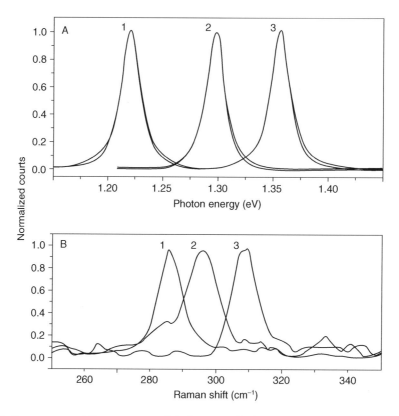

FIGURE 5.16 Fluorescence and Raman spectra of single SWNTs [78]. (A) Fluorescence spectra were detected for three different sample positions, labeled 1 to 3. Solid lines are fits with single Lorentzian line-shape functions. (B) Raman spectra obtained from the positions described in (A). The Raman shift corresponds to scattering from the nanotube RBM. (Reprinted with permission from A. Hartschuh et al., Science, 301, 1354, ©2003 AAAS.)

TABLE 5.1   Spectral Data and Structural Assignments for SWNTs

λ_{Em} (nm)	$h\nu_{Em}$ (eV)	$h\nu_{Em}$ (lit.) (eV)	Observed $\nu_{RBM} \pm 5$ (cm^{-1})	Predicted ν_{RBM} (cm^{-1})	(n,m) Structure
1023	1.212 (1.220[1])	1.212	286[1]	281.9	(7,5)
976	1.270	1.272	—	307.4	(6,5)
955	1.298 (1.298[2])	1.302	296[2]	298.1	(8,3)
915	1.355 (1.357[3])	1.359	309[3]	307.4	(9,1)
881	1.407	1.420	—	335.2	(6,4)

Note: λ_{Em} is the measured average emission wavelength. $h\nu_{Em}$ is the measured average emission energy. Superscripts correspond to the spectra (1, 2, and 3) marked in Figure 5.16. Raman spectra were not observed because the excitation energy was nonresonant. (Reprinted with permission from A. Hartschuh et al., Science, 301, 1354, ©2003 AAAS.)

actual measure of the amount of charge transfer has even been measured using Raman spectroscopy and found that in the case of the Br-intercalated SWNTs that the charge transfer was ~1 free hole per 19 carbon atoms [67].

Although the discussion up to this point has focused primarily on SWNTs, it should be mentioned that Raman spectroscopy has also been used for characterizing MWNTs [81–84]. The spectrum for MWNTs is quite different compared with SWNTs and consists of the disorder and tangential bands that appear at ~1300 cm^{-1} and 1584 cm^{-1}, respectively. While the dispersive D-band appears in approximately the same region of the spectrum as it does in graphite and SWNTs, the tangential band shows no splitting as seen in SWNTs and appears very much like the G-band in graphite. Other notable differences in

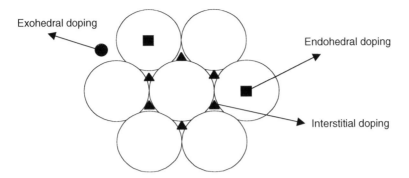

FIGURE 5.17 Available doping sites in a SWNT bundle.

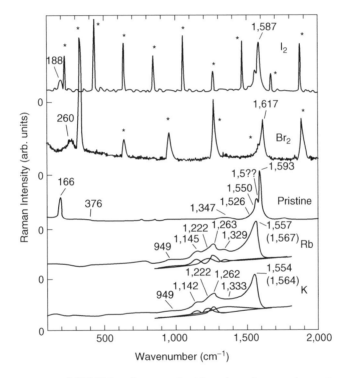

FIGURE 5.18 Raman spectra of SWNT bundles reacted with various donor and acceptor reagents. A Raman spectrum of pristine SWNT bundles is included for comparison [67]. The asterisks label the harmonic series of peaks associated with the fundamental stretching frequency for I_2 (220 cm⁻¹) and Br_2 (324 cm⁻¹).

MWNT Raman spectra compared with SWNT spectra are the absence of a RBM and the lack of resonance enhancement. These changes can be associated with the layered structure of the MWNT. With the many concentric cylinders having a different mode frequency, these different modes can interfere with each other, causing the RBM to be extinguished. These conflicting mismatches between mode frequencies can also lead to the resonance enhancement being destroyed.

5.8 X-Ray Diffraction

X-ray diffraction (XRD) has had a long and successful history of providing structural information for an infinite number of crystals. However, this technique has found relatively limited success for carbon nanotube research. In particular, it has found some degree of usefulness in investigating SWNT bundles

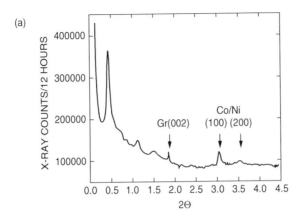

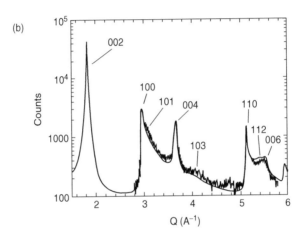

FIGURE 5.19 XRD spectra for (a) SWNT bundles [15] and (b) MWNTs [87]. (Figure 19a reprinted with permission from A. Thess et al., Science, 273, 483, ©1996 AAAS; Figure 5.19b reprinted with permission from D. Reznik, Phys. Rev. B, 52, 116, 1995. ©1995 by the American Physical Society.)

[15,85] and MWNTs [86,87]. Figure 5.19 shows examples of XRD data taken from both SWNTs and MWNTs. In the case of the XRD data taken from the SWNT bundles, there are five peaks that are seen, with the strongest one appearing at $Q=0.44 Å^{-1}$ [15]. This peak, along with the four weaker peaks, is characteristic of the triangular lattice the SWNTs make up when they are bundled. Peaks associated with the Co/Ni catalysts used in the growth of SWNT bundles are indicated in the figure along with the (002) graphite peak at 1.85 $Å^{-1}$, which is attributed to the presence of amorphous carbon in the sample. Notably absent is the graphite (100) peak that is seen in the XRD data of MWNTs. This study also found that the van der Waals gap between tubes was ~0.315 nm and that the SWNT diameter is ~1.38 nm [15]. The XRD data are quite different for the MWNTs, which exhibits a dominant peak at ~1.85$Å^{-1}$ attributed to the reflection from the layered structure of the MWNT. The perturbation of the line shape for the (100) peak in the MWNT XRD spectrum is due to the curvature of the graphene sheets that make up the MWNTs.

5.9   Summary

The primary objective of this chapter is to introduce a variety of characterization techniques that have been used extensively in carbon nanotube research. Other characterization techniques, such as the electron spin resonance [88,89] and nuclear magnetic resonance measurements [90–92], have also been used to some extent to study nanotube properties and are not discussed. Readers interested in a specific characterization technique should refer to the original research articles referred here for an in-depth evaluation of that technique.

References

1. S. Bandow et al. J. Phys. Chem., 101, 8839 (1997).
2. R. Andrews et al. Chem. Phys. Lett., 303, 467 (1999).
3. R. Andrews et al. Carbon, 39, 1681 (2001).
4. S. Iijima, Nature, 354, 56 (1991).
5. C.J. Lee et al. Chem. Phys. Lett., 359, 115 (2002).
6. Z.L. Wang, P. Poncharal, and W.A. de Heer, J. Phys. Chem. Solids, 61, 1025 (2000).
7. Y. Murakami et al. Chem. Phys. Lett., 374, 53 (2003).
8. D.E. Luzzi and B.W. Smith, Carbon, 38, 1751 (2000).
9. K. Hernadi et al. Chem. Phys. Lett., 367, 475 (2003).
10. Y. Feng et al. Chem. Phys. Lett., 375, 645 (2003).
11. J.-Y.Chang et al. Chem. Phys. Lett., 363, 583 (2002).
12. J. Sloan et al. Inorganica Chim. Acta, 330, 1 (2001).
13. N.A. Kiselev et al. Carbon, 37, 1093 (1999).
14. C. Kiang et al. Carbon, 33, 903 (1995).
15. A. Thess et al. Science, 273, 483 (1996).
16. J.M. Cowley et al. Chem. Phys. Lett., 265, 379 (1997).
17. M. Terrones et al. Appl. Phys. Lett., 75, 3932 (1999); B. Sadanadan et al. J. of Nanosci. Nanotech., 3, 99 (2003).
18. X.Z. Liao et al. Appl. Phys. Lett., 82, 2694 (2003).
19. M. Terrones et al. Phys. Rev. Lett., 89, 075505 (2002).
20. J. Hafner et al. J. Phys. Chem. B, 105, 743 (2001).
21. J. Kong et al. Nature, 395, 878 (1998).
22. J. Kong et al. Appl. Phys. A, 69, 305 (1999).
23. J. Chen et al. J. Phys. Chem. B, 105, 2525-2528 (2001).
24. H. Dai et al. Nature, 384, 147 (1996).
25. P. Kim et al. Carbon, 38, 1741 (2000).
26. R. Czerw et al. Nano Lett., 1, 457 (2001).
27. M. Ouyang, J.-L. Huang, and C.M. Lieber, Annu. Rev. Phys. Chem., 53, 201 (2002).
28. Ph. Lambin et al. Carbon, 38, 1713 (2000).
29. V. Meunier and Ph. Lambin, Carbon, 38, 1729 (2000).
30. K.F. Kelly et al. Chem. Phys. Lett., 313, 445 (1999).
31. Z. Osvath et al. Mat. Sci. Eng. C, 23 561 (2003).
32. K. Ichimura et al. Physica B, 323, 230 (2002).
33. Y. Shingaya, T. Nakayama, and M. Aono, Physica B, 323, 153 (2002).
34. M.S. Dresselhaus et al. Indian J. Phys., 77(B), 75 (2003).
35. Ph. Avouris, Chem. Phys., 281, 429 (2002).
36. W. Choi et al. Appl. Phys. Lett., 75, 3129 (1999).
37. M. Grujicic, G. Cao, and R. Singh, Appl. Surface Sci., 211, 166 (2003).
38. H.-F. Hu, Y.-B. Li, and H.-B. He, Diamond and Related Materials, 10, 1818 (2001).
39. Y. I. Prylutskyy et al. Synth. Metals, 121, 1209 (2001).
40. E. Jouguelet, C. Mathis, and P. Petit, Chem. Phys. Lett., 318, 561 (2000).
41. T. Pichler et al. Solid State Comm., 109, 721 (1999).
42. L. Valentini et al. Composites Sci. Technol., 64, 23 (2004).
43. R. Byron Pipes et al. Composites Sci. and Technol., 63, 1349 (2003).
44. Z. Ounaies et al. Composites Sci. Technol., 63, 1637 (2003).

45. C.-W. Nan, Z. Shi, and Y. Lin, Chem. Phys. Lett., 375, 666 (2003).
46. D.W. Schaefer et al. Chem. Phys. Lett., 375, 369 (2003).
47. J. Sandler et al. Polymer, 40, 5967 (1999).
48. Ph. Avouris et al. Physica B, 323, 6 (2002).
49. J. Appenzeller et al. Microelectronic Eng., 64, 391 (2002).
50. J. Appenzeller et al. Microelectronic Eng., 56, 213 (2001).
51. Ph. Avouris et al. Appl. Surf. Sci., 141, 201 (1999).
52. H. Hongo et al. Physica B, 323, 244 (2002).
53. T. Hayashida, L. Pan, and Y. Nakayama, Physica B, 323, 352 (2002).
54. J.P. Small, L. Shi, and P. Kim, Solid State Comm., 127, 181 (2003).
55. P.G. Collins et al. Science, 287, 1801 (2000).
56. K. Bradley et al. Phys. Rev. Lett. 85, 4361 (2000).
57. H.E. Romero et al. Phys. Rev. B, 65, 205410-1 (2002).
58. G.U. Sumanasekera et al. Phys. Rev. Lett., 85, 1096 (2000).
59. P.C. Eklund, Private communication.
60. B.Sadanadan et al. J. Nanosci. Nanotechnol., 3, 99 (2003).
61. K. McGuire et al. Phys. Rev. B, (submitted).
62. M.S. Dresselhaus and G. Dresselhaus, Adv. Phys., 30, 139-326 (1981).
63. M.S. Dresselhaus, G. Dresselhaus, and P.C. Eklund, *Science of Fullerenes and Carbon Nanotubes*, Academic Press (1996).
64. A.M. Rao et al. Science, 259, 955 (1993).
65. R. Saito et al. Physica B, 323, 100 (2002).
66. A.M. Rao et al. Science, 275, 187 (1997).
67. A.M. Rao et al. Nature, 388, 257 (1997).
68. P. Corio et al. Chem. Phys. Lett., 370, 675 (2003).
69. M.S. Dresselhaus et al. Physica B, 323, 15 (2002).
70. A.G. Souza Filho et al. Chem. Phys. Lett., 354, 62 (2002).
71. L.E. McNeil et al. Carbon, 24, 73 (1986).
72. V.W. Brar et al. Phys. Rev. B, 66, 155418 (2002).
73. A.G. Souza Filho et al. Phys. Rev. B, 65, 085417 (2002).
74. H. Kataura et al. Synth. Metals, 103, 2555 (1999).
75. M.A. Pimenta et al. Phys. Rev. B, 58, R16016 (1998).
76. M.J. O'Connell et al. Science, 297, 593 (2002).
77. S.M. Bachilo et al. Science, 298, 2361 (2002).
78. A. Hartschuh et al. Science, 301, 1354 (2003).
79. R.S. Lee et al. Nature, 388, 255 (1997).
80. L. Grigorian et al. Phys. Rev. Lett., 80, 5560 (1998).
81. A.M. Rao et al. Phys. Rev. Lett., 84, 1820 (2000).
82. X. Zhang et al. Chem. Phys. Lett., 372, 497 (2003).
83. G. Maurin et al. Carbon, 39, 1273 (2001).
84. M.S. Dresselhaus, A.M. Rao, and G. Dresselhaus, in *Encyclopedia of Nanoscience and Nanotechnology*, H. S. Nalwa, Ed., American Scientific Publishers, Los Angeles (2004).
85. A. Bougrine et al. Synth. Metals, 103, 2480 (1999).
86. Gy. Onyestyák et al. Carbon, 41, 1241 (2003).
87. D. Reznik et al. Phys. Rev. B, 52, 116 (1995).
88. M. Kosaka et al. Chem. Phys. Lett., 225, 161 (1994).
89. M. Kosaka et al. Chem. Phys. Lett., 233, 47 (1995).
90. I. Yu, J. Lee, and S. Lee, Physica B, 329, 421 (2003).
91. M. Schmid et al. Synth. Metals, 135, 727 (2003).
92. H. Yoshioka, J. Phys. Chem. Solids, 63, 1281 (2002).

Applications in Scanning Probe Microscopy

Cattien V. Nguyen
ELORET Corp.
NASA Ames Research Center

6.1   Introduction

Scanning probe microscopy (SPM) and the subset scanning-force microscopy (SFM) have seen tremendous development and progress in the past two decades since Binnng et al. [1] introduced the atomic-force microscope (AFM) in 1986. AFM has become an essential scientific research tool, particularly in the field of nanoscale science and technology. Because of its versatility, SPM has emerged as one of the techniques of choice for the investigation of single molecule phenomena in areas of scientific research from molecular biology to nanoscale fabrication. SPM is also playing an increasing role as a surface characterization technique for industrial applications; this is particularly important in semiconductor industry as devices approach length scales below the 100-nm regime. Some of the examples for industrial applications are (1) magnetic force microscopy (MFM) as applied in the data storage industry for the characterization of magnetic domains, (2) scanning capacitance microscopy for the characterization of gate dopant density, and (3) as a general surface roughness characterization of ultrathin films. With further understanding of these existing techniques and with more novel variations currently under development, SPM will become an even more important tool in scientific research and nanotechnology applications.

At the heart of SPM is the interaction between the tip of a scanning probe and a sample surface. The geometry and material properties of the tip on the scanning probe ultimately determine the performance and resolution of the instrument. The focus of this chapter is on the development of scanning probe technologies, culminating in a discussion of carbon nanotube (CNT) tips for advanced scanning probe technology and the applications of CNT scanning probes in SPM. First, a general discussion of how the

AFM works is presented. Then, a chronological development of CNT for the scanning probe applications is discussed followed by a highlight of the most recent research and developments in the field.

6.2 Development of the Atomic Force Microscope and the Role of the Scanning Probe

For an in-depth explanation on the functioning of the AFM, the reader is referred to text books on this topic [2,3]. In the most basic terms, the AFM can be described as a stylus profilometer (similar to that in a record player) used as a surface characterization instrument. All the SPM techniques function by measuring a local property — such as height, optical absorption, chemical forces, or magnetism — with the tip held very close to and interacting locally with the sample. The sub-nanometer probe-sample separation makes it possible to take measurements over a very small area. The microscope raster scans the probe over the sample while measuring a local property, generating a three-dimensional data representation of the surface. Because of the precise positional control and sensitive measuring capabilities of the AFM technology, atomic resolution data can be achieved. Although there are other well-established techniques such as scanning electron and transmission electron microscopies (SEM, TEM) for high-resolution imaging, these have certain drawbacks. For example, insulating samples require a conducting coating before SEM imaging. The sample preparation for TEM is often tedious. Electron beam microscopy and scanning tunneling microscopy (STM) require a carefully controlled vacuum environment and rely on specific sample limitations. Because the AFM relies on a mechanical response for its data acquisition, it is uniquely suited to work in ambient conditions and is not limited to conductive samples. AFM's ability to achieve atomic-scale resolution is based on a number of refinements: (1) flexible cantilevers, (2) sensitive detection, (3) high-resolution tip-sample positioning, (4) tip sharpness, and (5) force feedback.

6.2.1 Flexible Cantilevers

The probe consists of a sharp tip fabricated on one end of a long and narrow cantilever beam (see Figure 6.1). As the tip scans over the sample surface, the attractive or repulsive forces between the tip and the sample is measured through the flexing of the cantilever. At each point on the X-Y surface, the flexing of the cantilever gives rise to the data for the Z-coordinate and thus results in a three-dimensional map of the tip-sample interaction. For example, as the tip touches the surface and scans over a raised feature on the surface, the cantilever flexes vertically. Direct measurement of the vertical deflection signal of the cantilever gives the local sample height. Moreover, in this mode where the tip is in physical contact with the surface, often referred to as contact mode, the frictional force may also be studied by measuring the lateral deflection of the cantilever. For most advanced SPM techniques, a feedback mechanism is used to maintain a constant force of interaction between the tip and the substrate. This is achieved through adjusting the tip-sample relative position. The data of the tip-sample position gives rise to the topography (or height) image of the surface.

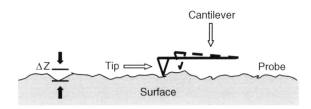

FIGURE 6.1 Schematic representation of the mechanism of the scanning probe in the AFM and its interaction with a surface.

Because the flexing of the cantilevers directly gives rise to the detected signal, the sensitivity of the detection obviously depends on the force constant of the cantilever. In the development of the first AFM, Binning et al. used a probe consisting of a cantilever that was made of a gold foil with a glued-on diamond tip [1]. The original cantilevers had to be cut by hand from thin metal foil or formed from fine wires, and then diamond fragments were attached to the these cantilevers for use as tips. This was later improved by Albrecht et al. [4] in which conventional microchip fabrication techniques were used to fabricate silicon-based probes. A process flow that includes microlithography and a combination of wet and dry etching steps was used to fabricate cantilevers with an integrated pyramidal tip from single crystal silicon. This allowed for the batch production of probes, resulting in inexpensive probes and, more importantly, probes with highly reproducible physical properties. Probe fabrication is a key technology in the development of AFM and marks the turning point for the more widespread use of AFM, leading to the successful commercialization of the instrument and resulting in the proliferation of the AFM throughout scientific laboratories.

In order to maximize the deflection for a given force and therefore maximize the sensitivity of force detection, the spring constant of the cantilever is required to be as soft as possible. At the same time a stiff spring with high resonant frequency (>100 Hz) is necessary in order to minimize the sensitivity to vibration noise. Probe design for optimizing the resonance frequency, f_o, of a spring is based on the equation $f_o = (1/2\pi)(k/m_o)^{1/2}$, where k is the spring constant and m_o is the effective mass of the spring. For a highly sensitive soft spring, i.e., small k, m_o must be small in order to keep the ratio of (k/m_o) large for a high resonant frequency [5]. Typical Si cantilevers have physical dimensions approximately 100-μm length, 10-μm wide, and a thickness of a few μm. Depending on the dimensions of the cantilever, the resonant frequency can range from 10s to 100s of kHz, and the force constant k can be smaller than 1 nN/nm. This allows the instrument to sense tip-sample interaction forces in the pico-Newton (10^{-12}) range.

6.2.2 Sensitive Detection

Sensitive detection of the flexing of the cantilever is another development that was required for the realization of the AFM as an instrument for surface analysis. The first AFM used a tunneling junction of an STM for detection. An STM tip is held in close proximity to the cantilever and the tip of the AFM probe on the underside of the cantilever is in physical contact with the sample surface. In this approach, a tunneling current between the cantilever and the tunneling electrode above the cantilever is employed as the signal to monitor the dynamic movement of the AFM cantilever as well as for the feedback circuit for maintaining the force f_o at a constant level as the probe is scanned over surface features. In general, STM is limited to only conducting surfaces because it requires tunneling current between the tip and the sample as the means of detection as well as feedback mechanism. The integration of an STM tunneling mechanism for the detection of the cantilever motion in the first AFM enabled the investigation of both conducting and nonconducting surfaces. The flexing of cantilever as the new mechanism of probing the surface allows for the examination of all types of surfaces. Thus, the design of the first AFM was able to image sample surfaces regardless of conductivity of the sample and demonstrated the intent of Binnig et al. to develop a general scanning probe technique for the investigation of all types of surfaces.

Other detection methods, mostly based on optical detection using laser light, were subsequently successfully demonstrated for the detection of the AFM cantilever motion. A variety of sensor methods, such as capacitive and diode laser feedback detection, have been developed, but we will concentrate on two optical methods: (1) optical interferometry and (2) optical beam deflection. These two optical sensor techniques exhibit stability and high sensitivity and are simple, robust, and small enough to be integrated into the instrument. In addition, when compared with the tunneling probe, these two optical sensor methods are totally insensitive to thermal drifts in the optics, and hence the distance between the optical probe and the cantilever does not need to be accurately controlled. Furthermore, in the case where the cantilever is operated at its resonant frequency, the optical-detection techniques allow for the monitoring of the cantilever vibration over a wide range of frequencies and with amplitude in the range of tens of nanometers.

In 1987 Martin et al. [6] demonstrated the use of optical heterodyne interferometric detection to accurately measure the vibration of the tip. In their set-up, a tungsten tip at the end of an iron wire was mounted on a piezoelectric transducer. The wire, acting as the cantilever, was excited to vibrate at its resonance frequency by the piezoelectric transducer. A He-Ne laser beam was reflected off the surface of the wire cantilever, and the reflected light was detected by a sensitive laser heterodyne probe, which is capable of detecting vibration amplitude and phase of AC displacements as small as 10^{-4} A. Later, Stern et al. [7] fashioned another type of optical detection scheme using optical fiber interferometry. In this instrument set-up, a Ga-Al-As diode laser ($\lambda = 830$ nm) with a direct single-mode fiber output was used as a light source. A single-mode directional coupler was used to split and separate both the incident and the reflected light. The intensity of the optical interference was measured with a photodiode; and the output of the photodiode signal was used directly as the force microscope signal.

$$\Delta i / i_o = 4\pi V (\Delta d / \lambda) - 4\pi d V (\Delta \lambda / \lambda^2) \tag{6.1}$$

The change in signal intensity (Δi) relative to the change in fiber-to-reflector spacing, Δd, is described by Equation 6.1, where i_o is midpoint current ($i_o = (i_{max} + i_{min})/2$), V is the fringe visibility ($V = (i_{max} + i_{min}) (i_{max} - i_{min})$), and λ is the laser wavelength. At $d = 4$ μm, the noise associated with root mean square (rms) wavelength variation $\Delta \lambda$ was measured to be less than 0.01 nm in a 1000-Hz bandwidth, which was sufficient for measurement of a magnetic bit pattern on a Co-alloy disk [8].

Laser-beam deflection offers another convenient and sensitive method of measuring cantilever deflection (Figure 6.2), which is widely used now in most commercial instruments. This technique was first reported in 1988 as applied in AFM by two groups: (1) Meyer and Amer of IBM Thomas J. Watson Research Center [9], and (2) Alexander et al. of the University of California, Santa Barbara and Longmire and Gurley of Digital Instruments, Inc [10]. Figure 6.2 shows a schematic representation of the cantilever laser beam deflection system. In the first reported systems, a mirror was mounted on the cantilever; however, at present, most commercial cantilevers have a highly reflective Si surface that does not require a mirror system. The laser beam reflecting from the cantilever is measured with a position sensitive photodetector (PSD). The PSD consists of two side-by-side photodiodes, and the difference between the two photodiode signals indicates the position of the laser spot on the detector and thus the angular deflection of the cantilever. Because the cantilever-to-detector distance is generally thousands of times the length of the cantilever, the optical lever greatly magnifies motions of the tip. The design of the signal obtained from the PSD and the ~2000-fold magnified lever detection enable the detection of deflections much smaller than 0.1 nm. Meyer and Amer [9] noted that the sensitivity limit is determined by the amplitude of the thermal vibration (10^{-5} nm/(Hz)$^{1/2}$) and not by the minimum detectable displacement. For measuring cantilever deflection, the laser beam deflection method offers high sensitivity and large deflection range but without the cumbersome apparatus in the interferometric method. The superior technical performance coupled with the economy of manufacture and ease of use have made laser beam detection methods the technique of choice for most commercial AFMs.

6.2.3 High-Resolution Tip-Sample Placement

High-resolution tip-sample position and force control in AFM require the use of piezoelectric materials. Movement or positioning by a piezoelectric material gives a precision greater than 0.1 nm. This is controlled by applying a voltage gradient to a piezoelectric ceramic tube, causing the tube to expand or contract. Most scanning probe microscopes use tube-shaped piezoceramics because they combine a simple one-piece construction with high stability and large dynamic range. The piezoceramic tube is segmented into four quadrants, each connected to a separate electrode, and another electrode covers the inner area in contact with all four quadrants. Application of voltages to one or more of the electrodes

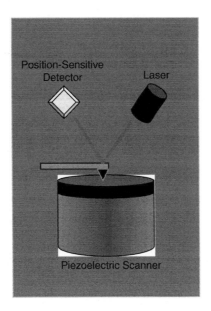

FIGURE 6.2 Laser beam deflection optical detection system in combination with the piezoelectric tube scanner for the AFM.

causes the tube to bend or stretch, thus moving the sample relative to the probe in three dimensions. For instance, when applying a voltage to one of the four outer quadrants it causes that quadrant to expand and the scanner to tilt away from the probe, resulting in the X-Y movement of the sample relative to the probe. Z movement or height position of the tip relative to the sample is achieved when applying a voltage to the inner electrode, resulting in the entire tube expanding or contracting. 87th Way

6.2.4 Sharp Tips

Sensitivity of the probe is mainly a function on the flexibility or softness of the cantilevers as described above, which gives the z-coordinate data for the three-dimensional image. However, spatial resolution or lateral resolution in the X-Y plane is mainly the function of the size and shape of the probe or *tip*. The scanning probe systems do not rely on optics, such as lenses and electromagnetic radiation, so the resolution is not limited to diffraction effects that limit resolution in optical microscopy. Rather, the determining factor for resolution in AFM is the physical size and geometry of the apex of the tip of the scanning probe. Conventional microfabricated Si probes have a pyramidal-shaped tip at the end of the cantilever, with tip apex radius of curvature of about 10 nm. The relatively inexpensive Si probes allow for routine high-resolution imaging; however, these probes reach certain limitations in advanced ultrahigh-resolution imaging applications. Figure 6.3 demonstrates how the tip geometry may govern the spatial resolution: (1) in the case where the tip diameter is larger than the imaging object, the topological detail and lateral resolution of the object will be lost due to a tip broadening effect; and (2) in the case where imaging deep and narrow features on the surface is required, a conventional Si scanning probe with its pyramidal shape will not be able to trace the surface morphology due to the angled sides and low aspect ratio of the Si probe. Various improvements in the probe fabrication process have been developed to address these problems. For example, focused ion beam (FIB) milling techniques have been used to fabricate a high aspect ratio stylus for imaging deep and narrow surface features (Figure 6.4). However, because of the intrinsically brittle nature of silicon at these dimensions; the robustness of the silicon-based stylus remains to be a problem. The stylus continuously wears out as it is used or occasionally experiences breakage as well. An alternative is to use a carbon nanotube that naturally has a high aspect ratio and robust mechanical properties, which will be discussed later.

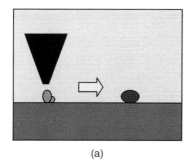
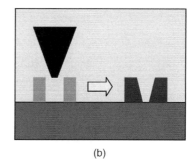

(a)   (b)

FIGURE 6.3 Tip shape and the effect on image. (a) Tip broadening with large radius of curvature tip; note that the smaller particle is not resolved in this case. (b) Tip-shape artifacts due to the interaction of the side of tip and high aspect ratio surface features.

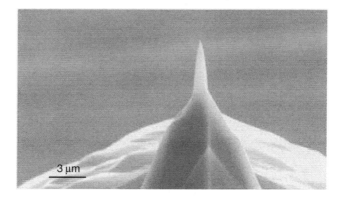

FIGURE 6.4 High aspect ratio Si probe fabricated by focused ion beam milling.

6.2.5   Force Feedback

The use of a feedback mechanism is another refinement that enabled the dramatic development of AFM into an instrument with widely ranging applications in research and industry. This also differentiated AFM from the older stylus-based instruments. The feedback in AFM regulates the force of interaction between the tip of the probe and the sample. A compensation network (controlled via a computer program) monitors the cantilever deflection and keeps the force of interaction between the tip of the probe and the local surface area of the sample at a predetermined set point. In the contact mode or the repulsive regime of the intermolecular force curve (see Figure 6.5), this is achieved by adjusting the applied voltage to the inner electrode of the piezoelectric tube, resulting in the change of the height position of the tip over the surface and thus keeping the deflection of the cantilever constant.

In more sophisticated modes, such as noncontact or tapping modes, the cantilever is oscillated at its resonant frequency (in the range of tens to hundreds of kilohertz) and positioned above the surface at a fixed distance. For tapping mode, which is a widely used method in AFM, the tip on the cantilever only taps the surface for a very small fraction of its oscillation period. The tip makes contact with the surface of the sample in the sense defined for the contact mode, but the very short time of contact in the oscillation cycle helps significantly reduce the lateral forces as the tip scans over the surface. Because of the lower lateral forces, tapping mode in general offers better imaging capability than contact mode. The feedback mechanism relies mainly on adjusting the voltage on the piezoelectric tube for the Z-axis so that the rms amplitude of the cantilever oscillation remains constant. There are several types of images that may be obtained from tapping mode: (1) amplitude signal, as there will be small variations in this oscillation amplitude due to the control electronics not responding instantaneously to changes on the

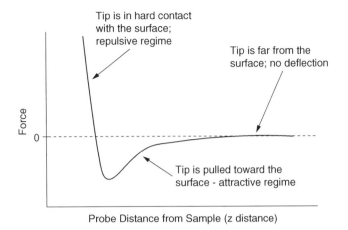

FIGURE 6.5 Potential energy curve (force vs. distance curve) of the scanning probe tip as a function of distance to the surface.

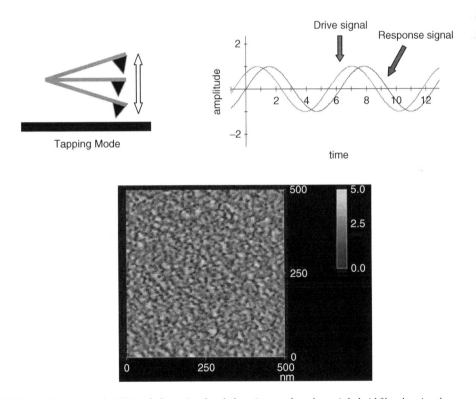

FIGURE 6.6 Tapping mode AFM and phase signal and phase image of a polymeric hybrid film showing the nanoscale phase-separated domains.

specimen surface; (2) Z-axis signal as the response for maintaining constant cantilever amplitude, which gives surface topography image; and (3) phase shift of the cantilever oscillation in response to the tip interaction with the surface. The phase-shift images give details of the cantilever energy dissipation through interaction of the surface. In phase-mode imaging, the phase shift of the oscillating cantilever relative to the driving signal is measured (see Figure 6.6). This phase shift can be correlated with specific material properties that affect the tip/sample interaction. The phase shift can be used to differentiate areas on a sample with such differing properties as friction, adhesion, and viscoelasticity. The technique is used in standard tapping mode, so topography can be measured as well.

Noncontact operation is another method for imaging with the AFM. Here, the cantilever oscillates at a defined distance above the surface of the sample. This is accomplished via the lift mode, in which the topography of the surface is first imaged and then the tip of the probe is lifted a predetermined distance above the surface to retrace the line following the contour of the surface. The tip-sample distance must be large enough in the intermolecular force curve, where the tip is not experiencing any repulsive force as well as any capillary bridge between the tip and the sample that may cause the tip to jump to contact.

In noncontact mode, which is ideal for use in situations where tip contact might physically alter the sample, the tip hovers 5 to 15 nm above the sample surface. Unfortunately the attractive forces from the tip to sample are substantially weaker than the forces used by contact or tapping mode. Therefore the tip must be given a small oscillation so that AC detection methods can be used to detect the small forces between the tip and the sample by measuring the change in amplitude, phase, or frequency of the oscillating cantilever in response to force gradients from the sample. For highest resolution, it is necessary to measure force gradients from van der Waals forces, which may extend only a nanometer from the sample surface. In general, the fluid contaminant layer, normally about 10 to 30 water monolayers thick under ambient conditions, is substantially thicker than the range of the van der Waals force gradient; therefore, attempts to image the true surface with noncontact AFM fail as the oscillating probe becomes trapped in the fluid layer or hovers beyond the effective range of the forces it attempts to measure. In noncontact mode, the tip-sample distance should be unaffected by topography, and an image can be built up by recording changes, which occur due to longer range force interactions. For instance, magnetic force imaging and electrostatic force images are obtained via noncontact force through mainly the phase shifting of the oscillation frequency of the cantilever.

6.3 Mechanical Properties of Carbon Nanotubes in the Context of SPM Applications

The previous section discusses the contribution of the various essential components that combine to develop and enhance the performance of the AFM. Because the tip-sample interaction is the fundamental element of SPM, it is suffice to say that the quality of the tip on the cantilever has the most direct influence of the imaging quality. The probe itself remains the most important component of the whole system that can benefit from new technological development. Although there has been great improvement in the fabrication technology for the probe, the materials for the probes, namely Si and Si-derived materials, have remained the same for some time. Because of the intrinsic nature of these materials, the tip of the probe tends to be brittle; therefore, the size and shape of the tip may change during scanning. The ability to control the size and shape and also rendering the tip more robust will enhance SPM as an analytical tool, thus expanding the usage of the AFM as an instrument. This section describes the use of CNTs as the tip for the AFM probe. The structural and mechanical properties of CNTs, specifically as related to SPM, are described in this section. A general and more detailed discussion on the mechanical properties of CNTs can be found in Chapter 1.

The physical dimension, namely the high aspect ratio with diameter on the nanometer length scale, renders CNT an excellent candidate for the application as the AFM tip. However, it is the mechanical properties of CNT that set it apart from the rest of nanomaterials, such as nanowires like Si, ZnO, etc., which have very similar physical dimensions. The Young modulus perpendicular to the plane of a graphite sheet is ~1.3 Tera pascal and, as expected, the Young modulus of single-walled nanotubes (SWNTs) is comparable at about 1 Tera pascal, making CNT one of the strongest materials.

At the most fundamental level, the unique mechanical as well as electrical properties of CNTs derive from the chemical bonds between the sp^2 hybridized carbon atoms of the graphitic material. The surface curvature and the arrangement of the carbon atoms in CNT may cause the graphitic carbon-carbon bonds to be nonplanar. This results in a less efficient overlapping of the pi-orbital between carbon-carbon

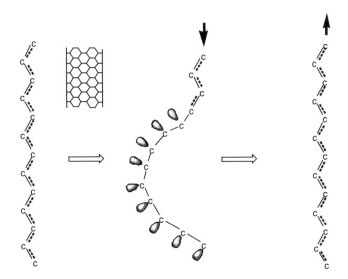

FIGURE 6.7 The elastic buckling of the CNT is derived from the chemical bonding energy from the sp^2 carbon atoms. The schematic in (a) is the double-bond characteristic of the graphitic carbons along the CNT; (b) external force, as presented by the black arrow, causes bending of the CNT and thus disruption of the pi-bonding of the carbon atoms. The carbon-carbon along the bend are more characteristic of single bond and have more sp^3-like hybridization. (c) The carbon atoms revert to their original configuration with graphitic sp^2 bonding when the external force is removed, represented by the arrow.

atoms, and hence the carbon atoms may not be perfectly sp^2 hybridized. Nevertheless the bonding of the carbon atoms in CNT maintains the same sp^2 hybridization characteristics, that is, the carbon-carbon bond as well as the bonding angle can be mechanically distorted in a reversible manner. Figure 6.7 schematically demonstrates the bending of sp^2 hybridized graphitic carbons when an external force is applied. The bonding angle may be twisted out of plane, with the central C atom becoming more sp^3 and thus losing most of the bonding contribution by the π-orbital, i.e., transforming from double-bondlike to more of a single-bondlike characteristic. When the strain is removed, the bonding of the C–C atoms would revert to its thermodynamically more-stable sp^2 configuration and thus retain the original nanotube structure.

The general chemical nature of the sp^2 carbons and the C–C bonds is the basis of the elastic buckling properties of CNTs. The elastic deformation renders the CNT structurally robust, that is, the CNT will not permanently change its structure when it experiences external forces. This highly desirable mechanical property, coupled with its nanoscale diameter and high aspect ratio, renders CNTs ideal for application as the tip of scanning probes. Therefore, it is worthwhile to explore some of the elastic responses of the CNT in more detail. The CNT tip is likely to encounter a force that causes it to bend while scanning. The bending force constant of a nanotube is given by

$$k_B = 3Y\pi(r_o{}^4 - r_i{}^4)/4L^3, \tag{6.2}$$

where Y is Young's modulus, r_o and r_i are the outer and inner diameters of the nanotubes, and L is the length of the CNT [11,12]. Elastic deformation also invokes a compression response. The compression force constant is given by

$$k_C = Y\pi(r_o{}^2 - r_i{}^2)/L. \tag{6.3}$$

It can be clearly seen from the two equations that the bending response is the larger component of the elastic deformation. Also the length of the CNT, with the length scale typically an order of magnitude larger than the radius, will dramatically influence the bending response of the CNT tip.

6.4   Fabrication of Carbon Nanotube Scanning Probes

In order to understand the fabrication of CNT probes, some knowledge of the CNT synthesis is required. Briefly, the minimal requirements for the growth of a CNT are a catalyst such as Fe, Ni, or Co either as particles or very thin film (a few nanometers) and a source of reactive carbon species. Energy from either thermal, plasma, arc discharge, or laser source is used to generate reactive carbon species. The active carbon species are proposed to diffuse through a diffusion process, crystallize on the surface of the catalyst to form the nanotubes. For example, in the thermal chemical vapor deposition (CVD) process for producing SWNT, methane gas is decomposed at 900°C on catalytic Fe particles deposited on a substrate. The carbon species diffuse into the catalyst particle and crystallizes into the structure of a nanotube protruding from the surface of the catalyst. It is likely that it is the size or the curvature of the surface of the catalyst particles that influences the size and the atomic arrangement of the CNT structure. For a more detailed discussion on the growth of CNTs, the reader is referred to Chapters 3 and 4 of this book.

6.4.1   Multiwalled Carbon Nanotube Probes

The first demonstration of CNT as the tip for SPM was accomplished with a multiwalled nanotube (MWNT) by Dai et al. [13]. Based on the chronology of development, and for no other reason, the discussion here will focus first on the fabrication of the scanning probe with a MWNT tip (referred to as MWNT probe from here on) and followed by that for the SWNT tip (referred from here on as SWNT probe). In this pioneering paper by Dai et al., the fabrication of the MWNT probe was accomplished by gluing the nanotube onto one of the sides of the pyramidal-shaped tip of a conventional silicon cantilever. A source of low-density MWNTs was created by dabbing the as-grown MWNTs onto a strip of adhesive tape. With an inverted optical microscope equipped with two X-Y-Z microtranslators/manipulators, one holding the Si cantilever and one for that of the MWNT tape, the adhesive from the tape was transferred onto the Si tip of a conventional cantilever. Next the MWNT from the tape, as either an individual tube or a bundle of tubes, was then transferred to the adhesive-coated Si tip. Using the force-distance mode in the AFM, the length of the MWNT extending past the tip of the Si probe was shortened by applying a voltage pulse that generated an arc between the MWNT tip and a conducting surface. A MWNT probe composed of a single 5-nm diameter MWNT tip, which extended 250 nm from about a 5-μm bundle of MWNTs, was employed to image (in tapping mode) a 400-nm wide, 800-nm deep, microfabricated high aspect ratio pattern on silicon wafer.

Dai et al. also mentioned that the attachment with adhesive allowed for the CNT to bend away from its point of attachment to the silicon tip whenever the tip crashed into a hard surface, but then the CNT would snap back to its original position. In this work, the signal for physical buckling of the CNT was demonstrated for the first time in the amplitude vs. distance force curve. It was correlated to the Euler buckling force, F_{EULER}:

$$F_{EULER} = \pi^3 Y r^4 / 4L^2, \qquad (6.4)$$

where Y is the Young's modulus (estimated to be ~1 Tpa), r is the radius of the nanotube, and L is the length of the nanotube. In the case where the nanotube's long axis is close to the normal of the surface, no tube bending was observed until the force exceeded the F_{EULER}. For the particular 250-nm length and 5-nm diameter tube, F_{EULER} is about 5 nN as calculated from the experimental data where the buckling occurred as the MWNT tip was pushed 20 nm past the initial point of contact with the surface. It was reasoned that the highest force from the MWNT to the surface would not exceed the F_{EULER} because the MWNT probe would serve as a compliant spring that moderated the impact of the tip tapping the surface.

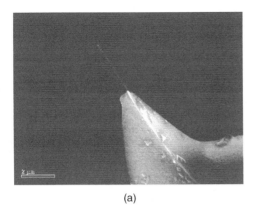
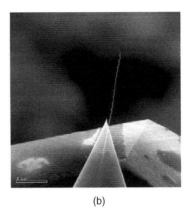

(a)   (b)

FIGURE 6.8 SEMs of MWNT probes composed of (a) an ~3-μm straight tube and (b) an ~5-μm coiled tube. Notice that the probe contains only a single MWNT as the tip, which was made possible by having a long, individually separated, and low-density MWNT source.

This translated to minimal damage to samples, which is very important for imaging biological and soft materials.

Several steps and materials in the original fabrication of MWNT probe were subsequently modified in a number of laboratories. Most probes fabricated using the original type of CNT source, i.e., adhesive tape dabbed with CNTs, tended to have tips with a bundle of MWNTs and with a very high likelihood of not having just a single MWNT protruding from the end of the bundle. The bundle of MWNTs gave a higher-effective tip diameter; therefore, the gain in lateral resolution was not being realized for these probes. Moloni et al. [14] invented a sharpening technique based on the same procedure for shortening nanotube probes. The method made use of microfabricated nanoscale V-groove patterns as the substrate on which the bundle of MWNTs may be sharpened, thus producing a tip with a single protruding MWNT.

Stevens et al. [15] enhanced the optical microscopic manual method of attachment as well as improved the point of attachment by employing an applied DC field between the MWNT and a cobalt metal-coated Si probe for the manual attachment process. The electric field in this modified technique provided an effective means to attract and orient the MWNT for better alignment onto the Si tip. A small DC field (less than 5 V) was efficient in the process of attracting and aligning the MWNT, and at this point, increasing the DC field to larger than 20 V induced breakage at a roughly predetermined nanotube-nanotube junction, such as a bending point along its length. The MWNT probes were used to image silacetein protein filament, with the results showing a more intrinsically accurate image, which had none of the imaging artifacts compared with the images obtained with conventional silicon probes.

An improvement in the nanotube source, which consisted of low-density, long, and individually separated MWNT, was reported in Reference 16 for the fabrication of MWNT probes. A viscous catalyst solution (composed of iron, tri-block copolymer P123, and $Si(OCH_2CH_3)_4$ in methanol) was coated on a Pt wire, and thermal CVD was employed to directly grow relatively low-density MWNTs on the Pt wire [16]. Some of the MWNTs had lengths greater than 10 μm and were physically well separated. This allowed for easier sorting and optimal alignment (i.e., manipulation) of individual MWNTs by optical microscopy and hence provided a very efficient means for attaching a single MWNT to a silicon probe (see Figure 6.8). This improvement of the MWNT source coupled with the electric field-enhanced attachment method enables an efficient and easier fabrication of the MWNT probes.

It is important to note that a scanning probe with an individual MWNT, rather than a bundle as the tip, enables tracking of deep and narrow feature at sub–100-nm length scale, which will be discussed in more detail in the next section on applications. These probes also have high lateral resolution, with tip diameters as small as 10 nm; more importantly, they were demonstrated to be mechanically very robust [17]. No degradation in the image quality was observed after continuous scanning for 10+ hours. Larson et al. [18] also demonstrated the robustness of MWNT probes derived from the fabrication technique reported by Dai et al [13]. In this work, after 1000 scans of a 2×2 μm area (equivalent to a linear length of two meters); minimal loss in resolution was observed in the imaging of a polysilicon surface. The authors based their finding on the rms surface roughness data for the polysilicon surface as well as direct SEM data, which also showed no observable change in the 220-nm length. The robustness of the MWNT probes points to the structural integration of the CNT graphitic structures derived from the nature of the carbon-carbon double bonds.

The point of attachment between the CNT and the surface of the Si probe also contributes greatly to the stability of the MWNT probe. The alignment of the MWNT with respect to the Si probe may be altered, or catastrophically fail in the worst case, particularly when the point of attachment is weak and the MWNT encounters excessive force (for example, during the approach routine). It is therefore highly desirable to have a strong point of attachment for reliable and stable probes. The interfacial junction between the MWNT and the surface of the Si probe may be welded to the surface of the probe by locally generated Joule heating (via electrical current) from a DC bias applied between the MWNT source and the Si probe. The welding at the interface may be strengthened with an ultrathin film of a metal such as Ni or other metals known to form metal-Si bond and metal-C bond.

In the work of Nakayama and coworkers, a low-density MWNT aligned on a knife edge [19,20] was produced by solution electrophoretic method [21]. The process involved placing two knife edges, with a 500-μm gap, in a solution of isopropryl alcohol containing dispersed MWNTs derived by an arc-discharge synthesis. Upon applying an AC electric field of 5 MHz and 1.8 kV/cm, the MWNTs moved onto the knife edges and aligned almost parallel to the electric field. This produced knife edges with a density of fewer than one MWNT per micrometer, with some of the nanotubes reported to protrude as much as 1.3 μm from the knife edge. Because of the short protruding length, the MWNTs were not observable with optical microscopy; therefore, the MWNT had to be manipulated using a scanning SEM. An individual MWNT was transferred from the knife edge to the Si tip by applying a DC bias voltage between the knife edge and the Si probe in an SEM. Similar to the original method reported by Dai et al., two independent translation stages, one for the MWNT source and the other for the Si tip, were used for the attachment process in the SEM. In addition, a DC field was applied between the MWNT source and the Si probe to induce electrostatic force for attracting and transferring a MWNT to the Si tip. *In situ* SEM amorphous carbon deposition by electron-beam dissociation of hydrocarbon contaminants was used to strengthen the point of MWNT attachment on the surface of the Si probe [19], and the attachment force was determined to be greater than 80 nN [22]. Alternatively, Nakayama and coworkers also reported a welding process, similar to the one described above, for strengthening the attachment between the MWNT and the Si probe. Here a subsecond pulse of less than 1 mA was introduced onto the contact, resulting in the MWNT fusing to the surface of Si or an Os-coated surface of Si_3N_4 probes [19]. The resolution of these MWNT probes was demonstrated by imaging double-stranded DNA molecule, the width of which was measured ranging from 7 to 8 nm, thus giving an estimated tip radius of curvature of about 3.5 nm.

Nakayama and coworkers also reported that the MWNT probes with lengths greater than 500 nm and diameters greater than 10 nm exhibited a thermal vibration with amplitude greater than 0.5 nm at the very tip and produced blurred images. In order to adjust the length of the MWNT, *in situ* SEM electron bombardment method was employed as a means to control the length of the MWNT [23]. This method utilized a counter nanotube as a field-emission source, and the emitted electrons collided with the nanotube tip on the probe, which functioned as the extraction electrode. At a nanotube-nanotube gap ranging from 0.5 to 2 μm and at pressure of 10^{-4} Pa in an SEM, a ramp in bias from 0 to 200 V caused the MWNT probe to be shortened by about 100 nm. A maximum current of about 100 μA was detected in the process. The mechanism for the shortening process was concluded to be thermally assisted field evaporation, which can be described as an induced heating of the nanotube tip due to the dissipation of kinetic energy from the emitted electrons, thus resulting in the field evaporation of carbon atoms at the tip.

The manipulation of MWNTs for the fabrication of scanning probes is made possible due to a long nanotube extending its length from substrate. In other words, a MWNT that is a few micrometers in length can physically support itself to grow out from the surface without bending. This is mainly due to the large radius of the tube and is clearly evident from Equation (6.4) describing the r^4 dependence of the bending force constant. In contrast, SWNT and even MWNTs with very small diameter have smaller bending force constant due to their smaller diameter. Hence manual manipulation as a method of probe fabrication with individual SWNT as the tip is therefore not possible. The fabrication of probes containing individual SWNT as tips must be approached differently and this is the topic of the next section.

6.4.2   SWNT Probes

After the first demonstration of MWNT as a tip for scanning probes, there was a lot of excitement in anticipation of SWNT probes, mainly because of the promise of ultrahigh lateral resolution of such probes due to the smaller inherent diameter (0.5 to 2 nm) of SWNT. Wong et al. [24,25] were the first to demonstrate the high lateral resolution capability of SWNTs in 1998. In their work, the probe was fabricated by manual attachment of a SWNT rope to the Si probe by the same technique used for the first reported MWNT probe. The tip of the SWNT probe was shortened and sharpened by applying a voltage between the tip and a sputtered Nb substrate. Direct imaging data of 5-nm diameter gold particles and amyloid-β 1-40 (Aβ40) fibrils render SWNT tips with radii of curvatures of 3.4 and 2.6 nm, respectively. The authors noted that this can be improved with probes composed of a single SWNT as the tip, in which case a tip with radius of curvatures as small as 0.5 nm may be possible.

The ability to grow SWNTs by CVD on surfaces, as discussed in Chapter 4, opens the opportunity for the direct growth of CNT on a Si probe and also mass production of such probes. The first direct growth of CNT probe was reported by Hafner et al. [26], which was achieved by first etching (via an HF anodization process) the surface of a Si tip that has been mechanically flattened, generating 60-nm diameter nanopores that aligned along the tip axis. Iron catalyst was electrodeposited into the porous structure from $FeSO_4$ solution, and the SWNTs were grown from these porous structures by CVD using ethylene-hydrogen gas mixture at 750°C. The authors indicated that the porous structure aided in the growth direction of the SWNT for optimal imaging. A growth time of 10 minutes produced SWNTs with lengths too long for imaging, thereby requiring the nanotube to be shortened, via the normal technique of applying an electric field between the tip and a conducting substrate. This fabrication technique was reported to produce probes composed of a single SWNT tip with radius of curvature ranging from 3 to 6 nm.

Hafner et al. also reported another fabrication technique for SWNT probe by direct growth of the nanotube on a Si probe. This approach relied on the fact that nanotubes with smaller diameter, such as the SWNT and small-diameter MWNT, prefer to grow along a surface of the substrate. Because the bending response scales to the fourth power of radius as in Equation 6.2, a small radius nanotube would therefore have a small bending force constant. In this case, the attractive force between nanotube-surface interaction overcomes the force required for bending, particularly for a long nanotube. Taking advantage of the surface alignment properties of SWNT, Hafner et al. [27] electrodeposited Fe containing catalyst onto the surface of Si tip and directly grew SWNT by CVD synthesis. SWNT, grown with a 1:200:300 C_2H_4:H_2:Ar at 800°C for 3 minutes, was demonstrated to align along the face of the Si probe and protrude from the tip apex. For the growth mechanism, the authors proposed that "nanotubes prepared from catalyst deposited on a pyramid AFM tip will grow along the surfaces until they reach the pyramid edges, then some will be directed toward the tip apex along the edges. At the pyramid end the nanotubes will protrude straight from the apex to create an ideal tip. In this instance, the strain energy for bending the nanotube is compensated by the nanotube-surface interactions." It was also reported that this process produced tips with an effective tip radius of 3 nm or less.

Fabrication of small-diameter SWNT probe by direct CVD approach was also demonstrated in References 17 and 28. The process involved transferring an island of catalytic metal particles to the inner base of the pyramid by dipping the very end of a cantilever into a solution of catalysts (1.2 g $AlCl_3$, 0.57 mL $SiCl_4$, 90 mg $FeCl_3$, 6 mg $MoCl_2O_2$, and 1 g of P123 block copolymer in 15 mL of ethanol). Figure 6.9 shows the schematic representation and actual SEM micrographs of SWNT growing from the

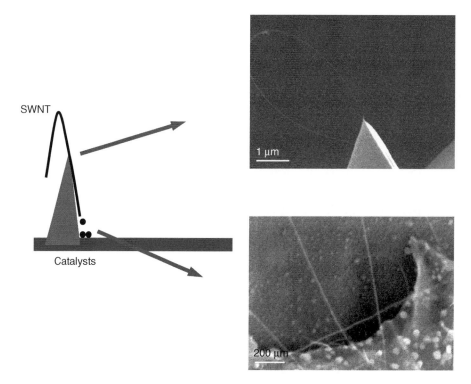

FIGURE 6.9 Direct growth of SWNT for the fabrication of CNT probe. The schematic shows the AFM cantilever with catalyst particles deposited on the base of the pyramid and SWNT grown from the catalyst island. The two SEM images on the right show the alignment of SWNTs on the Si surface and extending from the apex of the Si probe.

catalyst island at the base of the Si pyramid. Like other fabrication methods, this also produced SWNT too long to be practical; therefore, it too required shortening. The high lateral resolution of the SWNT fabricated by this technique was demonstrated by imaging surfaces of ultrathin films of Ir and Si_3N_4, in which the grain sizes of the films were resolved to be as small as 2 nm.

Hafner et al. [29] also reported another technique for the fabrication of SWNT probes, which was described by the authors as a "hybrid CVD/assembly method for the facile production of individual SWNT tips with essentially a 100% yield." This method is based on a conventional Si probe *pick-up* of vertically aligned SWNT off a planar surface during imaging to create a well-aligned SWNT probe. It was reported that a nanotube pick-up event occurred approximately once every 8×8 μm² scan when imaging a surface consisting of isolated, vertically aligned SWNT. The picking up of a nanotube from the surface was signified by a jump in the Z-position, which corresponded to an increase in the tip length. Evaluating the direct imaging data of a 5.2-nm Au particle, a full width obtained with the SWNT probe was only 7.2 nm, thus suggesting an impressive tip radius of curvature of 1 nm for the SWNT probe.

Although the high resolution using a probe with a single SWNT as the tip is impressive, the direct growth by CVD process due to the lack of control of the density of catalyst particles deposited on the Si probe sometimes produces more than one SWNT for the tip. This results in several unwanted tips simultaneously. Moreover, the SWNT growth process based mainly on the attractive force of nanotube-surface interaction produces nanotubes with random growth direction. In many of the probes, the SWNT tip may have a nonvertical orientation relative to the substrate because of the random growth direction. Snow et al. [30,31], with a SWNT probe fabricated by the hybrid CVD/assembly method, reported a detailed quantitative analysis of imaging phenomenon when the SWNT was not at the optimal vertical orientation. They deduced that the imaging force may cause a nonvertically oriented SWNT to bend and jump into contact with the surface, resulting from a large contact area between the surface of the side walls of the CNT and the sample [30]. In turn, this produces unstable feedback and causes the edge blurring image artifact. This problem is reduced to some degree by decreasing the cantilever vibration amplitude, thus reducing the imaging forces. Another way to solve the jumping into contact problem associated with nonvertical alignment of the SWNT is to have a shorter and thus stiffer SWNT, which would result in less bending of the nanotube.

6.5   Applications of Carbon Nanotube Probes

The nanoscale dimension, namely the small diameter and high aspect ratio, of the CNT probe is very desirable for a host of applications. As discussed above, the most advantageous factor for the CNT probe is the mechanical property, which renders the CNT tip very robust. The ability to use the same tip, with the radius of curvature remaining constant after many scans, permits reliable analytical assessment of the surface morphology. This is important in the evaluation of deposition processes for thin films in state-of-the-art microelectronics.

6.5.1   Applications of CNT Probes in Information Technology

The metrology section of the International Technology Roadmap for Semiconductors [32] covers challenges ahead in microscopy, critical dimension (CD) overlay, film thickness and profile, and dopant profile among other issues. These include imaging of high aspect ratio surface features such as photoresists, vias, and deep Direct Random Access Memory (DRAM) trenches. In the past, characterization by SEM required ultralow voltage electron beams to overcome image degradation due to charging and radiation damage to photoresist and other surfaces. The 193 nm generation of resists, composed of aliphatic polymers such as polynorbornene, has been recently reported to suffer from electron beam–induced damage, in which the physical shrinking of the resist features was observed while being characterized by SEM [33]. Moreover, as the feature size of devices continues to shrink, the application of CD-SEM to nonconducting photoresist may reach a physical limit because of surface charging and thus exacerbate the edge brightness effect. Atomic force microscopy has emerged as a viable alternative to SEM.

6.5.1.1   Imaging High Aspect Ratio Features

The conventional pyramidal tip of microfabricated Si cantilever with its 30° cone angle is unable to trace the deep and narrow features often required for characterization of photoresists. Generally, images of surface features with steep and deep patterns obtained with Si probes contain artifacts introduced by the interaction of the side of the pyramidal tip (see Figure 6.3 showing the effect of tip shape on imaging of deep surface feature). High aspect ratio Si probe fabricated by FIB milling, as seen in Figure 6.4, offers improvement over that of the conventional probe. However, the FIB-milled probe still suffers from the wear problem associated with Si probes. The high aspect ratio of the CNT tip may offer the best solution, particularly the MWNT with length of about 1 μm or longer and diameter as small as 10 nm. This is depicted in Figure 6.10, with an SEM image of a MWNT probe and a schematic representation for the high aspect ratio MWNT tip reaching the bottom of a deep and narrow feature.

MWNT tips, with many layers of concentric nanotubes and larger diameters, experience less thermal vibrations than the SWNT tip. The vibration amplitude of the CNT tip, x_t, in relation to thermal energy, is closely approximated by

$$k_B x_t^2 = k_b T, \tag{6.5}$$

where k_B is the force constant for the bending response of the CNT (as discussed previously with Equation (6.2)), k_b is Boltzmann constant, and T is temperature. Calculation based on Equations (6.2) and (6.5) gives a tip vibration amplitude, x_t, less than 0.5 nm for a CNT with a length of 1 μm and a diameter of 10 nm [34]. The mechanical stiffness of the MWNT tip allows for the probe to trace deep and narrow features. In addition, the mechanical stiffness also circumvents the problem of undesired interaction between the sidewall of the MWNT and the edge of tall features of the surface, which may cause the nanotube to bend, resulting in imaging difficulty.

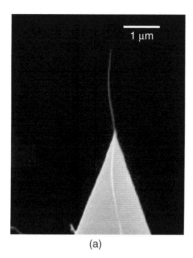
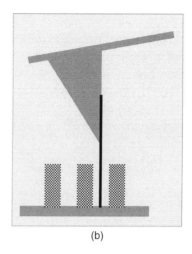

(a)   (b)

FIGURE 6.10 (a) SEM image of a MWNT probe and (b) schematic representation of the ability of a high-aspect ratio MWNT tip to reach the bottom of the trench of a line and space patterned photoresist film.

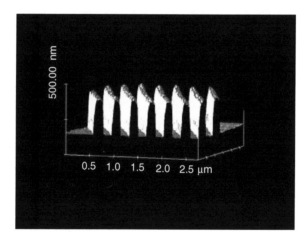

FIGURE 6.11 Three-dimensional image of 90-nm critical dimension line and space photoresist pattern obtained with a scanning probe consisting of a single MWNT as the tip.

Figure 6.11 shows a three-dimensional image of a line and space array of a photoresist pattern generated by a 193-nm lithographic tool. The 90-nm CD with a depth of 250-nm array is clearly resolved. The vertical sidewalls of almost 90° show no imaging artifact, thus suggesting that the MWNT probe does not undergo any bending response when interacting with the sidewall of the resist. It should be noted that the extremely narrow feature could be resolved only with a probe consisting of a single MWNT tip, which is possible and mainly the result of the improved low-density MWNT source [16]. The small diameter (~20 nm) and micron length of the MWNT enable the tip to reach the bottom and trace along the bottom surface of the trench. In contrast, if the probe consists of a bundle of CNTs, the diameter may be larger than the width of the trench, in which case the probe would not be able trace the bottom of the narrow trench. It is also noteworthy that the MWNT probe and AFM-based metrology circumvents the e-beam damage reported for the CD-SEM technique and therefore can be a viable alternative to SEM as a critical-dimension metrology tool.

In comparing SWNT and MWNT probes, the smaller diameter of SWNT translates to a smaller force constant for the bending response. Using Equation (6.5) to describe the thermal vibration amplitude along with Equation (6.2) for k_B, a SWNT with a diameter of 2 nm requires its length to be shorter than 100 nm in order to keep the tip vibration amplitude smaller than 0.5 nm [29]. Such a short length and the small bending force of SWNT make application of the SWNT probe for CD metrology intrinsically not feasible.

6.5.1.2 High Lateral Resolution Imaging

High lateral resolution application with the AFM for the characterization of ultrathin films is becoming very important in the IT industry. For example, future silicon devices require the integration of ultrathin film (thickness on the order of a few atomic layers) high-k gate dielectrics ($k > 10$). Among the materials considered for this application is HfO_2, which is deposited by CVD. Film characteristics such as surface roughness and uniformity of coverage need to be determined for process control. The nondestructive, high-resolution, and high-throughput nature of the SPM makes it a highly desirable characterization technique in many different sectors of the IT industry. In addition to high lateral resolution, a probe that is highly wear resistant, i.e., maintaining constant radius of curvature over extended periods of usage, would also be highly desirable for the characterization of thin films. Robust, small tip diameter probes enable self-consistent analytic assessment of multiple samples, providing precise data for the optimization of process parameters.

Clearly, for high lateral resolution imaging applications, the smaller diameter of SWNT is more attractive, as illustrated in Figure 6.12 (see color insert following page 146). Unfortunately, the low force constant for the bending response and the nonvertical orientation of SWNT, as a direct result of the fabrication process involving surface alignment as the mechanism of attachment and growth, pose many challenges for the routine utilization of SWNT probes in industrial applications. The stability of the point of attachment of the SWNT to the surface of the Si probe presents a great problem. The interaction between the SWNT and the Si surface is purely by van der Walls force. Because of this relatively weak noncovalent interaction, the SWNT may change its surface alignment configuration and orientation with respect to the Si tip. It has been observed that for some of the SWNT probes, the nanotube tips may disappear (as detected by the buckling signature in the tip amplitude vs. distance curve) after the SWNT probe has been in storage, and then it may appear again at a later time. This observation suggested that the configuration/orientation of the SWNT is meta-stable, and it is likely that there are multiple SWNT alignment configurations with similar energy states that allow the SWNT to shift its orientation. A nonuniform surface or a surface contaminant on the Si probe can also contribute to the instability of the SWNT probes.

Removing the outer shells of a MWNT probe also produce smaller-diameter nanotubes as an alternative to SWNT for high lateral resolution application. It has been demonstrated that the outer layers of a MWNT can be selectively removed with electrical current in the presence of oxygen, rendering a nanotube with fewer shells and a smaller diameter [35,36]. In addition, it has also been demonstrated that in a TEM chamber, one end of a MWNT can be sharpened by locally removing the outer layers through a field evaporation mechanism [37]. A similar, albeit a more practical, *in situ* AFM method for sharpening the tip of a MWNT probe has also been demonstrated [38]. The process, very similar to the shortening of a CNT tip, involves using the AFM force calibration mode to bring the tip into intermittent soft contact with the surface of an electrode (Figure 6.13). A DC field is applied between the tip and the metal electrode, and the resulting Joule heating at the MWNT tip-electrode surface junction causes selective degradation of the graphite layers at the tip. It should be noted that the applied voltage for the sharpening process is much smaller than that required for the shortening procedure (2 to 3 V vs. greater than 7 V).

The sharpening process is highly reproducible for CNT probes composed of only a single MWNT tip. The process generally reduces the size of the MWNT tip to less than 5-nm radius of curvature, as directly analyzed from AFM data obtained from the imaging of a 5-nm gold particles dispersion on mica (see Figure 6.14). The stability of this sharpened MWNT tip has also been demonstrated by scanning the surface of a HfO_2 film for more than 20 hours and with multiple reapproaches. The RMS surface roughness values of the two images obtained at the first and the 56th scans are about the same, 0.564 nm and 0.562 nm (Figure 6.15; see color insert following page 146). The very little change in the roughness

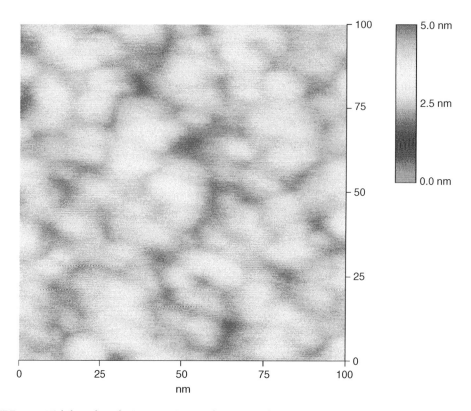

FIGURE 6.12 High lateral resolution AFM image of a 2-nm Ir film deposited on mica substrate obtained with a SWNT probe. Grain size as small as 2 nm was resolved. [**See color insert following page 146.**]

data strongly suggests that the sharpened CNT tip structure has not degraded. Comparing SEM micrographs of the MWNT probe before and after the experiment also confirmed that the tapered graphitic structure of the sharpened tip did not exhibit any observable change. The general durability of the MWNT probe (derived from the fabrication method in References 16 and 17) and the robustness in the structure of the sharpened CNT tip are clearly demonstrated from these sets of experiments. Control experiments performed with silicon probes on the HfO$_2$ surface exhibit an initial RMS roughness value of 0.42 nm for a fresh Si probe and a value of 0.32 nm after 10 scans. This is indicative of the well-known tip degradation of Si probes, which is undesirable in the analytical evaluation of the surface morphology of thin films.

A detailed statistical analysis of the data obtained with MWNT probes and Si probes has been reported for sputtered Si films on Si (100) surface [39]. AFM data collected for a series of Si films with thickness ranging from 53.5 nm to 6420 nm demonstrated the superior performance of the MWNT probe over that of conventional Si and Si$_3$N$_4$ probes. The data for the roughness exponent $\alpha = 0.61$ for a MWNT probe, compared with $\alpha = 0.832$ for both Si and Si$_3$N$_4$ probes, demonstrates the ability of the MWNT probe to more precisely track the surface. The degradation of the probe tip can also be correlated to the value of the roughness exponent, where the value of α increases as the tip degrades and the diameter becomes larger. In addition, the lateral correlation length, ζ, vs. sample thickness is significantly and consistently lower as derived from the MWNT probe images when compared with the Si probe images. The value for the MWNT probe correlates well at short growth times (films with thickness less than 800 nm), whereas the Si probe shows flattening, or saturation curves. This effect supports the notion that the MWNT probe is able to resolve smaller features and therefore shorter-length correlations.

6.5.1.3 Magnetic Force Microscopy

The stability of the MWNT probe with its high aspect ratio and high lateral resolution of sharpened MWNT probes has much potential for industrial applications. In the magnetic storage industry, there is

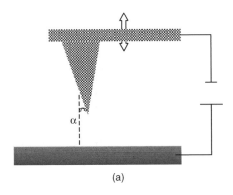

(a)

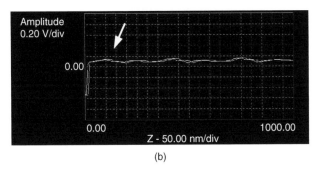

(b)

FIGURE 6.13 Schematic diagram for the MWNT probe sharpening process. (a) Picture depicting the contact of the MWNT tip with a conducting substrate through the use of the force calibration mode in the AFM. (b) Amplitude vs. distance plot in the AFM force calibration plot to monitor the placement of the tip relative to the electrode substrate. The white arrow indicates the point where the tip of the MWNT makes contact with the substrate, at which point the amplitude of the cantilever becomes smaller.

a great need for a characterization technique to resolve nanoscale magnetic domains. MFM, one of the modes in an SPM, is widely used in the storage industry for imaging magnetization patterns with high resolution and minimal sample preparation. MFM responds to forces that range over a long distance. The lateral resolution in imaging using the long-range forces depends on the actual tip geometry characterized not only by the radius of curvature of the tip but also by the half angle of the conical taper, the tip length, and the tip-sample spacing.

An ideal probe in MFM would consist of a single-domain ferromagnetic nanoparticle located at the very end of a nonmagnetic tip. To meet this need, Nakayama and coworkers, using their nano-manipulation technique in an SEM, were able to fabricate a MWNT probe with a Ni particle at the end of the CNT [40]. The 35-nm diameter Ni particle at the tip of a 20-nm diameter MWNT was demonstrated to resolve 750-nm spaced magnetic tracks of a magnetic recording medium. An individual Fe-alloy-capped MWNT probe was also reported for MFM [41]. Here the diameter of the MWNT was 30 nm and the diameter of the Fe-alloy particle was about 50 nm. The dipole moment of the Fe-alloy-capped MWNT probe was estimated to be 1.9×10^{-12} emu, as compared with 1.1×10^{-12} emu for a 60-nm diameter Ni-capped MWNT probe. Although the diameter of the Fe alloy is smaller than that of the Ni particle, it has twice the dipole moment and is on the same order as that of conventional MFM probes.

Simple coating of a MWNT probe with a Co film is also effective for MFM. MFM imaging results with a 30-nm Co-coated MWNT probe are shown in Figure 6.16. Compared with other commercial Si probes for MFM, the resolution is clearly much higher with the MWNT probe. This is presumably due to the long MWNT with a small radius of curvature, which minimizes the fringe effect often associated with the pyramidal shape of conventional Si probes. It should be noted that optimization of tip parameters such as length and radius as well as thickness of the magnetic film coating would yield better data than this initial result.

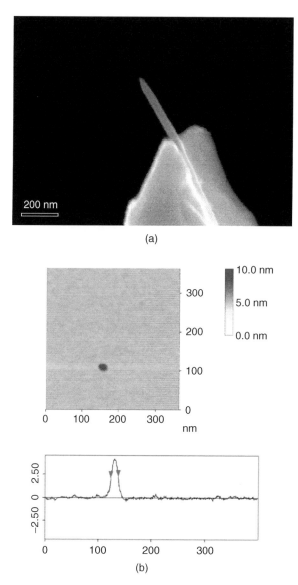

FIGURE 6.14 (a) SEM image of a tapered MWNT tip obtained by the AFM sharpening method. (b) AFM image and cross-section data of a Au particle on mica surface obtained with a sharpened MWNT probe. Deconvolution of a cross-section data 13.8 nm at full width and half-max of the 4.8-nm Au particle gives a 4.2-nm radius of curvature tip.

6.5.1.4 Scanning Probe Lithography

Lithography by a scanning probe is an emerging nanoscale patterning technique. One of the most popular methods is to utilize a scanning probe as a source for electron emission and chemically modify a substrate in the nanometer regime [42]. In general, scanning probe lithography (SPL) produces low-energy electrons (<50 eV) compared with those in conventional electron-beam lithography (EBL; 300 eV to 100 keV) [43]. SPL has the advantage of operating in ambient environment, whereas EBL requires a high-vacuum system. The low-energy electrons of SPL avoid the effect of electron back-scattering, thereby virtually eliminating proximity effects and resulting in greater resolution and superior reproducible patterns. Moreover, because of the lower energy of the emitted electrons, SPL is less sensitive to dose variations compared with EBL. This also translates to SPL requiring a higher dose to write the same feature size. The low-energy electrons in SPL have a mean free path below 2 nm and reach through the

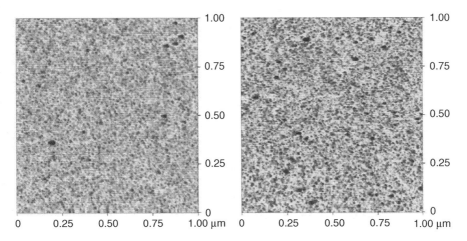

FIGURE 6.15 AFM image of HfO$_2$ film demonstrating the stability of the sharpened tip structure of MWNT probe. The RMS roughness values show little change, 0.564 nm for the initial image (left) and 0.562 nm for the 56th image collected after 20+ hours of continuous scanning. [**See color insert following page 146.**]

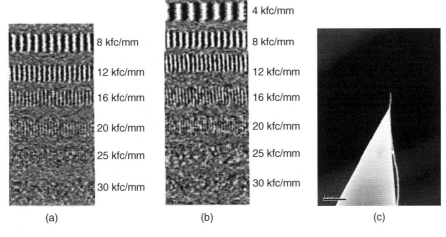

FIGURE 6.16 Magnetic force microscopic images of a magnetic recording medium obtained with (a) a standard MFM probe and (b) a 30-nm Co coating on a MWNT probe; (c) the MFM MWNT probe with 30-nm Co coating. (MFM images courtesy of B. Terris, L. Folks, and M. Best of IBM Almaden Research Center.)

resist under the influence of an electric field. In the process, the electron also undergoes a number of scattering events before transversing through to the resist/substrate interface. SPL also suffers a serious tip-wear problem when using conventional Si probes.

The nanometer dimension and mechanical properties of CNT coupled with high electrical conductivity render MWNT probe ideal for SPL. Particularly the CNT field-emission property with a low turn on voltage (2 to 3 V/μm) may be an advantage to using it as the probe in SPL. The topic of CNT field emission is covered in detail in Chapter 8. Another advantage of CNT tip is that field evaporation of atoms will not be a problem under a strong applied field because of the strong binding energy of carbon atoms (14.5 eV and 12.5 eV for hexagon and pentagon, respectively). The anodization of hydrogen-passivated Si (100) surfaces in air by SPL [43] with a MWNT probe has been demonstrated [17,44]. Figure 6.17 (see color insert following page 146) shows a SiO$_2$ pattern generated on a Si substrate by SPL with a MWNT probe at a –6 V applied bias and a scan rate of 0.5 μm/sec. The SiO$_2$ features have dimensions of about 10-nm width and about 2-nm thickness.

FIGURE 6.17 AFM image of SiO_2 pattern generated on H-passivated Si surface by SPL with a MWNT probe, with line width of about 10 nm. (Image courtesy of Keith Jones of Veeco Metrology.) [**See color insert following page 146.**]

SPL with MWNT probes employing a variety of polymeric resists has been demonstrated. A comparison study of a gold-coated Si probe with a MWNT probe was reported [45]. In this study, a 10- to 15-nm poly(methylphenylsilane) film was used as the resist layer. The 20-nm MWNT tip, at a -20 V bias and scan speed of 10 μm/sec, directly produced a 60-nm wide groove, whereas a gold-coated Si probe produced a 130-nm wide groove at –10 V bias and 0.25 μm/sec scan rate. The poly(methylphenylsilane) resist did not require a liquid development step because the emitted electrons directly induced physical change in the polymer. It was found that grooves are not formed under humidity less than 20% and also that the width of the groove increased with an increase in ambient humidity. An electrochemical reaction with moisture was proposed to be responsible for the Si-Si bond scission, which generated volatile species as the by-product [46]. SPL with tapping mode appears to have better performance compared with a contact mode operation. The smaller SPL features were a direct result of the smaller tip curvature of the MWNT probe. The smaller radius of curvature of the tip also generated a higher electric field, resulting in higher energy electrons and a higher current density, which enable faster-scan SPL. The lifetime of the MWNT tip was also longer than that of the Au-coated Si probe because of the stronger binding energy of the carbon atoms in CNT, whereas the Si probe with a metal coating may wear away during the SPL process.

The major technical challenge in SPL is the problem of tip-wear with the conventional Si probe. To a large degree this problem can be alleviated with scanning probes composed of MWNT tip, as demonstrated in the examples discussed above. Improvement in terms of generating smaller features has also been demonstrated with the MWNT probe. Further improvements in the generation of smaller lithographic patterns may be possible with better understanding of the mechanism of the chemical reaction. At present, SPL has been performed on resists that are not designed specifically for this application. The designing and engineering of a new class of materials specifically for use as SPL resists require a better understanding of the underlying mechanism for pattern generation in SPL. In light of the problem of proximity effects of EBL and the need for techniques for generating ever-increasing density of patterns for nanodevices, SPL has the potential of becoming the best alternative method for nanolithography and nanofabrication for the general development of nanoscience and nanotechnology.

6.5.2   Applications of CNT Probes in Biological Sciences

The inherent mechanical flexibility of the CNT tip prevents the transfer of large forces from the probe to the sample; thus, it is a very attractive feature for the imaging of soft materials such as biological samples. As discussed earlier, the buckling force of the CNT is about 5 nN as calculated from Equation (6.4) for a 250-nm long, 5-nm diameter MWNT probe [13]. This suggests that a sample will not encounter a force greater than the Euler buckling force because the CNT acts as a compliant spring that moderates the impact of the tip tapping the sample. It is generally agreed that high-frequency, large-force constant cantilevers (high Q factor) are the most desirable for obtaining high-resolution images; however, using such high force probes may cause damage to soft samples. In theory, with a CNT tip mounted on a high-frequency cantilever, the problem of sample damage may be mitigated to a large degree by the transfer of imaging force to the buckling of the CNT. This may be a great advantage for high-resolution imaging of soft biological samples.

Because of the ability of AFM to resolve single molecules, there is a great interest in the molecular biology community to use this technique to investigate the structure as well as the activity of biological molecules, such as protein-protein or protein-DNA interactions. Impressive resolution of SWNT probes for investigation of biomolecules in the dried state have been demonstrated [24]. Nakayama and coworkers also used MWNT probes, fabricated by their *in situ* SEM attachment method, to investigate the structure of DNA molecules on mica substrate. In ambient atmosphere, a nanotube with diameter of 11 nm afforded DNA image with a diameter of about 7 to 8 nm [47]. This was reported to be a factor-of-2 improvement in resolution over that of conventional Si tip. By noncontact AFM techniques in high vacuum ($<8 \times 10^{-11}$ Torr), a MWNT probe with ~5-nm tip diameter afforded images of DNA molecules with a full-width and half-maximum of about 2.7 nm and a height measurement of 0.5 nm [48]. A periodic structure at an interval of 3 to 4 nm on the DNA was reproducibly resolved. This periodic structure exhibited right-hand turns with a spacing distribution maximum at 3.5 nm, suggesting the structure of the 3.4 nm spacing major grooves in B-form DNA.

More important in biological applications is the ability of AFM to image in a fluid. This allows for real-time investigation of biochemical activity and measurement of biomechanical elements. A probe composed of a bundle of MWNTs has been used to obtain the image of DNA at a solid-liquid interface [49]. It is interesting to note that the hydrophobic nature of the CNT graphitic sidewall is chemically incompatible with the aqueous solutions in some biological imaging applications. This may result in the instability of the CNT tip, as demonstrated in Figure 6.18. The MWNT probe in this case was fabricated by current-induced welding of the CNT to the surface of the Ni-coated Si probe. After immersion of the MWNT probe in water, it was observed that the CNT bent backwards and permanently attached itself to the surface of the Si probe. This failure may be attributed to unfavorable chemical interaction of hydrophobic CNT sidewall and the surface tension of water as CNT penetrates the liquid-air interface. The unfavorable interaction may be altered by chemical modification the CNT sidewall to render it less hydrophobic. Indeed, by exposing the MWNT probe to ethylene diamine ($H_2NCH_2CH_2NH_2$), a molecule that is known to adsorb onto the sidewall of the CNT, the failure of the MWNT tip is no longer observed when the probe is immersed in water [50]. Therefore, manipulating the CNT probe in a liquid environment must be performed with great care to ensure that the CNT tip has not changed its configuration in the liquid.

6.6 Summary

CNT probes in AFM have become prominent recently, and there is sufficient data in the literature to demonstrate the enhanced resolution as well as performance of CNT probes. Impressive lateral resolution has been demonstrated with the SWNT probe, where a tip with a radius of curvature as small as 1 nm has been demonstrated. However, the stability of SWNT probe still needs further refinement. Fabrication techniques that rely mainly on the surface alignment and van der Walls force for attaching the SWNT to the Si probe must be improved in order to enhance stability.

The MWNT probes have been demonstrated for high aspect ratio imaging applications, showing that deep and narrow features with CD in the sub–100-nm range can be resolved. This is possible because of the large force constant for the bending response of the MWNT. In contrast to the SWNT probe, the point of attachment of the MWNT to the surface of the Si probe is strengthened by current-induced

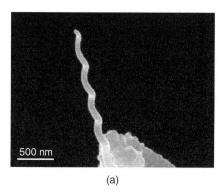
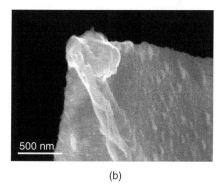

(a)   (b)

FIGURE 6.18 The effect of immersion of MWNT probe in liquid: SEM images of MWNT probe (A) before and (B) after immersion, showing the permanent bending of the CNT tip.

welding, thus dramatically enhancing the stability of the probe where no degradation in image quality is observed after a long period of scanning. In addition, a simple technique involving locally sharpening of the MWNT tip improves the lateral resolution of the MWNT probes, closely approaching that of SWNT probes. Taken together, the MWNT probes have universal applications; high lateral resolution; and high aspect ratio surface imaging, SPL, and MFM with a magnetic-coating. However, one area of the MWNT probe fabrication that needs improvement is the enhancement of the orientation. This is required for future metrology applications, in which trenches with over a micron depth and width less than 100 nm need to be characterized.

As for the commercialization of CNT probes, the growth of SWNTs on a patterned 4-inch Si wafer as a means for mass production of CNT scanning probes has been demonstrated [51]. In this case, the surface alignment mechanism is still the means of controlling orientation as well as fixing the SWNT to the Si substrate. In theory, similar CVD growth of MWNTs on a wafer scale can also be used for the mass production of MWNT probes. Ideally, the length and diameter of the CNTs must be uniform over the whole wafer, and more importantly the orientation must be uniform across the whole wafer as well. This remains the challenge in the growth process. Once these technical challenges are overcome, the CNT probes will be the probes of choice for many industrial and scientific applications. Moreover, with the development of multiprobes AFM on the horizon [52], SPL will be a very attractive nanofabrication technique due to the higher throughput of the multiprobe AFM. Coupling the multiprobe AFM with CNT probes will prove to be a great combination for nanolithography — a corner stone for nanofabrication and the development of nanoscience and nanotechnology.

References

1. G. Binning, C.F. Quate, and C. Gerber, Phys. Rev. Lett., 59, 930 (1986).
2. D. Sarid, *Scanning Force Microscopy: With Applications to Electric, Magnetic and Atomic Forces. Oxford Series in Optical and Imaging Science, No. 5*, Oxford University Press on Demand, Oxford (1994).
3. V.J. Morris, A.P. Gunning, and A.R. Kirby, *Atomic Force Microscopy for Biologists*, Imperial College Press, London (1999).
4. T.R. Albrecht et al. J. Vac. Sci. Tech. A, 8, 3386 (1990).
5. P. Hansma and D. Rugar, Physics Today, 33, 23 (1990).
6. Y. Martin, C.C. Williams, and H.K. Wickramasinghe, J. Appl. Phys., 61, 4723 (1987).
7. J.E. Stern et al. Appl. Phys. Lett., 53, 2717 (1988).
8. D. Rugar, H.J. Mamin, and P. Guethner, Appl. Phys. Lett., 55, 2588 (1989).
9. G. Meyer and N.M. Amer, Appl. Phys. Lett., 53, 1045 (1988).
10. S. Alexander et al. J. Appl. Phys., 65, 164 (1989).

11. R.S. Ruoff and D.C. Lorents, Carbon, 33, 925 (1995).

12. E.R. Wong, P.E. Sheehan, and C.M. Lieber, Science, 277, 1971 (1997).

13. H. Dai et al. Nature, 384, 147 (1996).

14. K. Moloni, A. Lal, and M.G. Lagally, in Optical Devices and Diagnostics in Materials Science, Proceedings of SPIE, Vol. 4098 (2000).

15. R.M. Stevens et al. Nanotechnology, 11, 1 (2000).

16. R.M. Stevens et al. Appl. Phys. Lett., 77, 3453 (2000).

17. C.V. Nguyen et al. Nanotechnology, 12, 363 (2001).

18. T. Larson et al. Appl. Phys. Lett., 80, 1996 (2003).

19. H. Nishijima et al. Appl. Phys. Lett., 74, 4061 (1999).

20. Y. Nakayama et al. J. Vac. Sci. Tech. B, 18, 661 (2000).

21. K. Yamamoto, S. Akita, and Y. Nakayama, Jpn. J. Appl. Phys., 35, L917 (1996).

22. S. Akita et al. J. Phys. D: Appl. Phys., 32, 1044 (1999).

23. S. Akita and Y. Nakayama, Jpn. J. Appl. Phys., 41, 4887 (2000).

24. S.S. Wong et al. J. Am. Chem. Soc., 120, 603 (1998).

25. S.S. Wong et al. Appl. Phys. Lett., 73, 3465 (1998).

26. J.H. Hafner, C.L. Cheung, and C.M. Lieber,, Nature, 398, 761 (1999).

27. J.H. Hafner, C.L. Cheung, and C.M. Lieber, J. Am. Chem. Soc., 121, 9730 (1999).

28. A. Cassell et al. J. Phys. IV France, 11, 401 (2001).

29. J.H. Hafner et al. J. Phys. Chem. B, 105, 743 (2001).

30. E.S. Snow, P.M. Campbell, and J.P. Novak, Appl. Phys. Lett., 80, 2002 (2002).

31. E.S. Snow, P.M. Campbell, and J.P. Novak, J. Vac. Sci. Technol. B, 20, 822 (2002).

32. See http://public.itrs.net/Files/2001 ITRS/Home.html for 2001 ITRS Roadmap.

33. Kudo, T. et al. Advances in Resist Technology and Processing XVIII, SPIE Proceedings, 4345, 179 (2001).

34. C.V. Nguyen et al. Appl. Phys. Lett., 81, 901 (2002).

35. P.G. Collins, M.S. Arnold, Ph. Avouris, Science, 292, 706 (2001).

36. P.G. Collins et al. Phys. Rev. Lett., 86, 3128 (2001).

37. J. Cumings, P.G. Collins, and A. Zettl, Nature, 406, 586 (2000).

38. C.V. Nguyen et al., J. Phys. Chem. B, 108, 2816 (2004).

39. Q.M. Hudspeth et al. Surface Sci., 515, 453 (2002).

40. T. Arie et al. J. Vac. Sci. Tech. B, 18, 104 (2002).

41. N. Yoshida et al. Jpn. J. Appl. Phys., 41, 5013 (2002).

42. M. J. Madou, *Fundamentals of Microfabrication The Science of Miniaturization,* 2nd Ed. CRC Press, Boca Raton (2002).

43. K. Wilder et al. J. Vac. Sci. Tech. B, 16, 3864 (1998).

44. H. Dai, N. Franklin, and J. Han, Appl. Phys. Lett., 73, 1508 (1998).

45. A. Okazaki et al. Jpn. J. Appl. Phys., 39, 3744 (2000).

46. A. Okazaki, S. Akita, and Y. Nakayama, Jpn. J. Appl. Phys., 41, 4973 (2002).

47. H. Nishijima et al. Appl. Phys. Lett., 74, 4061 (1999).

48. T. Uchihashi et al. Jpn. J. Appl. Phys., 39, L887 (2000).

49. J. Li, A. Cassell, and H. Dai, Surf. Interface Anal., 28, 8 (1999).

50. R.M. Stevens, C.V. Nguyen, and M. Meyyappan, IEEE Transactions on Nanobioscience, 3, 56 (2004).

51. E. Yenilmez et al. App. Phys. Lett., 80, 225 (2002).

52. S.C. Minne et al. App. Phys. Lett., 73, 1743 (1998).

7

Nanoelectronics
Applications

Toshishige Yamada

NASA Ames Research Center

Carbon nanotubes are either semiconducting or metallic depending on their structures [1–5] as discussed in Chapter 1. Usually, semiconductors are used for metal-semiconductor (Schottky) diodes, pn junction diodes, and field-effect transistors (FETs), whereas metals are used for single-electron tunneling transistors. This situation is more or less the same in nanotube (NT) electronics. In this chapter, semiconducting NT devices will be emphasized. Semiconducting NT characterization is discussed in Section 7.1, followed by doping in Section 7.2, NT FETs in Section 7.3, intermolecular NT Schottky junctions in Section 7.4, and NT pn junctions as Esaki diodes in Section 7.5. Coulomb blockade phenomena are observed in both semiconducting and metallic NTs. Coulomb oscillations and Coulomb diamonds are covered in Section 7.6. In Section 7.7, other semiconducting NT devices are examined, and some topics in metallic NT transport are briefly covered in Section 7.8. Finally, in Section 7.9, general remarks on nanoFETs, including NT FETs, are given.

7.1 Carrier Characterization

In electronics applications, semiconductors must be identified as p-type or n-type. In the NT field, the thermoelectric power measurement and the FET are used for this purpose.

0-8493-2111-5/05/$0.00+$1.50

7.1.1   Thermoelectric Power Measurement

In the thermoelectric power measurement [6], the doping nature is determined by the sign of Seebeck voltage generated by a temperature gradient [7] with a hot probe and a cold probe as shown in Figure 7.1. Because of the temperature gradient, the particle flow occurs from the hot probe to the cold probe. When the NT is p-type, the electric current flows in the same direction as the particle flow, from the hot to the cold probe inside the NT as shown in Figure 7.1a. When the NT is n-type, the electric current flows in the opposite direction, from the cold to the hot probe as in Figure 7.1b. A voltmeter is connected to form a closed circuit and measures the electric current direction. The voltmeter can be regarded as a load resistance. Because of the current conservation, the electric current must have a circular motion. Thus, in the p-type NT, the current flows from the cold probe, through the voltmeter (load resistance), to the hot probe. Therefore, the cold probe is higher and the hot probe is lower in voltage as in Figure 7.1a. In the n-type NT, the current flows from the hot probe, through the voltmeter, to the cold probe as in Figure 7.1b. Therefore, the hot probe is higher and the cold probe is lower in voltage. Figure 7.1c is a model to be used for later discussions.

There is an easy way to visualize which probe is positive in the n-type NT [7,8]. A hot probe will accelerate electron diffusion. Thus, there will be an electron depletion under the hot probe, and this region must be positively charged. The cold probe will decelerate electron diffusion. There will be an electron accumulation under the cold probe, and this region must be negatively charged. Thus, a dipole is created such that the hot probe is positive and the cold probe is negative. The voltmeter simply detects the potential difference V created by these charges.

Usually the voltmeter has an extremely large resistance and the current through it is negligibly small. In this case, the electric field is created so that the thermal diffusion is cancelled. The thermopower measurement detects the bulk properties of the NT. In the thermoelectric power experiment, an internal electric potential caused by the thermal gradient is compared with that of the standard metal such as lead, and the potential difference is measured without practically allowing a current to flow. As long as there is no significant change in the contact properties due to temperature, the contact potential modulation in air and vacuum, if any, will cancel at the hot and the cold probes and will not influence the thermoelectric power coefficient.

The thermoelectric power of semiconducting single-walled nanotubes (SWNTs) was measured in air and in vacuum [9–13], and it has been shown that the thermoelectric power coefficient is positive in air and negative in vacuum as shown in Figure 7.2 [9] and Figure 7.2 [13], respectively. In Figure 7.2, SWNTs have been left in air and are oxidized well. Then the samples are brought into a vacuum chamber at $T = 500$ K at time $t = 0$ and the oxygen molecules are removed. At $t = 0$, the thermoelectric power coefficient is positive, indicating p-type, but as the time passes ($t > 7$ hours) and the oxygen molecules are removed, the coefficient becomes negative, indicating n-type. In Figure 7.3, SWNTs have been left in vacuum and most of the oxygen molecules are removed from the SWNTs. At $t = 0$, the SWNTs are exposed to oxygen molecules at $T = 300$ K. At $t = 0$, the thermoelectric power coefficient is negative, indicating n-type, but as time passes ($t > 70$ hours), the coefficient is positive, indicating p-type. In both experiments, it is clearly demonstrated that the SWNTs in air are p-type and those in vacuum are n-type.

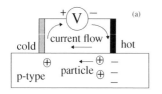

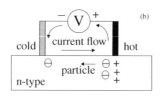

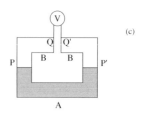

FIGURE 7.1 Schematic diagram of thermoelectric power measurement: (a) p-type case and (b) n-type case; (c) thermocoupler with a voltmeter.

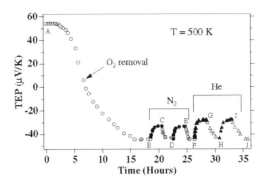

FIGURE 7.2 Thermoelectric power of an initially oxidized SWNT left in vacuum as a function of time. (Reprinted with permission from G. U. Sumanasekera et al. Phys. Rev. Lett. 85, 1096, 2000. ©2000 by the American Physical Society.)

Thermoelectric power is not related to the contact properties [6], and this is experimentally verified in the context of NT. In fact, applying the electrode contact pressure and changing the resultant contact resistance did not change the thermoelectric power coefficient [10]. This means that the thermoelectric power coefficient reflects the NT bulk properties and is not influenced by the electrode contact property change, which can be seen using the following analysis [6]. When there is no current flow, the electric field and the flow due to the temperature gradient must balance with a coefficient $e(T)$ so that:

$$dV/dx = e(T)dT/dx, \tag{7.1}$$

where x is a space coordinate. Introducing a function $\phi\{T(x)\}$ defined by $e(T) = d\phi/dT$,

$$e(T)dT/dx = d\phi/dT \; dT/dx = d\phi/dx. \tag{7.2}$$

Thus,

$$dV/dx = d\phi/dx. \tag{7.3}$$

By integrating this,

$$V(x) = \phi\{T(x)\} + \text{constant}. \tag{7.4}$$

If such an electric potential $V(x)$ is generated on a ring made of two pieces with the same material, the thermal diffusion is completely cancelled in each piece and no current flows. However, if the ring is created using two different materials A and B with junctions P and P′ as shown in Figure 7.1c, the voltage E caused by the thermal gradient is measurable. Assuming that the temperatures at Q and Q′ are the same,

$$E = V(Q) - V(Q') = \phi_A(T_Q) - \phi_A(T_P) + \phi_B(T_P) - \phi_B(T_{P'}) + \phi_A(T_{P'}) - \phi_A(T_{Q'})$$

$$= \{\phi_A(T_{P'}) - \phi_A(T_P)\} - \{\phi_B(T_{P'}) - \phi_B(T_P)\}. \tag{7.5}$$

This expression indicates that the thermoelectric power is due to the bulk properties of materials A and B through ϕ_A and ϕ_B and have nothing to do with contacts P and P′, although E is determined by the contact temperatures T_P and $T_{P'}$. The fact that E depends only on T_P and $T_{P'}$ does not mean that E is determined by the contact properties only. $\phi\{T(x)\}$ is given by an integration of $e(T)$, and $e(T)$ describes how the electric field is created to balance the temperature gradient as in Equation (7.4). This is nothing but the bulk properties of an NT.

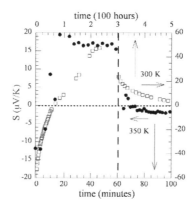

FIGURE 7.3 Thermoelectric power of an initially degassed SWNT left in air as a function of time. (Reprinted with permission from K. Bradley et al. Phys. Rev. Lett., 85, 4361, 2000. © by the American Physical Society.)

Another observation is that the thermoelectric power of A is determined *relative* to B. Usually, lead is chosen for material B, as the standard. The positive and the negative thermoelectric power coefficients simply mean that the thermopower of A is larger or smaller than that of B as seen in Equation (7.5), but in our case of the NTs, the magnitude of the thermoelectric power itself is so distinctively large (20 to 60 μV/K) that there is no ambiguity for the type of the semiconducting NT. Oxygen certainly changes the bulk properties of the NTs by inducing p-type doping. This does not necessarily mean that the oxygen molecules will remove electrons from the NT. The studies indicate either a chemisorption with a small charge transfer of about 0.1 of the unit charge [14,15] or physisorption with a negligible charge transfer in the NT-oxygen interaction [16–22]. The system consists of metallic (often gold) electrodes, oxygen molecules, and the NT, and charges can move around the system between the metal and the NT. Thus, the metal-oxygen interaction and the NT-oxygen interaction are equally important, and the system must be understood from this point of view [23].

7.1.2 Doping Characterization Using FETs

The type of a semiconducting SWNT can be clarified by building an NT FET and measuring the gate-voltage dependence [24–26]. The transistor structure and the corresponding energy-band diagram will be discussed in Section 7.3. In principle, if the transistor is normally-on, i.e., conducting with no gate voltage, V_G, a p-type is confirmed by the drain current increase with a negative V_G, and an n-type is confirmed by the drain current increase with a positive V_G. The current is influenced by the NT bulk properties and the source drain contact properties. Any difference in the NT FET characteristics in air and vacuum will be either due to the NT bulk property modulation or due to the source drain contact property modulation. In order to differentiate them, the V_G-characteristics of the NT FET need to be analyzed theoretically, including comprehensive device models.

In early experiments [27–29], the drain current was observed at $V_G = 0$ and increased at a negative V_G as shown in Figure 7.4a. This means that the NT FET had a p-channel and was normally-on in air. As V_G was increased, the NT was in the depletion mode, and the drain current decreased to a negligible level. However, a large positive V_G did not induce a finite drain current, and it was not possible to detect the inversion transport. This indicates that the source/drain contacts allow only holes to flow in and out and block out electrons.

In recent experiments with improved contacts [30–34], it is possible to observe a finite drain current in both positive and negative V_Gs as shown in Figure 7.4b. Thus, both the electron channel and the hole channel are confirmed. The NT FET is again normally-on, and the drain current increases with a negative V_G if the NT is left in air. The same measurement is repeated after the NT is annealed in vacuum and the NT is degassed. It is shown that the NT FET is at the edge of turn-on at $V_G = 0$, and the drain current

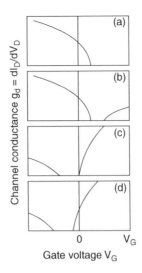

FIGURE 7.4 Schematic plots of channel conductance g_d as a function of gate voltage V_G in SWNT FET under various conditions: (a) characteristics in air at early stage of research, (b) with an improved contact in air, (c) with an improved contact in vacuum, (d) after potassium doping.

increases with a positive V_G as shown in Figure 7.4c. These results do not necessarily contradict the thermoelectric power experiments. In comparison, the characteristics for potassium doping is shown in Figure 7.4d. Alkali metals easily release electrons and dope a NT to be n-type. The NT FET is normally-on, and a positive V_G increases the drain current.

It is possible to determine a semiconductor type by examining the V_G dependence. It has to be noted, however, that the NT FET performance is influenced by the Schottky barrier modulation (contact property change) as well as the Fermi level modulation (bulk property change). Unlike the thermoelectric power measurement, the contact property change will also play an important role. For a thorough understanding, detailed modeling is warranted.

7.2 Doping Methods

In semiconductor device applications, doping and, hence, a control of the Fermi level is of great importance. Generally, there are two ways to dope a semiconductor [24]: substitutional doping and interstitial doping as schematically shown in Figure 7.5.

In substitutional doping, dopant atoms replace the lattice carbon atoms and form an sp^3 bonding in the bulk semiconductors with a diamond lattice and an sp^2 bonding in the NTs with a graphite lattice. In either case, the group III atoms are acceptors, and the group V atoms are donors. Substitutional doping has been done for an NT with boron (group III) or nitrogen (group V) recently [35,36], whereas the interstitial doping has been much more popular in the NT field, probably because of the experimental feasibility.

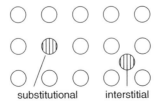

FIGURE 7.5 Schematic figure of substitutional and interstitial doping.

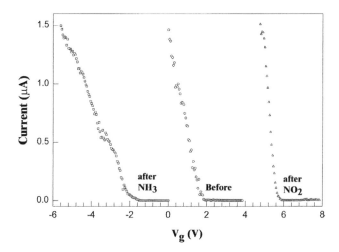

FIGURE 7.6 Drain current as a function of gate voltage for SWNT FETs: after NH_3 doping, initial, and after NO_2 doping. (Reprinted with permission from J. Kong et al. Science 287, 62, 2000. © 2000 AAAS.)

In interstitial doping, the NT lattice is unchanged, and newly introduced dopant atoms stay at the NT surface as adatoms. Dopant atoms may emit or absorb electrons depending on the relation between the highest occupied molecular orbital/lowest occupied molecular orbital (HOMO/LUMO) level of the dopant atom and the Fermi level of the NT, and it does not matter whether they are group III or group V. Materials having a deep LUMO will work as acceptors, and materials having a shallow HOMO will work as donors. It has been reported that O_2, NO_2, Br_2, and I_2 are acceptors, whereas alkali metals such as K, Cs and NH_3 are donors, and other chemicals do change the NT FET properties [37–43]. When an NT is exposed to these materials, the Fermi level rises in case of donors and lowers in case of acceptors. This was confirmed in the change in the threshold voltage of an NT FET or formation of a single-electron transistor.

As discussed in Section 7.1.2, two independent modulations, the NT bulk Fermi level modulation and the Schottky barrier modulation, contribute to the change in NT FET characteristics. One way to distinguish is to examine the channel conductance g_d vs. V_G characteristics, which is often used to determine the FET threshold voltage. The Fermi level modulation causes the horizontal shift of the entire g_d-V_G characteristics without changing the gradients of the onset points for a hole channel or an electron channel as in Figure 7.4. This can be described as a modulation of the FET threshold voltage. The Schottky barrier modulation can be seen as a change in the gradient of the onset branches in the g_d-V_G characteristics as in Figure 7.6, where the data is taken from Reference 37. Both effects are clearly seen. NH_3 has a smaller gradient, whereas NO_2 has a larger gradient in Figure 7.6. This means that the Schottky barrier for holes is higher with NH_3 than that with NO_2.

7.3   SWNT FETs

7.3.1   Basic FET Structure

Using a semiconducting SWNT, an FET has been built with significant gate modulation effects. In early experiments [27,28], a backgated structure was employed with the NT placed on a silicon substrate where the substrate surface was oxidized. Source and drain electrodes (Au or Au-Pt) were placed on the NT as shown in Figure 7.7. When the drain voltage was fixed, the drain current has a strong dependence on the applied gate voltage V_G. This is a gate modulation effect leading to signal gain.

Some fundamental properties have been studied to date, such as subband formation in the SWNT channel [44], a long-channel NT FET behavior [45], transport in a semiconducting channel [46], an effect of a defect in the NT channel [47], the use of electrolyte gate [48], and the use of large diameter NTs [49]. In addition, design principles are also discussed [50,51]. Recently, there are two areas of progress

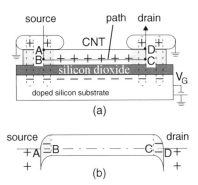

FIGURE 7.7 Backgated SWNT FET: (a) schematic structure and (b) band diagram.

in the NT FET structure. One involves placing an individual gate for each FET [31,52,53], and the other involves building a complementary FET with p- and n-channels [33,54,55].

7.3.2 Schottky Barrier at Electrode Contact

Because of the source/drain electrode contacts, there was a kink structure in the measured $g_d – V_G$ characteristics as shown in Figure 7.8 [27]. This can be explained based on the transport across the metal-semiconductor contact [56]. In the $g_d – V_G$ characteristics for a p-type NT FET, four operating points are shown with a band diagram. The drain current is negligible for (a) to (c) in the subthreshold region, and (d) is in the turn-on region. In Figure 7.8a, the position of the Fermi level is around the middle of the band gap, because the large positive V_G induces a band bending at the source and drain contacts. Because of the thermionic emission process, the drain current depends on V_G exponentially. As V_G is decreased, the SWNT band in the bulk shifts up. In Figure 7.8b, a flat band condition at the source contact is realized, and the drain current stays constant with V_G. This is the origin of the kink. Figure 7.8c corresponds to an onset of tunneling at the source contact, and the drain current starts to depend on V_G exponentially again. In Figure 7.8d, the FET is turned on. The band structure is shown with a broken line, because such a band structure can be realized only when there is inelastic scattering in the NT channel.

Recently, the SWNT FET characteristics have been studied in the context of the Schottky barrier modulation in oxidation [30,57–59]. According to Reference 54, the Schottky barrier height changes in oxidation. The g_d was observed to rise slowly at a negative V_G (V_A) and rapidly at a positive V_G (V_B) in vacuum as in Figure 7.9a, whereas in air, the g_d asymmetry flipped, i.e., g_d rose rapidly at a negative V_G (V_C) and slowly at a positive V_G (V_D) as in Figure 7.9b. This is attributed to the contact property change of the electrode and the NT in Figures 7.9c to f: at $V_G = 0$, the Schottky barrier for holes (Φ_{Bh}) was high and that for electrons (Φ_{Be}) was low in vacuum whereas Φ_{Bh} was low and Φ_{Be} was high in air. Then, g_d rose slowly at V_A because the holes started to tunnel through high Φ_{Bh}, and g_d rose rapidly at V_B because electrons started to tunnel through low Φ_{Be}. Similarly, g_d rose rapidly at V_C because of low Φ_{Bh} and g_d rose slowly at V_D because of high Φ_{Be}. In this $g_d – V_G$ experiment, the contact property modulation of the NT in oxidation was much more influential than the bulk property change, and the experimental findings were consistently explained through the Schottky barrier modulation.

7.3.3 Drain Current as a Function of Drain Voltage

In a standard metal-oxide-semiconductor FET (MOSFET) in Figure 7.10, the current saturation occurs because of the pinch-off point formation [60]. If x is the distance measured from the source to the drain, then the local carrier density is given by $q(x) = C_{ox}[V_G – V(x) – V_T]$, where C_{ox} is an oxide capacitance and V_T is an FET threshold voltage. The voltage monotonically changes from 0 at the source to V_D at the

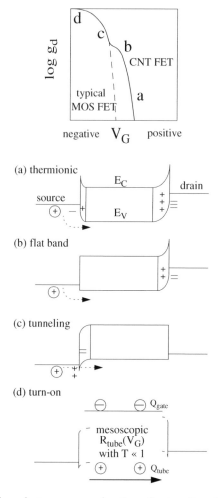

FIGURE 7.8 Schematic channel conductance g_d as a function of gate voltage V_G in SWNT FET with a kink as observed by S.J. Tans et al. [27] compared with a smooth kinkless g_d for MOSFET. (a) to (d) are band diagrams for corresponding operating points. (From T. Yamada, Appl. Phys. Lett., 76, 628, 2000. With permission.)

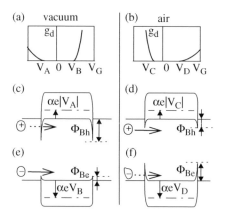

FIGURE 7.9 Schottky barrier behavior in vacuum and in air. (a) and (b) are experimental tendency of g_d as a function of V_G. (c) to (f) are band diagrams at V_A, V_B, V_C, and V_D, respectively.

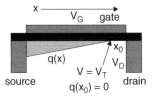

FIGURE 7.10 Pinch-off point formation in an FET.

drain. Thus, when V_D is larger than $V_G - V_T$, there exists a point X_0 where the carrier density is negligible, and this is the pinch-off point. Once this is formed, the drain current saturates, because the current magnitude is determined by the process in the inversion region between the source and the pinch-off point, which is about the channel length. Carriers passing the pinch-off point are simply swept to the drain, and there is no physical process to restrict the current flow. The saturated drain current depends on gate voltage in a quadratic way $(V_G - V_T)^2$, because both the carrier density and velocity are proportional to the gate voltage. This occurs for a long-channel FET where there is frequent carrier-carrier Coulomb scattering for the pinch-off point formation.

As the channel length reduces and becomes intermediate, the electric field between the source and the drain increases for the same drain voltage. Thus, it is possible that the threshold electric field is reached where carriers have a saturation velocity due to increasing phonon scattering before the pinch-off point is formed. This is how another type of drain current saturation occurs. The saturated drain current depends only on $(V_G - V_T)$, unlike the long-channel saturation case. Here, the carrier density depends on $(V_G - V_T)$, but the carrier velocity is the saturated velocity, which is constant and independent of the gate voltage.

When the channel length reduces further and becomes short, carrier-carrier Coulomb scattering is absent (no pinch-off point formation) and phonon scattering is absent (no carrier velocity saturation). Carriers suffer at most impurity scattering or do not suffer any scattering when traveling from the source to the drain. In this situation, the transmission picture prevails and there is no physical mechanism to cause the drain current saturation. The drain current does not saturate in these short-channel transistors.

It is experimentally confirmed that in a SWNT FET with a long-channel length of 10 μm [45], the drain current shows a saturation with a pinch-off point formation. The saturated drain current is a quadratic function of the gate voltage $(V_G - V_T)^2$, and this is the evidence of the pinch-off point formation. This means that the carrier-carrier Coulomb scattering occurs in an NT with a length of 10 μm. Such long channels are created to study basic physics of NT FETs and are certainly not suitable for nanoelectronics. In some NT FET experiments with a short-channel length of ~0.1 μm, the drain current as a function of drain voltage does not show a saturation. Thus, in these short-channel NT FETs, phonon scattering and carrier-carrier Coulomb scattering are absent. In many cases, the drain current I_D is a linear function of drain voltage V_D, but the gradient dI_D/dV_D decreases with increasing V_D in some cases [58]. Such a weak saturation often disappears when source and drain terminals are switched. This indicates that the bulk NT does not contribute to the weak saturation. From an examination of temperature dependence, it is shown that the Schottky barrier at the source contact is responsible for this. The weakly saturated current varies linearly with $(V_G - V_T)$.

7.3.4   Meaning of Drain Current Saturation in Digital Applications

Drain current saturation is important in digital applications [26]. Consider an operation of a resistor loaded inverter in Figure 7.11a. An ideal inverter will have a vertical voltage transfer characteristics as shown in Figure 7.11b, indicating that only low and high V_{out} values are possible. This is the essence of the digital applications. Because only the two signals, low and high, are allowed in the output, the inverter is quite resistant to the noise. In fact, even if there is a small deviation from the ideal low or high output,

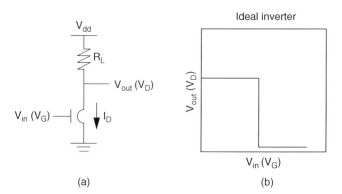

FIGURE 7.11 Resister loaded inverter: (a) circuit configuration and (b) ideal voltage transfer characteristics.

the signal is reshaped after going through another inverter. Each time the signal goes through an inverter, the unwanted noise component is removed, and this is the strength of digital circuitry.

Whether the inverter can have ideal characteristics or not depends on the performance of the FET. There are three different FETs: FET 1 in Figure 7.12a shows traditional saturation behavior, and the saturated drain current is proportional to $(V_G - V_T)^2$; FET 2 in Figure 7.12b shows a saturation, but the saturated current value is proportional to $V_G - V_T$; FET 3 in Figure 7.12c does not show any drain current saturation at all. Long-channel ($\sim 10~\mu m$) NT FETs or individual gate NT FETs with a pinch-off point formation are like FET 1, whereas short-channel ($\sim 0.1~\mu m$) NT FETs with at most elastic scattering are often like FET 3. Middle-channel NT FETs where the velocity saturation is relevant are like FET 2. Using a standard load line analysis, the following voltage transfer characteristics can be presented.

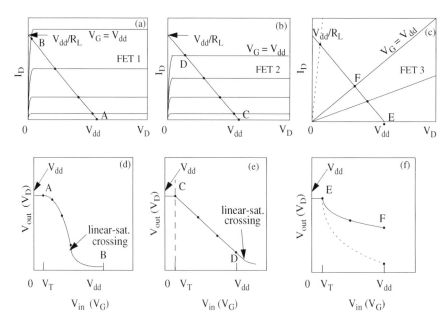

FIGURE 7.12 FET characteristics and resistor-loaded inverter performance: (a) to (c) are various FET drain current (I_D) – drain voltage (V_D) characteristics and (d) to (f) are corresponding voltage transfer characteristics.

When there is a saturation with the pinch-off point formation in Figure 7.12a, the saturated drain current lines are dense for small V_Gs but are sparse for large V_Gs reflecting the quadratic dependence, $(V_G - V_T)^2$. This is crucial in digital applications. The small change in the saturated drain current causes the convex shape of V_{out} around $V_{in} \sim V_T$ near operating point A as shown in Figure 7.12d. The saturated

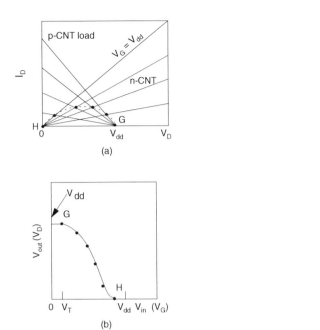

FIGURE 7.13 FET characteristics and complementary inverter performance: (a) p-channel and n-channel SWNT FET characteristics and (b) voltage transfer characteristics.

drain current increases more rapidly as V_G increases, and this brings about a rapid vertical drop in the $V_{out} - V_{in}$ voltage transfer characteristics. Near operating point B, there is a transition of the FET from the saturation mode to a linear mode, and this creates a kinklike concave-type change in the $V_{out} - V_{in}$ characteristics. The above voltage transfer characteristics are close to ideal. In Figure 7.12b, the drain current saturation is present, but the saturated drain current depends linearly on $V_G - V_T$. Although the load line intersects the drain current curve mostly in the saturation current mode like Figure 7.12a, the transition portion in the $V_{out} - V_{in}$ characteristics is mostly straight with a 45° gradient in Figure 7.12e. Thus, the inverter operational margin is narrow. Finally, when there is no drain current saturation as in Figure 7.12c, V_{out} cannot be low, and the inverter does not work appropriately as shown in Figure 7.12e.

By adopting a different circuit scheme, however, FET 2 or FET 3 can be used to build an inverter; this is necessary because micron-scale source-drain lengths are not realistic. If a complementary circuit [20,21] using a p-NT FET and an n-NT FET [39,49] is created, the inverter can have a reasonable margin, although each FET does not show an ideal drain current saturation with a pinch-off point formation. This is illustrated graphically in Figure 7.13. Performing circuit analysis with the p-NT FET as a load, the operating point will form a dotted curve like a half circle as in Figure 7.13a. This means that near the threshold voltage of either the p-NT FET (operating point G) or the n-NT FET (operating point H) the V_{out} changes slowly. Thus, a convex shape near operating point G and concave shape near operating point H are created as in Figure 7.13b, and we can achieve a reasonable inverter margin.

Experimentally fabricated complementary SWNT FETs show a more vertical transition line in the voltage transfer characteristics than a simple resistance of 45° gradient. Many nanodevices, including the NT devices, show ballistic transport in the channel and may not have a drain current saturation, but an inverter can be built with a finite operating margin. This is encouraging for the future nanoelectronics.

7.4   Intermolecular Metal-Semiconductor SWNT Heterojunctions

By fusing metallic and semiconducting SWNTs of different chiralities, intermolecular metal-semiconductor heterojunction diodes can be created [61–66]. The diodes are expected to have a kink at the junction [65]. Indeed, rectifying current-voltage characteristics were observed for such a kink-shaped NT fused diode [66], where V_G was applied to change the carrier density in the diode and the rectifying characteristics were modulated.

This device corresponds to a gate-controlled diode known in the semiconductor electronics. The gate bias modifies the carrier density in the diode and changes the rectifying characteristics. This property itself is useful in electronics applications, but the device has been used rather in the fundamental study of the semiconductor surface potential [67,68]. For example, MOSFET energy-band structures have been studied using a gate-controlled diode [67] because of their geometrical resemblance. NT gate-controlled diodes will play the same role with respect to NT FETs. As will be shown below, the gate modulation occurs essentially the same way in these devices, resulting in the same surface potential behavior at the junction. The only difference is that NT doping is p-type in the FET whereas it is n-type in the diode.

The NT was placed on Ti-Au electrodes on a SiO_2/doped-Si substrate (backgate) as in Figure 7.14a and V_G was applied to the backgate with electrode 3 grounded [66] at 100 K. Although the circuit between electrodes 0 and 1 showed linear characteristics (110 kΩ) without noticeable V_G dependence, the circuit between electrodes 1 and 3 across electrode 2 showed rectifying characteristics with appreciable V_G dependence as in Figure 7.14b. Therefore, the right NT between electrodes 2 and 3 had to be semiconducting.

The operation mechanism can be analyzed using a one-dimensional transport modeling [69,70]. A band-diagram modeling is used here to understand how V_G modulates the diode characteristics [71]. Consider an equivalent circuit with drain current I_D and voltage V_D at electrode 1 with a linear resistor R_1 and a capacitor C_{NT} with respect to the substrate. The metallic and the semiconducting NTs meet at kink 2, and an MS junction (J_2) is formed. The semiconducting NT reaches electrode 3, and a semiconductor-metal (SM) junction (J_3) is formed. For rectification to take place, either J_2 or J_3 should be a Schottky diode and the other should be a resistive element.

The forward direction current transport occurred when $V_D > 0$ [66]. Thus, two equivalent circuits are possible [71]: J_2 is a Schottky diode with an n-type NT and J_3 is a resistor as in Figure 7.15a, or J_2 is a resistor and J_3 is a Schottky diode with a p-type NT as in Figure 7.16b. Forward and reverse turn-on voltages are introduced for a diode, V_{onF} and V_{onR}, respectively. The experimental V_G dependence is such that increasing V_G shifts both V_{onF} and V_{onR} in the positive V_D direction. Such V_G dependence is possible

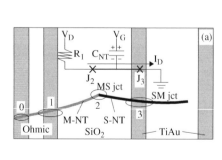

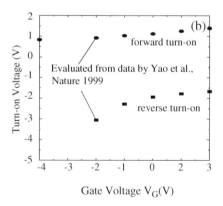

Gate Voltage V_G(V)

FIGURE 7.14   (a) Experimental setup of Z. Yao et al. [66] and its equivalent circuit; (b) forward and reverse on voltages (see text for definition) between electrodes 1 and 3 with V_G as a parameter at 100 K with data evaluated from Z. Yao et al. [66]. (Figure 7.14a from T. Yamada, Appl Phys. Lett., 80, 4027, 2002. With permission.)

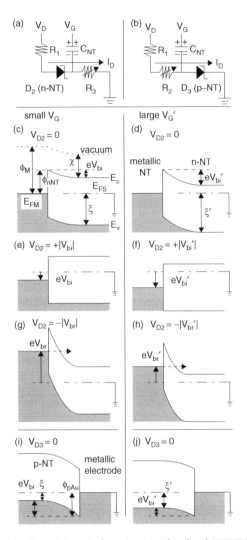

FIGURE 7.15 Rectification mechanisms: (a) equivalent circuit with a fused SWNT MS junction D_2 of n-type and an Ohmic contact R_3 with a capacitor C_{NT}, (b) equivalent circuit with an Ohmic contact R_2 and SWNT-electrode SM junction D_3 of p-type, (c) to (h) energy-band diagram for selected V_{D2}s in the n-SWNT scenario of (a), (i) and (j) energy-band diagrams for $V_{D3} = 0$ in the p-SWNT scenario of (b). Small V_G (left) and large V_G (right) cases are examined. (From T. Yamada, Appl. Phys. Lett., 80, 4027, 2002. With permission.)

with an n-NT but not with a p-NT. The band diagrams for Schottky diode D_2 of n-type in Figure 7.15a are shown in Figures 7.15c to h for selected D_2 voltages (V_{D2}s), and those for D_3 of p-type in Figure 7.15b are shown in Figure 7.15i and j for null D_3 voltage (V_{D3}). We compare small V_G (left) and large V_G (right) cases. ϕ_M is a metallic NT work function. ϕ_{nNT} and ϕ_{pAu} are Schottky barriers for electrons at D_2 and holes at D_3. E_{FM} and E_{FS} are electrochemical potentials (Fermi levels) in the metallic and the semiconducting NTs. E_c and E_v are conduction and valence band-edges with a band gap, e.g., ξ is a chemical potential $E_{FS} - E_v$ and χ is an electron affinity. V_{bi} (> 0) is a built-in voltage and V_{br} (< 0) is a breakdown voltage. e (> 0) is the unit charge.

In the n-NT scenario in Figure 7.15a, increasing V_G results in higher electron density, and ξ increases. Thus, $\xi < \xi'$, where a prime indicates a quantity at V'_G ($> V_G$). Because ϕ_{nNT} is independent of V_G, $V_{bi} < V'_{bi}$ as shown in Figures 7.15c and d. In the thermionic emission [24,25], $V_{onF} \sim V_{bi}$. Therefore, $V_{onF} < V'_{onF}$, as in Figures 7.15e and 7.15f. This is consistent with the experiment. The reverse turn-on occurs when $V_D \sim V_{br} = -|V_{br}|$. This is the beginning of the tunneling breakdown. The effective doping is larger

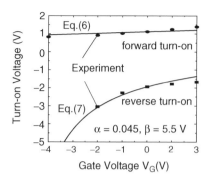

FIGURE 7.16 Comparison of modelling results Equation (7.6) and (7.7) to an experiment by Z. Yao et al. (From T. Yamada, *Appl. Phys. Lett.*, 80, 4027, 2002. With permisson.)

for larger V_G, leading to the thinner Schottky barrier as in Figures 7.15g and h. Thus, $-|V_{br}| < -|V'_{br}|$ and $V_{onR} < V'_{onR}$. This is also consistent with the experiment. However, neither trends for V_{onF} and V_{onR} are explained by the p-NT scenario in Figure 7.15b. Again $\xi < \xi'$ but $V_{bi} > V'_{bi}$ for holes as shown in Figures 7.15i and j. Thus, $V_{onF} > V'_{onF}$. In the reverse direction, $-|V_{br}| > -|V'_{br}|$ and $V_{onR} > V'_{onR}$. Both are against the experiment. Therefore, we conclude that the rectification occurred at D_2 and the NT must be n-type.

V_{onF} and V_{onR} can be expressed as a function of V_G based on the above view. V_G causes a linear change in ξ, such that $\xi(V_G) = \xi(0) + \alpha e V_G$. The coefficient α is related to the NT state density and NT [71]. By the band diagram in Figure 7.15c, $eV_{bi} = \phi_{nNT} - [E_g - \xi(V_G)]$, where e is the unit charge. The forward turn-on is achieved by applying $V_{D2} = V_{bi}$ and $\Delta V_{bi}(V_G) = \alpha \Delta V_G$. The turn-on voltage modulation by V_G including the R_3 contribution in the vacuum-gap mode [72] is given [71] by

$$V_{onF}(V_G) = V_{onF}(0) + \alpha V_G, \tag{7.6}$$

$$V_{onR}(V_G) = V_{onR}(0) + \alpha V_G + (V_{onF}(0) + |V_{onR}(0)| - E_g/e)V_G/(\beta + V_G), \tag{7.7}$$

where $V_G = -\beta$ is a voltage such that electrons are completely repelled, and the planar junction theory [24,25] is assumed. The choice of $\alpha = 0.045$ and $\beta = 5.5$ V recovers voltagethe experimental $V_{onR}(V_G)$ quite well as in Figure 7.16. Quasi–one-dimensional junction field [73], Fermi level pinning [74], and image potential [24,25] effects would not be relevant and are not included in our model, but they could be necessary in the analysis for finite I_D.

The gate-controlled MS diode measurements are studied, and it is shown that the rectification occurred at the kink of the NT junction and the carriers involved in the transport must be electrons, not holes. NT FETs with gold electrodes in air have p-type channels in that the transistors are already on with $V_G = 0$, and the drain current increases with a larger negative gate voltage. These two are not necessarily contradictory, because the doping effect and the Schottky barrier effect can co-exist and both influence the final device behavior. In the gate-controlled NT diode, the Schottky barrier is formed with respect to the metallic NT and is lower for electrons, whereas in the NT FETs, the Schottky barrier is formed with respect to the gold electrode and is lower for holes. For the further NT FET fundamental research, various gate-controlled NT diodes need to be studied actively, as in the silicon MOSFET research [67].

7.5   SWNT pn Junction as Esaki Diode

In the history of semiconductor electronics, people focused on tunnel diodes exhibiting negative differential resistance (NDR) as shown in the current-voltage characteristics in Figure 7.17, when three-terminal transistors were not fabricated reliably. With a device having an NDR and a gyrator allowing a signal to flow in a designated circulation direction (like a traffic rotary), it is possible to create a signal

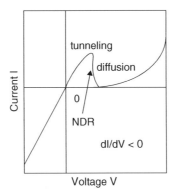

FIGURE 7.17 Current-voltage characteristics with a negative differential resistance (NDR).

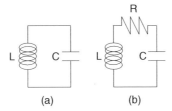

FIGURE 7.18 (a) LC and (b) LCR circuits.

amplifier [75]. The operation of such an amplifier is equivalent to that of a transistor. Another application is to use it for an oscillator circuit. An *LC* circuit with an inductance *L* and a capacitance *C* in Figure 7.18a has a resonance frequency of $\omega = (LC)^{-1/2}$. If a resistor *R* is added to the loop and an *LCR* circuit is formed as in Figure 7.18b, the resonance frequency is modulated to $\omega = (R^2 - 4L/C)^{1/2}/2L$ and the oscillation damps by $exp(-Rt/2L)$ with *t* the time. If $0 < R < 2(L/C)^{1/2}$, the oscillation will decay exponentially with time. If $-2(L/C)^{1/2} < R < 0$, however, it is expected that the oscillation grows in time. This is an oscillator generating an AC signal with a DC bias and such a negative *R* can be achieved with a device with the NDR. Because of these two major applications, NDR devices have been attracting a lot of attention historically in electronics.

An Esaki diode [76–78] belongs to this tunnel diode family with an NDR. The structure is simply a highly doped pn junction, denoted by p^+n^+, and because of this feasibility in fabrication. The NDR occurs in the forward direction. There is a bias point where the tunneling becomes maximum, and these characteristics are superimposed for a usual pn junction forward bias characteristics — larger diffusion current for increasing the bias voltage, as shown in Figure 7.17. The physical mechanism for this behavior is visualized in the band structure below.

A SWNT Esaki diode has been fabricated experimentally [43]. An NT is placed on a SiO_2/Si substrate. After placing source and drain electrodes, the drain current is measured as a function of the gate voltage. The entire NT is unintentionally doped to be p-type in air, and this is confirmed in the normally-on characteristics and the increasing drain current with a negative gate voltage. Then a half of the NT in the source side is covered by polymethylmethacrylate (PMMA), and a half of the NT in the drain side is exposed. The exposed side is doped with potassium. The potassium has a very shallow work function and dopes the NT to be n-type. Because of this initial doping difference, the source side consistently has a higher hole density and a lower electron density than the drain side, regardless of the gate voltage. The drain current as a function of the gate voltage shows that a finite drain current flows when $-12\ V < V_G < -7\ V$ and $-1\ V < V_G$, as shown in Figure 7.19A. The drain current is zero elsewhere. This is because the band structure of p^+n junction is effectively achieved for $V_G < -12\ V$, p^+n^+ for $-12\ V < V_G < -7\ V$,

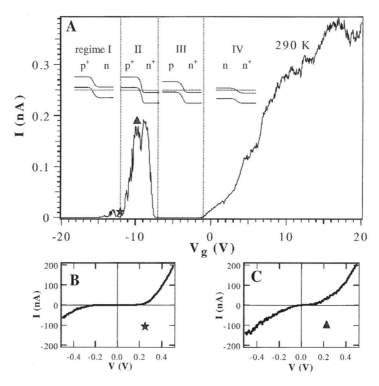

FIGURE 7.19 Experimental SWNT Esaki diode drain current vs. gate voltage: I-IV are corresponding band diagrams. (Reprinted with permission from C. Zhou et al. Science 290, 1552, 2000. ©2000 AAAS.)

pn$^+$ for -7 V $< V_G$, and nn$^+$ for -1 V $< V_G$. The p$^+$n$^+$ junction is conductive. The Fermi level in the p$^+$ is below the valence band top, and the Fermi level in the n$^+$ is above the conduction band bottom. The electrons can tunnel either from p$^+$ to n$^+$ or from n$^+$ to p$^+$ because they can find post-tunneling states in either case. The junction nn$^+$ is also conductive because it is Ohmic without any rectification.

It is interesting to note that after the potassium doping, the source side of the junction is no longer p-type but n-type, when $V_G = 0$. When the potassium doping is ample, there are so many electrons introduced in the NT system that some electrons are introduced even in the PMMA covered region, too. The uniform NT in air is most likely p-type, but this does not mean if the NT is partially doped n-type that the rest can remain p-type. Thus, it is critical that the experiment uses the backgated structure to manipulate the Fermi level of the NT system. Without the backgate, the resulting NT junction would be simply nn$^+$, which is nothing but an ohmic junction and does not show the Esaki behavior of negative differential conductance.

The NT p$^+$n$^+$ junction works as a standard silicon Esaki diode, showing an NDR as in Figure 7.20A. Electrons tunnel through from the grounded source of p$^+$ region to the drain of n$^+$ region in the reverse direction with a negative drain voltage, where the built-in voltage is increased, as in Figure 7.20B. In the forward direction with a positive drain voltage, the built-in voltage is reduced and electrons tunnel from the n$^+$ region drain to the p$^+$ region source, as in Figure 7.20C. When the drain voltage is increased, the built-in voltage is further reduced and electrons in the drain cannot find states to which they can tunnel, or the Fermi level in the drain corresponds to the band gap of the source, as in Figure 7.20D. This unavailability of post-tunneling states is the origin of the negative differential conductance and has been experimentally observed using the NT p$^+$n$^+$ junction. The built-in voltage with the further increase in drain voltage is so small that it cannot prevent an electron diffusion from the source and the hole diffusion from the drain, as in Figure 7.20E. This is the scenario of the Esaki diode and has been clearly observed in the experiment.

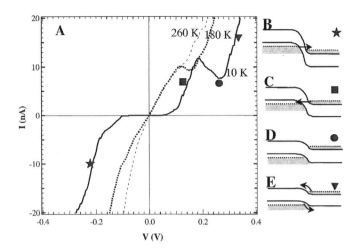

FIGURE 7.20 Experimental SWNT Esaki diode drain current vs. drain voltage characteristics. (Reprinted with permission from C. Zhou et al. Science 290, 1552, 2000. ©2000 AAAS.)

7.6 Single-Electron Tunneling Devices Using SWNTs

Single electron tunneling (SET) devices [79–82] have been studied from physics and electronics points of view. NTs have a narrow diameter, and when contacted to an electrode the overlap area is naturally small without any specific microfabrication, resulting in an extremely small capacitance C (e/C typically ranges from ~10 mV to ~1 V, with e the unit charge). This is a significant advantage of NTs in the SET device applications.

In this section, an emphasis is placed on how the current-voltage characteristics change with the device component parameters. At an early stage, two-terminal SET devices had been examined in the study of correlated tunneling in time or in space. Recently, three-terminal SET devices have been studied, where the SET characteristics can be modulated with the gate bias. There are two important phenomena related to the gate-controlled SET operation: the Coulomb oscillation and the Coulomb diamond formation. The familiar source, drain, and gate are still used to distinguish the terminals in the SET devices. In applications, an SET device is brought to a certain operation point by applying a designated bias or inducing a designated charge on an appropriate part of the device. An input is applied as the source-gate voltage and an output is obtained as the source-drain voltage, which is the same as the conventional FETs, although the current-voltage characteristics are different from those of the conventional FETs.

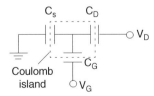

FIGURE 7.21 Equivalent circuit with a tunneling source capacitor, a tunneling drain capacitor, and a usual insulating gate capacitor.

7.6.1 Coulomb Oscillation

When a device structure is small so that the charging energy is comparable with the thermal temperature, the Coulomb blockade [79–82] often occurs. SWNT FET structures operating at a low temperature sometimes exhibit a periodic oscillation of the drain current as a function of gate voltage, known as

Coulomb oscillation [82], a form of the Coulomb blockade phenomena. An idealized circuit for Coulomb oscillation is given in Figure 7.21. Two capacitors in series represent the source and the drain contacts, and the central island is coupled to the gate capacitance through gate modulation effect. The energy level in the central island is continuous in a long NT but is discrete in a short NT. We will see a conductance peak condition using a method developed in Reference 81.

The probability $P(N)$ for the central island to have N particles is given by

$$P(N) = \text{const} \times \exp(-[F(N) - NE_F]/k_B T), \qquad (7.8)$$

where $F(N)$ is the free energy of the central island and T is the temperature [81]. If only one value of N minimizes the thermodynamic potential $\Omega(N) = F(N) - NE_F$, then the system will prefer that single N, and there will be no conductance because the number of particles in the central island cannot change. However, if two values, i.e., N and $N+1$, minimize the thermodynamic potential, $\Omega(N) = \Omega(N-1)$, then the system can allow a finite current flow with a small applied voltage because the central island can have states N, $N-1$, N, $N-1$, N, $N-1$,..., etc. $\Omega(N) = \Omega(N-1)$ gives

$$F(N) - NE_F = F(N-1) - (N-1)E_F. \qquad (7.9)$$

In the low-temperature limit, the free energy $F(N)$ is the ground state energy $U(N)$ of the island. Thus, the current peak appears when

$$U(N) - U(N-1) = E_F \qquad (7.10)$$

for some N. The role of the gate voltage is to change the functional form of U continuously and change the N value satisfying this relation. $U(N)$ is given by integrating $\phi(Q) = -Q/C + \phi_{ext}$ from 0 to $-NE$.

When the discrete energy separation in the island is much smaller than the nanotube charging energy, which is usually a good approximation for metallic islands with a reasonable size, we have

$$U(N) = (Ne)^2/2C - Ne\phi_{ext}. \qquad (7.11)$$

The gate voltage is related to ϕ_{ext} by

$$\phi_{ext} = \text{const} + \alpha V_G, \qquad (7.12)$$

where α is a capacitance ratio of the system. Now, the current peak condition can be rewritten by

$$(2N-1)e^2/2C = E_F + e\phi_{ext}. \qquad (7.13)$$

Thus, the period in V_G is $e^2/\alpha C$.

When the discrete energy separation is comparable with the Coulomb-charging energy, the discreteness of the energy comes into the expression $U(N)$.

$$U(N) = \Sigma E_p + (Ne)^2/2C - Ne\phi_{ext}, \qquad (7.14)$$

where the first term is a summation of the N single-electron energies in the ascending order measured from the bottom of the conduction band. Thus, the current peak condition is rewritten as

$$E_p|_{p=N} + (2N-1)e^2/2C = E_F + e\phi_{ext}. \qquad (7.15)$$

Thus, the period in V_G has a doublet structure, with a spacing alternating between $e^2/\alpha C$ and $(\Delta E_N + e^2/C)/2$, where ΔE_N is the relevant discrete energy separation. There are two remarks. The spin degeneray is lifted by the charging. Equation (7.15) predicts the current peak locations, but cannot predict the conductance, which requires information of the tunneling rates.

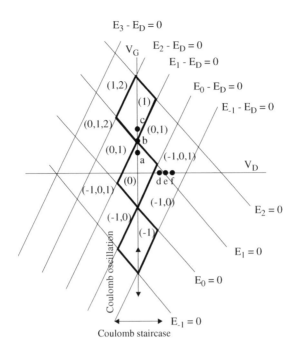

FIGURE 7.22 Coulomb blockade conductance plotted in V_G-V_D plane. Thick parallelograms are Coulomb diamonds with null conductance inside.

7.6.2 Coulomb Diamonds

In some cases, a plot is made for V_D-V_G plane, and the drain current or conductance is shown for this two-dimensional plane, as in Figure 7.22. We will disregard the discreteness of energy levels in the island for simplicity, but it will be incorporated in the same way as above. The boundaries for a finite current or conductance and zero current or conductance are parallelogram and are often called the Coulomb diamond. We will evaluate the ground-state system energy gain when an electron tunnels from electrode i (i = S, D, or G) to the island, where the number of electrons change from N-1 to N [83,84]. The battery loses eV_i; the island gains $e \, \Sigma C_j V_j / C$ with a summation taken for S, D, and G and $C = C_S + C_D + C_G$; and the Coulomb energy gain is $U(N)$-$U(N$-$1)$. The resultant ground-state system energy difference $\Delta(N, N$-$1)$ is given by

$$\Delta(N, N\text{-}1) = U(N) - U(N\text{-}1) - e \, \Sigma C_j V_j / C + eV_i. \tag{7.18}$$

By introducing new variables

$$E_N = U(N) - U(N\text{-}1) - e \, \Sigma C_j V_j / C, \tag{7.19}$$

$$E_i = -eV_i, \tag{7.20}$$

we have

$$\Delta(N, N\text{-}1) = E_N - E_i. \tag{7.21}$$

If this is negative, then tunneling from electrode i to the island certainly reduces the system energy and is preferred. If positive, oppositely tunneling from the island to electrode i reduces the system energy and is preferred. In the low-temperature limit, the ground-state system energy is equivalent to the free energy and tunneling absolutely occurs when this energy is negative, and the tunneling does not occur when this energy is positive. As usual, we take the source to be grounded ($V_S = 0$). In the $V_D - V_G$ plane, we have a series of boundary lines, which are $E_N - E_D = 0$ and $E_N = 0$, as shown in Figure 7.22.

The thick boundaries are the so-called Coulomb diamonds. Inside the diamonds, the conductance is 0. By taking $V_D \sim 0$ and scanning V_G, we will see a Coulomb oscillation. In fact, the drain current is finite only at the corners of the diamonds on the V_G axis and is 0 elsewhere. The oscillation period is e/C_G. Comparing with the previous discussion, $\alpha = C_G/C$. By fixing V_G and scanning V_D, we will see Coulomb staircase. We will see each case more in detail, considering a few operating points in the $V_D - V_G$ plane.

Operating point (a) is inside the diamond with a preferred number of electrons being 0. In fact, if the number of electrons is different from 0, the lacking or excess electrons will tunnel into or out of the island and the preferred number of electrons is reached as indicated in Figure 7.23a. For example, if the initial number is -2, then incoming tunneling is preferred as an arrow indicates, either from the source or from the drain, because $E_{-1} < 0$ and $E_{-1} - E_D < 0$. Thus, the number is -1, and incoming tunneling is still preferred as an arrow indicates because $E_0 < 0$ and $E_0 - E_D < 0$. When the number becomes 0, then it is a steady state. $E_0 < 0$, $E_0 - E_D < 0$, $E_1 > 0$, and $E_1 - E_D > 0$ means no incoming or outgoing tunneling is preferred. A similar procedure is possible if the initial number is positive.

Operating point (b) is shared by a diamond with a preferred number 0 and another diamond with a preferred number 1. Thus, the number of electrons can fluctuate between 0 and 1 and the current flows. This situation is explained in Figure 7.23b, where $V_D = 0^+$ is assumed. Now, let us assume that the initial number of electrons is -1. Then the incoming tunneling is preferred, and an electron is added to the island either from the source or from the drain. Now the number is 0. The incoming tunneling from the source is preferred because $E_1 = 0^-$, although the incoming tunneling from the drain is not since $E_1 - E_D = 0^+$. Thus, an electron tunnels from the source and the number is 1. Then, the outgoing tunneling from the drain is preferred because $E_1 - E_D = 0^+$. The electron will not tunnel to the source because $E_1 = 0^-$. Thus, we have an oscillation of the number of electrons 0, 1, 0, 1…, etc. This fluctuation gives a finite current.

Operating point (c) is similar to operating point (a). The only difference is that the preferred number is 1. Again, whatever the initial number is, the necessary number of electrons tunnel in or out through the source and the drain, and the final preferred number of 1 is reached, as in Figure 7.23c. There is no current flow.

Thus, by changing V_G at $V_D = 0$ and moving operating points from (a) through (b) to (c), we have a finite drain current only at point (b). A similar procedure repeats every time we cross a corner of a parallelogram, whereas the current is zero inside the parallelogram. This means a periodic oscillation of the drain current as a function of gate voltage and is known as the Coulomb oscillation.

By comparing operating points (d), (e), and (f), we can see how the Coulomb staircase occurs. At point (d), the preferred numbers of electrons are -1 and 0, as shown in Figure 7.23d. Let us start with no electrons on the island. An electron tunnels out from the island to the drain, and the number of electrons is -1. Then, an electron tunnels in from the source to the island, and the number of electrons is 0 again. This cycle is repeated.

At points (e) and (f), double the drain current flows because two electrons flow at once, as shown in Figures 7.23e and f. Let us again start with -1 electrons on the island. An electron tunnels in from the source to the island. Now the number of electrons is 0. This is not a stable state. Another electron tunnels in from the source of the island, and the number of electrons is 1. Then, an electron tunnels out from the island to the drain, and the number of electrons is 0. This is not a stable state, either. Another electron further tunnels out from the island to the drain, and the number of electrons is -1. This cycle is repeated. Comparing with point (d), double the drain current flows because two electrons tunnel successively. As the drain voltage increases, the drain current increases as a step, because each time the operating point crosses a parallelogram boundary the number of tunneling electrons increases by one. Depending on the value of gate voltage, the drain current may not be constant, but may change with the drain voltage on a step of the Coulomb staircase.

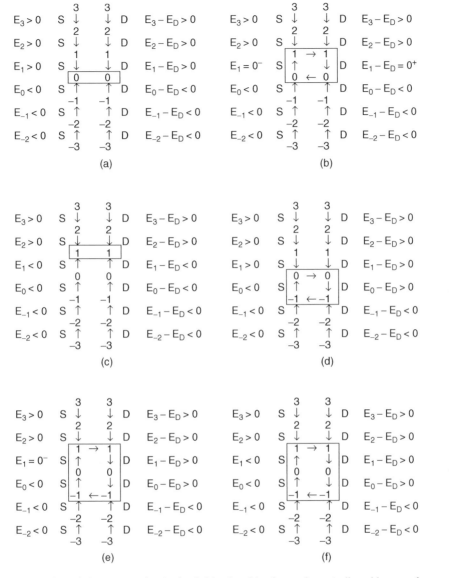

FIGURE 7.23 Number of electrons on the Coulomb island and its thermodynamically stable states for operating points (a) to (f).

7.6.3 SET Device Experiments Using SWNTs

There are many reports on the SET devices using SWNTs [34,41,43,85–96]. SWNT FETs at low temperature exhibit Coulomb diamonds in the $V_G - V_D$ characteristics, but experimentally observed diamonds often have a sawtooth structure [34,43], with a superposition of many small parallelograms. This indicates that there are multiple Coulomb islands connected in series in the channel, and multiple periods are superimposed. Similar irregularity has been observed in an NT pn junction structure [41]. We expect that there is one large NT island between the source and the drain. For various reasons, there will be random potential barriers and the large NT island is further divided. If this happens, then each island will create Coulomb diamonds. Because these islands are connected in series and no current flows inside the diamonds, the resultant Coulomb diamonds for the entire circuit will be a union of these diamonds. Thus, we will see a sawtooth structure.

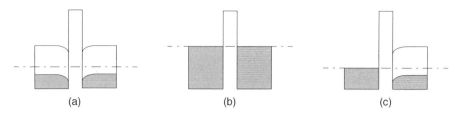

FIGURE 7.24 Band diagrams for (a) semiconducting SWNT-semiconducting SWNT (SS), (b) metallic SWNT-metallic SWNT (MM), and (c) metallic SWNT-semiconducting SWNT.

There are a few possible physical mechanisms for the formation of multiple islands. One mechanism is an impurity-induced potential barrier [43]. When dopants such as alkali metal atoms are introduced to the NT, they will not be uniformly distributed. A potential barrier will be created where the dopants are accumulated, and this will divide the NT into multiple islands. Another mechanism is the formation of pn junction near an electrode contact. In air, the Schottky barrier at an electrode contact is high for electrons and low for holes. This tends to let an end of the NT be p-type through the Schottky barrier modulation in air and oxygen doping. With an application of a large positive gate voltage, the central part of the NT is n-type. Thus, pn junctions are formed at both ends, with small p-islands. Depending on whether the NT is far enough to the electrode, the p-islands will or will not work as Coulomb islands. In Reference 41, the source and the drain contacts are not symmetric, and only one side of the p-region works as a Coulomb island. At a large positive gate voltage, the authors observed superimposed sawtooth Coulomb islands with two different Coulomb oscillation periods, whereas at a negative gate voltage they observed usual Coulomb islands with a single period, because the entire NT remained in p-type and there was one large island. A sawtooth structure was dominant in both p-type and n-type cases, and this suggests that a small Coulomb island was made probably because of the impurity barriers.

Regular Coulomb diamonds have been observed using metallic NTs. Reference 96 reports an extremely large Coulomb-charging energy in the NT field. A pair of potential barriers are introduced on a metallic NT by creating artificial defects with a chemical process, and the NT is placed on the SiO_2/Si substrate. The defects create a Coulomb island of 1~2 nm feature size, corresponding to the Coulomb energy of 400 meV or 5000 K. Thus, the room temperature operation is quite stable, and nearly ideal Coulomb islands are observed. There was an experiment with a metallic NT using the same source drain electrode geometry of an NT FET, and regular diamonds with a single Coulomb oscillation periodicity are observed. A Coulomb island is made between the source and the drain electrodes, and the period is consistent with it.

7.7 Other Semiconducting SWNT Devices

SWNT junctions have been fabricated by crossing two independent SWNTs [97]. The conductance is large in homojunctions — 0.01 to 0.06 e^2/h for semiconductor-semiconductor (SS) junctions and 0.086 to 0.26 e^2/h for metal-metal (MM) junctions — but is small in heterojunctions — 2×10^{-4} e^2/h for metal-semiconductor (MS) junctions where h is the Planck constant. The two NTs are slightly separated via a van der Waals interaction (0.34 nm), and this gap creates a potential barrier. Thus, this is a vacuum gap mode. The difference in conductance can be understood as follows. The conductance is large in homojunctions because the states before tunneling and the states after tunneling are available at the same energy level as shown in Figures 7.24a and 7.24b for the SS and the MM cases, respectively. This makes tunneling easier. The conductance is small in heterojunctions because there is a band bending in the semiconducting side, and the states for tunneling are not available at the same energy level as shown in Figure 7.24c. Rectification is observed in the current-voltage characteristics.

An NT FET may exhibit a hysteresis behavior in the $I_d - V_G$ characteristics, and this property can be used for a memory application [97,98]. Figure 7.25 schematically shows the origin of this process and the motion of positive charges. In Figure 7.25a, $V_G - V_{th}$ is 0 and the holes are just starting to be

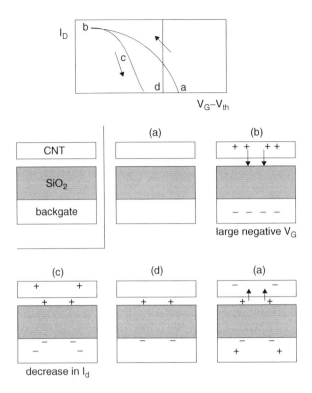

FIGURE 7.25 Mechanism for hysteresis in I_D-V_G characteristics of a SWNT FET. (a) to (d) are a real space charge distribution corresponding to the operating points.

accumulated. In Figure 7.25b, large negative V_G is applied, and a number of positive charges are induced in the NT. There are water molecules [99] between the NT and the SiO_2. In Figure 7.25c, because of the high electric field between the NT and the SiO_2, positive charges in the NT will be transferred to the water molecules. In Figure 7.25d, the positive charge remains between the NT and the SiO_2, and this increases the FET threshold voltage. The onset of hole accumulation starts at Figure 7.25a.

The silicon flash memory [100] has been quite popular recently because of its simple double-gated structure and compact size. It has a control gate and a floating gate as shown in Figure 7.26a. By charging or discharging the floating gate electronically, the FET threshold voltage is changed. There has been an effort to build an NT memory based on this flash memory mechanism, where the NT is used as an FET channel. Figure 7.26 explains the basic operation principle of the flash memory. In writing "1" in Figure 7.26a, the source is grounded, the drain is biased at 5 V, and the control gate is biased at 12 V. The FET has a pinch-off and hot electrons are created beyond the pinch-off point. These electrons can tunnel to the floating gate, and this is how the floating gate is charged. Because of this charge, the FET has a higher threshold voltage. In writing "0" in Figure 7.26b, the source is biased at 12 V, the control gate is biased at 0 V, and the drain is floating. Electrons on the floating gate will tunnel to the source, and the floating gate is discharged. This decreases the threshold voltage of the FET. In reading, V_{CGR} is applied to the control gate. Depending on whether there are electrons on the floating gate ("1") or not ("0"), the FET will have negligible I_D("1") or finite I_D("0"). In the experimental NT flash-memory operation, an appreciable threshold modulation was observed [101].

Two different modes for contact, the vacuum-gap mode and the touching mode, are proposed for a metal-semiconducting NT junction [72] realized in a metallic scanning tunneling microscope (STM) tip and a semiconducting NT [102,103]. With the tip grounded, the tunneling in vacuum-gap mode would produce a large conductance with a positive bias $V > 0$ and a small conductance with $V < 0$ for either an n-type or p-type NT, where the gap ΔV exists as shown in Figure 7.27a. The Schottky mechanism in

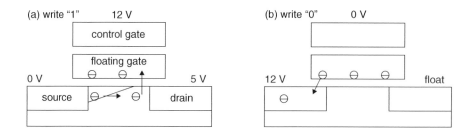

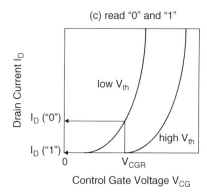

FIGURE 7.26 The principle of flash memory: (a) write "1;" (b) write "0;" (c) read.

the touching mode would result in rectifying characteristics, where the current flows only with $V < 0$ as in Figure 7.27b. Thus, the NT must be n-type. The vacuum-gap mode is schematically shown at left and the touching mode is shown at right in Figure 7.27c. These current-voltage characteristics are explained by the band diagrams. For the vacuum-gap mode, (d) is the valence band tunneling ($V < 0$), (e) is the equilibrium ($V = 0$), and (f) is the conduction band tunneling ($V > 0$). For the touching mode, (g) is the Schottky forward transport ($V < 0$), (h) is the equilibrium ($V = 0$), and (i) is the Schottky reverse transport ($V > 0$). The two observed current-voltage characteristics are entirely explained by a tip-NT contact of the two types with the above band diagrams.

In the NT production process, it is often the case that metallic and semiconducting NTs are obtained. An elegant way to obtain semiconducting NTs is reported [104]. By conducting an electric current, it is possible to burn metallic NT and obtain semiconducting NT selectively.

7.8   Transport in Metallic SWNTs

Metallic SWNTs are quasi–one-dimensional conductors. A conductance of metallic NT up to 4 μm was measured replacing a scanning probe microscope tip with an NT fiber, which was lowered into a liquid metal to establish a gentle electrical contact [105,106]. The conductance was quantized with $G_0 = 2e^2/h$ at a room temperature. Although the lowest subband was doubly degenerate in metallic NTs theoretically [1–3], the observed lowest conductance was G_0, indicating that the degeneracy was lifted. In the statistical data of conductance, the observed conductance had a strict cutoff at an integer multiple of G_0. This is because the transmission probability is limited to at most 100%, and it is impossible to go beyond it. Thus, there is a natural cutoff in the statistical data of the measured conductance at an integer multiple of G_0.

If there are scattering centers in a metallic NT, we can estimate a conductance using the Landauer-Butticker formula [107–113] and find a quantitative agreement with the experiment above. The coherent transport, including resonance tunneling, is possible in metallic NTs [114–120]. This is thanks to a good Ohmic contact between the metallic NT and the metallic electrodes. In case of semiconducting NTs, an

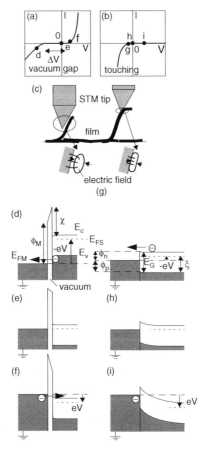

FIGURE 7.27 Vacuum gap and touching modes in the metallic tip — semiconducting SWNT. (a) and (b) the current-voltage characteristics for the vacuum-gap mode and the touching mode, respectively. (c) is the schematic showing the vacuum-gap (left) and the touching (right) mode. (d) to (f) are the band diagrams for the vacuum-gap mode. (g) to (i) are the band diagrams for the touching mode. (From T. Yamada, Appl. Phys. Lett., 78, 1739, 2001. With permission.)

Ohmic contact is not achieved easily and a Schottky barrier is often formed. In this situation, the effect of the Schottky barrier is much more influential in the current-voltage characteristics.

It is reported that when NTs are crossed, there are two different modes for crossing each other: the contact mode and the separate mode. The contact mode can conduct a lot of current, whereas the separate mode can conduct much less current [121]. The difference in a mechanical contact property is converted to a electronic transport difference and will be useful in device applications.

Metallic NTs are often discussed in the context of strongly correlated electrons [122–125], and the interplay of the many-body effects and the Coulomb interaction is discussed in Reference 125, but they are often studied in view of fundamental physics rather than device applications. They are not discussed further.

7.9 General Remarks on NanoFETs

SWNT FETs are some of the most actively studied nanoFETs and share many properties with other nanoFETs. Thus, nanoFETs in general are discussed below with the emphasis on the device-level problems. The problems related to future highly integrated circuitry with nanoFETs, such as an expected large heat dissipation from a chip beyond the present cooling technology [126], are not discussed here.

7.9.1   Properties of NanoFETs in Comparison with Macroscopic FETs

In the present silicon-based technology, FETs play a central role in electronics. For this reason, most recent nanodevice research, regardless of the channel conduction material, focuses on the FET scheme. Here a comparison of the macroscopic FETs and nanoFETs is drawn in view of how the channel material will influence the device characteristics. Because the FETs are genuine three-terminal devices, the characteristics in response to the drain voltage and the gate voltage need to be discussed independently.

The drain-voltage characteristics show significantly different behaviors. The characteristics are strongly dependent on the channel material in macroscopic FETs, whereas the characteristics are mostly independent of the channel material in nanoFETs. In a macroscopic FET, the drain current as a function of drain voltage is proportional to the carrier mobility [60]. The mobility is proportional to the mean free time of a carrier traveling without being scattered and the effective mass of the carrier. One of the main scatterers is phonon, and the phonon properties are different from material to material, reflecting the bulk material property difference. The effective mass represents the band structure of the bulk material and is again material dependent. Adopting a different channel material will certainly result in different transport characteristics. However, in nanodevices with an ideal electrode contact, the drain current as a function of drain voltage is determined by a transmission coefficient of an electron flux flowing from the source to the drain in the Landauer-Butticker view [101–107]. In the limit of an ideal nanoFET where carriers can run ballistically from source to drain, the transmission coefficient is unity, and there is no difference among different channel materials. In reality, it is not trivial to have ideal electrode contacts, and the transmission coefficient is reduced at the source drain contacts. Depending on the channel material, the feasibility of taking an electrode-channel contact differs, and in this sense, the characteristics will depend on the contacts. It has to be noted, however, that the bulk channel material properties do not play any role in this discussion.

The gate-voltage characteristics show similar behaviors for the macroscopic FETs and nanoFETs. In both cases, the problem is reduced to how effectively carriers are induced in the channel using the gate capacitance. Thus, the characteristics are determined by the thickness and the dielectric constant of the insulating layer, which are related to the environment around the channel. Ironically, this has no direct connection to the channel material itself. Depending on the choice of the channel material, the feasibility of preparing the insulation layer may differ, but there is no essential difference for the macroscopic FETs and nanoFETs.

In nanoFETs, the properties of the bulk channel material do not influence the FET performance directly. The performance, however, is still channel-material dependent in the sense that the feasibility in creating an ohmic contact or placing an insulating layer is different from material to material. Sometimes, this indirect influence is significant. As seen in Section 7.6, the narrow geometry of nanowire channels does not need a microfabrication to create an extremely small capacitor and is highly advantageous in SET applications. These situations are unique compared with macroscopic FETs, where the bulk channel material properties are everything in the performance.

7.9.2   Forgotten Benefit of NanoFETs

When Shockley wrote a paper on an FET [60], he demanded that the FET channel be very thin compared with the channel length, so that the variation of the electric field along the channel is much less than the corresponding variation perpendicular to the channel. In other words, the dominant electrostatic problem is in the perpendicular direction to the channel, and carriers are securely confined to the thin channel so that the distance to the gate electrode is minimal and the on state and the off state of the transistor are very well distinguishable. A largest possible current ratio for on and off states is one of the basic requirements for FETs. In order to achieve this, carriers need to be placed right below the gate electrode. Confined in a thin channel layer, carriers are minimally distant from the gate electrode, and their concentration is most effectively controlled by the gate voltage. This will result in excellent transistor performance.

As the channel length shrinks, this situation does not hold any more [24,26], and the electric field variation near the source and the drain cannot be negligible. The carriers are no longer confined to a thin layer below the gate electrode and are distributed deeper into the substrate near the source and the drain. The significant portion of the carriers is away from the gate electrode. This will cause the so-called short-channel effects. In this situation, the gate voltage cannot effectively control the carrier concentration and the transistor does not shut off well. Even if a gate voltage is applied to shut off the transistor, there is still significant drain current, and this degrades the transistor performance. This is often referred to as the reduction of threshold voltage. This is one of the serious problems the current silicon complementary MOS (CMOS) technology faces, and we are forced to engineer the channel doping to suppress these unwanted short-channel effects, but such doping control is highly challenging.

The fundamental solution is to adopt a transistor channel material that is, from the beginning, two-dimensional or one-dimensional so that the channel thickness is minimal. Nanodevices based on molecules, nanotubes, or DNAs are, from the beginning, quasi–one-dimensional and will be potentially quite advantageous in inherently suppressing the short-channel effects. In fact, nanotube FETs have shown excellent performance, which is already comparable with that of the state-of-the-art complementary MOS (CMOS) technology [59], and this is largely due to this "thin layer effect" caused by the adoption of a thin nanotube for the channel.

Nanodevices are inherently free from the short-channel-effects, and we will never face this serious problem in the future. Despite many anticipated fabrication problems in the future, it is quite meaningful to pursue nanotechnology in this context, in addition to the apparent benefit of the ultimate small size.

7.9.3   Two-Terminal vs. Three-Terminal

In electronics applications, special attention is paid to input and output electrodes. When a device has electrically independent input and output electrodes it is called three-terminal, having an input electrode, an output electrode, and a ground electrode. The electric isolation of the input and the output electrodes is essential, because otherwise the input and the output signals are mixed. Because of this isolation, in series connection of these devices, the signal can transmit only to the designed direction. Let us suppose a series connection of device A, device B, and device C such that the output of A drives the input of B and the output of B drives the input of C. In this case, the output signal of B transmits only to C, and it will never bounce back to A because of the input-output isolation. Therefore, the signal flow from A to B to C is established. A two-terminal device has only two electrodes. In the series connection of device A, device B, and device C, if the output signal of B will transmit to A as well as C because of the lack of the input-output isolation. The signal flow is not well established, and this is a serious problem in circuit applications. As is obvious here, even if a device has three electrodes, if the input and the output electrodes are not electrically isolated, it is not three-terminal.

A MOSFET is a three-terminal device, and in a most basic inverter circuit application, the gate electrode receives an input signal, the drain electrode gives an output signal, and the source electrode is connected to the ground. The gate electrode and the drain electrode are electrically isolated via a silicon dioxide layer, which is an excellent insulator, and there is no possibility for the input-output signal mixture. A Schottky diode is two-terminal, and one terminal is used for input/output and the other is used for ground. A Josephson junction diode is another example of two-terminal devices. In these diode examples, special circuitry has to be built to prevent a reverse, unwanted signal flow (from device B to device A in the example above).

Extensive circuit libraries suitable for three-terminal devices have been created; in fact, the current silicon CMOS technology relies on them. Thus, circuitry with any new devices can be immediately created as long as they are three-terminal. The new devices can be dropped in for the present MOSFETs. The situation is not this straight for two-terminal devices, because different circuit libraries than what we are using now must be developed. However, starting from scratch may not be necessary. Historically, new circuit libraries were developed for two-terminal devices [127,128]. In the 1950s, when semiconductor three-terminal devices were not very reliable but semiconductor diode devices were, "diode logic" was proposed and provided a circuit scheme based on the two-terminal devices only. In the 1960s, tunnel diodes were studied actively, and again suitable circuit schemes were considered. In the 1980s, Josephson junction diodes were studied. In these studies, it has been shown that "quasi-three-terminal devices" having independent input and output electrodes would be created by connecting diode elements.

In nanodevices, NTs, molecules, or DNAs are used for the device channels. All these structures are one-dimensional having two ends. Thus, without any special effort to place a third electrode, the device is inherently two-terminal. The proposal [129] to place a foundation in the diode logic scheme in the future nanoelectronics is based on this understanding. A nanosale molecular-switch 8 × 8 memory cell has been demonstrated toward this direction, using a cross-bar geometry memory scheme [130]. New two-terminal circuit schemes need to be developed, or three-terminal devices using nanotechnology need to be created.

Acknowledgment

The author acknowledges discussions with M. Meyyappan, A. Ricca, C. Bauschlicher, A. Cassel, W. Fan, Jing Li, Jun Li, J. Kong, H. T. Ng, Q. Ye, M. P. Anantram, D. Srivastava, M. A. Osman, and T. R. Govindan.

References

1. J.W. Mintmire, B.I. Dunlap, and C.T. White, Phys. Rev. Lett., 68, 631 (1992).
2. N. Hamada, S. Sawada, and A. Oshiyama, Phys. Rev. Lett., 68, 1579 (1992).
3. R. Saito et al. Phys. Rev. B, 46, 1804 (1992).
4. M.S. Dresselhaus, G. Dresselhaus, and P. C. Eklund, *Science of Fullerenes and Carbon Nanotubes*, Academic, San Diego (1996).
5. R. Saito, G. Dresselhaus, and M.S. Dresselhaus, *Physical Properties of Carbon Nanotubes*, Imperial, London (1998).
6. H. Takahashi, *Electromagnetism*, Shyokabo, Tokyo (1979).
7. D.K. Schroder, *Semiconductor Material and Device Characterization*, Wiley, New York (1990).
8. R.F. Pierret, Semiconductor fundamantals, in *Modular Series on Solid State Devices Vol. 1*, R.F. Pierret and G.W. Neudeck, Eds. Wesley, Menlo Park, CA (1988).
9. G.U. Sumanasekera et al. Phys. Rev. Lett., 85, 1096 (2000).
10. H.E. Romero et al. Phys. Rev. B, 65, 205410 (2002).
11. J. Hone et al. Phys. Rev. Lett., 80, 1042 (1998).
12. P. G. Collins et al. Science, 287, 1801 (2000).
13. K. Bradley et al. Phys. Rev. Lett., 85, 4361 (2000).
14. S.-H. Jhi, S.G. Louie, and M.L. Cohen, Phys. Rev. Lett., 85, 1710 (2000).
15. J. Zhao et al. Nanotechnology, 13, 195 (2002).
16. D.C. Sorescu, K.D. Jordan, and Ph. Avouris, J. Phys. Chem., 105, 11227 (2001).
17. A. Ricca and J. A. Drocco, Chem. Phys. Lett., 362, 217 (2002).
18. A. Ricca, C.W. Bauschlicher, Jr., and A. Maiti, Phys. Rev. B, 68, 035433 (2003).
19. P. Giannozzi, R. Car, and G. Scoles, J. Chem. Phys., 118, 1003 (2003).
20. S. Dag et al. Phys. Rev. B, 67, 165242 (2003).
21. H. Ulbricht, G. Moos, and T. Hertel, Phys. Rev. B, 66, 075404 (2002).
22. X.Cui et al. Nano Lett., 3, 783 (2003).
23. T. Yamada, Phys. Rev. B, 69, 125408 (2004).
24. S.M. Sze, *Physics of Semiconductor Devices*, Wiley, New York (1981).
25. R.S. Muller and T.I. Kamins, *Device Electronics for Integrated Circuits*, Wiley, New York (1986).
26. Y. Taur and T.H. Ning, *Fundamentals of Modern VLSI Devices*, Cambridge, New York (1998).
27. S.J. Tans, A.R.M. Verschueren, and C. Dekker, Nature (London), 393, 49 (1998).
28. R. Martel et al. Appl. Phys. Lett., 73, 2447 (1998).

29. C. Zhou, J. Kong, and H. Dai, Appl. Phys. Lett., 76, 1597 (2000).
30. R. Martel et al. Phys. Rev. Lett., 87, 256805 (2001).
31. A. Bachtold et al. Science, 294, 1317 (2001).
32. M. Radosavljevic et al. Nano Lett., 2, 761 (2002).
33. A. Javey et al. Nano Lett., 2, 929 (2002).
34. B. Babic, M. Iqbal, and C. Schonenberger, Nanotechnology, 14, 327 (2003).
35. K. Liu et al. Phys. Rev. B, 63, 161404 (2001).
36. R. Czew et al. Nano Lett., 1, 457 (2001).
37. J. Kong et al. Science, 287, 622 (2000).
38. J. Kong et al. Appl. Phys. Lett., 77, 3977 (2000).
39. R.S. Lee et al. Phys. Rev. B, 61, 4526 (2000).
40. M. Bockrath et al. Phys. Rev. B, 61, 10606 (2000).
41. J. Park and P. L. McEuen et al. Appl. Phys. Lett., 79, 1363 (2001).
42. T. Someya et al. Nano Lett., 3, 877 (2003).
43. C. Zhou et al. Science, 290, 1552 (2000).
44. R.D. Antonov and A.T. Johnson, Phys. Rev. Lett., 83, 3274 (1999).
45. H. Dai et al. J. Phys. Chem. B, 103, 1246 (1999).
46. A. Bachtold et al. Phys. Rev. Lett., 84, 6082 (2000).
47. M. Freitag et al. Phys. Rev. Lett., 89, 216801 (2002).
48. S. Rosenblatt et al. Nano Lett., 2, 869-872 (2002).
49. A. Javey, M. Shim, and H. Dai, Appl. Phys. Lett., 80, 1064 (2002).
50. J. Guo et al. Appl. Phys. Lett., 81, 1486 (2002).
51. J. Guo, M. Lundstrom, and S. Datta, Appl. Phys. Lett., 80, 3192 (2002).
52. A. Javey et al. Nano Lett., 2, 929 (2002).
53. F. Nihey et al. Jpn. J. Appl. Phys., Part 2, 41, 1049 (2002).
54. V. Derycke et al. Nano Lett., 1, 453 (2001).
55. X. Liu et al. Appl. Phys. Lett., 79, 3329 (2001).
56. T. Yamada, Appl. Phys. Lett., 76, 628 (2000).
57. S. Heinze et al. Phys. Rev. Lett., 89, 106801 (2002).
58. J. Appenzeller et al. Phys. Rev. Lett., 89, 126801 (2002).
59. S.J. Wind et al. Appl. Phys. Lett., 80, 3817 (2002).
60. W. Shockley, Proc. Inst. Radio Eng., 40, 1365 (1952).
61. J.-O. Lee et al. Appl. Phys. Lett., 79, 1351 (2001).
62. M. Freitag et al. Appl. Phys. Lett., 79, 3226 (2001).
63. M. Puyang et al. Science, 291, 997 (2001).
64. M. Ouyang et al. Acc. Chem. Res., 35, 1018 (2002).
65. L. Chico et al. Phys. Rev. Lett., 76, 971 (1996).
66. Z. Yao et al. Nature (London), 402, 273(1999).
67. A.S. Grove, *Physics and Technology of Semiconductor Devices*, Wiley, New York (1967).
68. J.H. Forster and H.S. Veloric, J. Appl. Phys., 30, 906 (1959).
69. A.A. Odintsov, Phys. Rev. Lett., 85, 150 (2000).
70. R. Tamura, Phys. Rev. B, 64, 201404 (2001).
71. T. Yamada, Appl. Phys. Lett., 80, 4027 (2002).
72. T. Yamada, Appl. Phys. Lett., 78, 1739 (2001).
73. F. Leonard and J. Tersoff, Phys. Rev. Lett., 83, 5174 (1999).
74. F. Leonard and J. Tersoff, Phys. Rev. Lett., 84, 4693 (2000).
75. W. Shockley and W. P. Mason, J. Appl. Phys., 25, 677 (1954).
76. L. Esaki Phys. Rev., 109, 603 (1958).
77. L. Esaki, Proc. IEEE, 62, 825 (1974).
78. L. Esaki, IEEE Trans. Electron Devices, ED-23, 644 (1976).
79. K.K. Likharev, IBM J. Res. Dev., 32, 144 (1988).

80. M.A. Kastner, Rev. Mod. Phys., 64, 849 (1992).

81. C.W.J. Beenakker, H. van Houten, A.A.M. Staring, D.K. Ferry, J.R. Barker, and C. Jacoboni, Eds. *Granular Nanoelectronics*, NATO ASI Series B, Physics, Vol. 251, Plenum, New York, 1990.

82. A.A.M. Staring, H. van Houten, and C.W.J. Beenakker, Phys. Rev. B, 45, 9222 (1992).

83. H. Tamura, S. Hasuo, and Y. Okabe, J. Appl. Phys., 62, 1909 (1987).

84. H. Tamura and S. Hasuo, J. Appl. Phys., 62, 3036 (1987).

85. M. Bockrath et al. Science, 275, 1922 (1997).

86. M. Ahlskog et al. Appl. Phys. Lett., 77, 4037 (2000).

87. A. Kanda et al. Appl. Phys. Lett., 79, 1354 (2001).

88. N. Yoneya et al. Appl. Phys. Lett., 79, 1465 (2001).

89. H.W. Ch. Postma et al. Science, 293, 76 (2001).

90. J.B. Cui et al. Nano Lett., 2, 117 (2002).

91. M.T. Woodside and P.L. McEuen, Science, 296, 1098 (2002).

92. Y.-H. Kim and K.J. Chang, Appl. Phys. Lett., 81, 2264 (2002).

93. J. Kong, J. Cao, and H. Dai, Appl. Phys. Lett., 80, 73 (2002).

94. J.W. Park, J.B. Choi, and K.-H. Yoo, Appl. Phys. Lett., 81, 2644 (2002).

95. K. Ishibashi et al. Appl. Phys. Lett., 82, 3307 (2003).

96. K. Matsumoto et al. Jpn. J. Appl. Phys., Part 1, 42, 2415 (2003).

97. M.S. Fuhrer et al. Science, 288, 494 (2000).

98. M.S. Fuhrer et al. Nano Lett., 2, 755 (2002).

99. W. Kim et al. Nano Lett., 3, 193 (2003).

100. B.G. Streetman and S. Banerjee, *Solid State Electronic Devices*, Prentice Hall, Upper Saddle River, NJ (2000).

101. W.B. Choi et al. Appl. Phys. Lett., 82, 275 (2003).

102. P.G. Collins et al. Science, 278, 100 (1997).

103. P.G. Collins, H. Bando, and A. Zettl, Nanotechnology, 9, 153 (1998).

104. P.G. Collins et al. Science, 292, 706 (2001).

105. S. Frank et al. Science, 280, 1744 (1998).

106. Ph. Poncharal et al. Science, 283, 1513 (1999).

107. R. Landauer, IBM J. Res. Dev., 1, 223 (1957).

108. M. Buttiker, Phys. Rev. Lett., 57, 1761 (1986).

109. Y. Imry, in *Directions in Condensed Matter Physics*, Memorial Volume to S.-K. Ma, G. Grinstein and G. Mazenko, Eds. World Scientific, Singapore (1986).

110. M. Buttiker, Transmission, reflection, and the resistance of small conductors, in *Electronic Properties of Multilayers and Low Dimensional Semiconductor Structures*, J.M. Chamberlain et al. Eds. Plenum, New York (1990).

111. S. Datta, *Electronic Transport in Mesoscpic Systems*, Cambridge, New York (1995).

112. D.K. Ferry and S.M. Goodnick, *Transport in Nanostructures*, Cambridge, New York (1997).

113. M.S. Lundstrom, *Fundamentals of Carrier Transport*, Cambridge, New York (2000).

114. P.L. McEuen et al. Phys. Rev. Lett., 83, 5098 (1999).

115. M. Terrones et al. Phys. Rev. Lett., 89, 075505 (2002).

116. M. Buongiorno Nardelli and J. Bernholc, Phys. Rev., B 60, 16338 (1999).

117. V. Meunier et al. Appl. Phys. Lett., 81, 5234 (2002).

118. A. Andriotis et al. Phys. Rev. Lett., 87, 066802 (2001).

119. A. Andriotis et al. Appl. Phys. Lett., 79, 266 (2001).

120. A. Maiti, A. Svezhenko, and M. P. Anantram, Phys. Rev. Lett., 88, 1268050 (2002).

121. T. Rueckes et al. Science, 289, 94 (2000).

122. R. Pati et al. Appl. Phys. Lett., 81, 2638 (2002).

123. M. Bockrath et al. Science, 291, 283 (2001).

124. W. Liang et al. Nature (London), 411, 665 (2001).

125. R. Tarkiainen et al. Phys. Rev. B, 64, 195412 (2001).

126. R.K. Cavin, III., Electron transport device scaling barriers, IEEE Nano 2003, Aug. 12–14, 2003, San Francisco, CA.
127. R. Landauer, Physica A, 168, 75 (1990).
128. R.W. Keyes, Science, 230, 138 (1985).
129. J. Birnbaum and R.S. Williams, Phys. Today, 53, 38 (2000).
130. Y. Chen et al. Nanotechnology, 14, 462 (2003).

8

Field Emission

Philippe Sarrazin

NASA Ames Research Center

This chapter discusses the application of carbon nanotubes (CNT) as field emission electron sources. After an introduction of the basics of field emission and the main parameters driving the quality of a field emitter, the possible configurations of CNT field emitters are discussed. Potential applications of CNT field emitters are then developed, with particular attention to the promising market of field emitter displays (FED) and the development of CNT cold cathodes for x-ray tubes.

8.1 Introduction

Modern technology makes a wide use of controlled propagation of electrons in vacuum. There are several examples, for instance, in CRT displays, vacuum electronics, microwave amplification, electron microscopy, x-ray generation, plasma processing, gas ionization, ion-beam neutralization, electron-beam evaporation, and electron-beam lithography. This broad range of applications finds its origins in the properties of free-flowing electrons: their energy (momentum) can be modified with ease using electric fields to accelerate or decelerate them, their trajectory can be controlled with electric or magnetic fields, and their generation and detection can be directly controlled with conventional electronics. No other type of particle radiation offers such latitude of control. At the basis of each system utilizing electrons in vacuum is an electron source. Several techniques are known to extract electrons from matter, but by far the most common one to date is thermionic emission. With thermionic emission, the source is heated at high temperature ($\sim 1000°C$) to give its free electrons the thermal energy required to overcome the surface potential barrier. Practically speaking, a thermionic source is usually a tungsten filament (Figure 8.1) or a porous tungsten matrix impregnated with a material allowing reduction of the work function (dispenser cathodes). During operation, the cathode temperature is raised by resistive or other means of heating, and emitted electrons are collected with an electric field. Although thermionic emission is a simple and well-proven technique, it does not provide good power efficiency because the high operating temperature necessarily comes with energy loss by radiation. Other disadvantages include the inertia of the emitter,

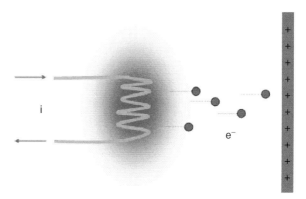

FIGURE 8.1 Principle of thermionic emitter: a metal is heated high enough to give the free electrons enough energy to overcome the surface potential barrier.

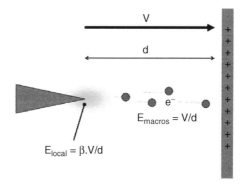

FIGURE 8.2 Schematic diagram of a field emitter; the electric field around the sharp tip is enhanced to the point electrons can tunnel through the surface potential barrier.

the dimensional changes from thermal expansion, the outgassing of species that contribute to vacuum degradation, and a limited lifetime from the thermal cycling.

Among alternative means of electron generation, field emission is regarded as one of the best choices in many applications. Field emission involves extraction of electrons from a conducting solid by an electric field. Unlike thermionic emission, no heat is required for obtaining field emission. Field emission can provide extremely high current densities (up to 10^7 A·cm^{-2} at the emitting site) and produces a low-energy spread of the emitted electrons. Extremely high electric fields (several kV/μm) are required for the electrons to tunnel through the surface potential barrier. Such high fields can practically be obtained using only the phenomenon of local field enhancement. A conducting solid shaped as sharp tips shows a geometrical enhancement of the electric field around the tip, the electric field lines being concentrated around the regions of small radius of curvature (Figure 8.2). When sharp cathodes are used for field emission, the macroscopic electric field required to induce a current is reduced to the point where macroscopic fields of a few V/μm can be sufficient.

Field emission is typically analyzed with the Fowler-Nordheim (FN) model, which describes the tunneling of electrons through a metal-vacuum potential barrier [1]. Figure 8.3 shows the expression of the FN model of a tip-shaped emitter. This model shows that maximizing the current output for a given applied voltage requires a material with a low workfunction that is shaped as sharp as possible to offer the highest field enhancement factor. The FN model is not rigorous in the description of field emission from CNT because it describes the emission from a flat metallic at 0 K. Although it is not strictly satisfactory in describing the physics of the emission from CNT, it is practical for analyzing field emission measurements. Consequently, field emission data are often presented in an FN representation (FN plots)

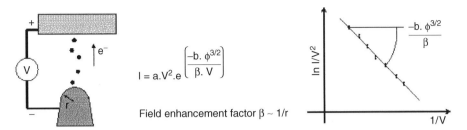

FIGURE 8.3 FN model of field emission from a metallic tip; typical FN plot in which field emission appears as a straight line with a negative slope. I, emitted current; V, applied voltage; φ, workfunction of the material; β, field enhancement factor.

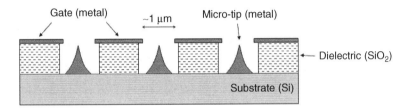

FIGURE 8.4 Schematic structure of a micromachined array of gated tips.

for which an FN regime is carried by a straight line with a negative slope. More-rigorous models describing the emission from CNT are being developed [2].

Very sharp field emitting microstructures were proposed and developed by Spindt in the form of micromachined miniature tips [3,4]. The structure of such a micromachined field emitter array (Spindt cathode) is presented in Figure 8.4. The emission relies on having very small radii of curvature of the tips. Consequently it is crucial to the preservation of the performance of the cathode that the tip sharpness remains unaltered. It has been observed that these tips are very sensitive, and sputtering of cation tends to dull the tip. Much effort was dedicated to the application of these Spindt cathodes, but this technology was proved to be expensive to scale and shows limited reliability in technical vacuum conditions.

A variety of field emitting thin films has been proposed as an alternative to the micromachined tip arrays for reduced fabrication cost and improved robustness. The potential of CNTs for field emission was first discussed in 1995 [5]. Increasing attention has ever since been devoted by the scientific and industrial communities to this possible application of nanotubes. At the time of writing, this technology is still under development, but some commercial applications are already being reported.

8.2 Structure and Microstructure of CNT Field Emitters

Carbon nanotubes have been shown to have excellent emission characteristics: emission has been observed at fields lower than 1 V/μm, and high current densities of over 1 A/cm^2 have been obtained. Detailed reviews of the properties of CNT as field emitters can found in the literature [6,7]. CNT emitters can be fabricated in a variety of configurations depending on the nature of the nanotubes, their microscopic arrangement, the preparation process, and the desired emitter structure.

8.2.1 Nature of the Nanotubes

Field emission has been observed from single-wall nanotubes (SWNTs), multiwall nanotubes (MWNTs), and nanofibers. The very broad range of configurations used for characterizing the CNT emitters (test geometries, gap and voltage range, vacuum conditions, etc.) make a quantitative comparison of the results published from different sources extremely difficult. Although there is a consensus that well-graphitized nanotubes give superior emission performance than do nanofibers because of the larger dimensions of the latter (hence lower field enhancement), it remains unclear if SWNT or MWNT is the most appropriate for field emission. Both SWNT and MWNT are, however, excellent field emitters. One would expect SWNTs to provide lower turn on voltages because of their sharper tips, but these often bundle into ropes, which limit this potential gain. MWNTs were shown to be more robust than SWNTs with their multiple-shell architecture providing a more stable structure upon ion bombardment and irradiation associated with field emission [8]. The choice of the nanotube type is often dependent upon the emitter fabrication process selected. MWNTs are often the primary choice for *in situ* CVD growth, whereas SWNTs are typically encountered with postprocessing emitter fabrication.

8.2.2   Importance of the Environment

It was shown earlier that field emission is a surface phenomenon that depends on the potential barrier at the solid-vacuum interface. Modification of the nature of this interface accordingly affects the emission. It has been shown that the adsorbate state of the emitting site can very significantly affect the emission. Water and oxygen are particularly important: H_2O adsorbates were shown to increase the emission current [9,10], whereas O_2 adsorbates decrease the emission dramatically [10,11]. These adsorbates were also shown to increase the instability of emission. The modifications in the emission characteristics resulting from the adsorption of H_2O or O_2 are reversible, and clean nanotube characteristics are recovered after desorption. Other gases such as N_2 have not shown significant effects on the field emission. This emphasizes the importance of ultrahigh vacuum (UHV) and the need for thermal processing (bake-out) of the cathode in order to stabilize its emission properties.

8.2.3   Microstructure of the Emitter

Single nanotube emitters have been fabricated primarily for fundamental research purposes [8]. Figure 8.5A illustrates this configuration: a single nanotube is mechanically mounted on a sharp tip (metal, Si). A technique for assembling such a structure with a MWNT is described in Chapter 6. Fabrication of such a structure with SWNTs would be much more delicate because of the tendency of SWNTs to bundle in ropes and the complexity of handling such a small structure. Maximum emission current over 100 µA has been reported from single MWNT emitters, giving extremely high current densities in the $10^7 A/cm^2$ range. One of the drawbacks of this single nanotube approach is that physical damage to the nanotube leads to the loss of emission. Possible applications of single nanotube emitters are limited to systems for which the sharpness of the electron source is of primary importance. High-resolution electron microscopy or lithography might, for instance, benefit from such sharp emitters. In most technical applications, however, a single nanotube is unlikely to provide the expected current and robustness.

Nanotube films have the largest potential for technical applications. They combine good emission characteristics, relative ease of fabrication, and possibility to scale for production. Two configurations of nanotube films are presented in Figure 8.5B and C. The fabrication of CNT film emitter is divided into two fundamentally different approaches: deposition or layering of pregrown tubes on a substrate, or *in situ* growth of the CNT. To date, there is no evidence for clear superiority of one approach over the other. The layering or transplanting approach tends to be favored when using SWNTs because their *in situ*

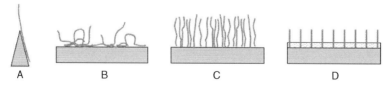

FIGURE 8.5 Microstructures of CNT emitters: (A) single nanotube, (B) random film, (C) vertically aligned nanotube film, (D) highly ordered array.

CVD growth requires high-temperature processes that pose significant constraints on the cathode substrates. Patterned films can be obtained with printing techniques using CNT-containing inks. Direct growth of the nanotube film on a cathode substrate using CVD processes offers the advantage of scalability for production. Patterned films can be obtained by controlling the catalyst deposition. Vertically aligned nanotube films are usually obtained with high densities with the nanotubes supporting each other in their vertical position, whereas low-density films tend to have randomly oriented tubes. CVD of nanotubes has been discussed in Chapter 4.

Highly ordered arrays of MWNTs can be fabricated using anodic aluminum oxide (AAO) templates [12]. The MWNTs are supported at their base by the aluminum oxide layer, as pictured in Figure 8.5D. This technique offers the advantage of controlling the distance between vertically aligned tubes. The process is, however, more complicated than regular film growth. Because of their larger dimensions, nanofibers can be self-supporting, and highly ordered arrays of vertical carbon nanofibers have been fabricated [13].

8.2.4   Importance of the Screening Effect

The aspect ratio and tip sharpness are not the only factors that determine whether a CNT film will have good emission characteristics. Indeed, an important factor affecting the field emission properties is field screening. The environment of a CNT tip can be such that it is shielded from the macroscopic electric field, preventing it from emitting to the extent it would if placed alone in the field. When considering the current density of emission from a film or array, it is important to take into account the screening from one tube to another. Although increasing the nanotube density implies the availability of more emitting sites per unit area, the emitted current density can actually be lowered because of higher screening between tubes. The compromise between emitter density and field screening points toward an optimum density of nanotubes. A theoretical investigation of the field screening in nanotube arrays concluded that the optimum distance between neighbor emitters is equal to two times their height [14], whereas an experimental study using AAO template arrays showed an optimum for intertube distances equal to the nanotube height [15]. The ideal density for a particular application will depend on what criteria the emitter has to be optimized for. Indeed, different nanotube density optima will be obtained depending on the emphasis: the lowest possible field at moderate current density or the highest current density regardless of the electric field required to obtain such current. In any case, optimizing the nanotube density of a CNT film or array is crucial in the design of CNT field emitters.

8.2.5   Geometry of the Emitter

As shown by the FN model, the current drawn from a field emitter is controlled by the macroscopic electric field E. Electron sources are often used in high-voltage devices. In such situations, a simple diode structure can be utilized (see Figure 8.6A) where a high voltage applied between the cathode and the anode provides the extraction field. The distance between the electrodes has to be adapted accordingly. The diode structure is very convenient for research purposes because of its simplicity. However, it is not necessarily the best choice in practical devices because the emission current must be controlled either by regulating the high voltage, which affects the energy of the electrons at the anode and requires complex power supplies, or by changing the gap distance, which requires mechanical controls.

Field emitters are often used in triode geometry where the control of the extraction field on the emitting surface is independent of the accelerating voltage. Figure 8.6B illustrates a triode configuration where the extraction field is controlled by the voltage on a transmission grid placed above the CNT film. The voltage on the gate (the grid) can be reduced by placing it very close to the emitter, typically within a few micrometers to a few hundred micrometers. This allows a simple electronics for the emission current regulation. With a transmission gate such as pictured in Figure 8.6B, a portion of the emitted current is collected by the grid. Other gate designs allow reducing the gate current compared with a simple grid.

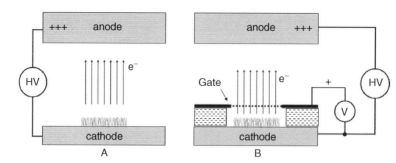

FIGURE 8.6 Emitter structure: (A) diode structure that requires regulation of the high voltage to control the emission current; (B) triode structure (gated emitter) with transmission gate illustrated, where the current is controlled by the gate voltage, independent of the acceleration voltage.

8.3   Applications of CNT Emitters

The range of potential applications for CNT-based field emission is very large. Any system that uses an electron source could potentially host a CNT field emission device. CNT field emitters are particularly suited when high efficiency is necessary, such as for most space and portable applications, when fast switching or ultrahigh frequency modulation are required and also, when very high current densities are desirable. Several applications for CNT emitters are discussed below.

8.3.1   Vacuum Microelectronics

Vacuum electronics is based on the controlled propagation of electrons in a vacuum to obtain a signal gain. Vacuum tubes were widely used until integrated solid-state electronics (transistors) became common, offering a cheaper, faster, and more robust alternative. Miniaturized vacuum electronic devices could overcome the major drawbacks responsible for the abandoning of vacuum tubes. One of the technical advances that may enable future vacuum microelectronics is the possibility of replacing the thermionic emitter found in most classic vacuum tubes by a cold cathode. Because field emitters do not release heat and can be made very small, a miniature field emission cathode could be integrated into a microfabricated vacuum device.

Vacuum microelectronics would be fast and efficient while providing some definite advantages over solid-state electronics. Indeed, the latter is based on semiconductors that are sensitive to radiation and high temperature. Conversely, vacuum electronics is immune to both. For these reasons, vacuum electronics is still being used in some specific applications (especially military) where improved resistance to temperature variation and high radiation is required. CNTs, with their excellent field emission properties, are very likely to find applications in future vacuum microelectronics systems. This, however, requires significant progress in the control of *in-situ* synthesis of CNT in microdevices.

8.3.2   Microwave Amplifiers

Long-range telecommunication relies primarily on microwave links. To achieve ultrahigh frequency communication, high-power transmitters are used on satellites and ground stations. The growth of telecommunication has lead to a saturation of the present bands and an increasing demand for new bands at higher frequencies. The use of higher frequencies (10s to 100s of GHz) is, however, limited by the current amplifier technology.

The amplifiers used for high-power microwave transmitters cannot be based on solid state electronics because conventional transistors are limited by the speed at which charge carriers move through the semiconductor. This limitation does not apply to vacuum tubes; hence, microwave amplifier tubes are typically used for this application. The basic layout of a traveling wave tube (TWT) is presented in

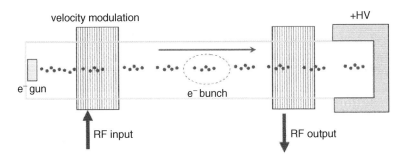

velocity modulation +HV

e⁻ gun e⁻ bunch

RF input RF output

FIGURE 8.7 Typical layout of a TWT for microwave amplification: current technology applies thermionic sources combined to a velocity modulation system.

Figure 8.7. A thermionic electron source has been typically used in the past to generate the electrons. It is, of course, impossible to modulate the emission with a thermionic source at high frequency; therefore, this modulation has normally been obtained by a technique called *velocity modulation*, which delays and accelerates electrons in a section of their path along the tube in order to generate electron bunches.

Cold cathodes could replace the thermionic source in TWTs. Because field emission offers ultrafast switching capabilities, a triode emitter could be used as an electron buncher in place of the velocity modulator. A field emitter capable of producing very high current densities (>10 A/cm²) could significantly improve the performance of a radio frequency power amplifier tube.

These very high current densities are difficult to achieve with micromachined arrays but could be reached with a CNT cathode. Hence, CNT field emitters would allow operation of high-power TWTs at higher frequencies while reducing cost and improving the efficiency compared with the current technology. Research groups around the world are pushing the limit of current density of CNT cathodes. For instance, current densities over 10 A/cm² have been reported by Applied Nanothec Inc., Agere Systems was reported to be developing a CNT-based cathode for high-power TWT, and many other groups worldwide are currently pursuing similar projects for this promising application.

8.3.3 Space Applications: Example of Electric Propulsion

The improvements CNT field emitters promise over the thermionic source is, of course, of particular interest for space applications. A successful application of CNT field emitters to improve miniature x-ray tubes for planetary exploration will be discussed later, but development of electric thrusters is discussed first.

Electric thrusters are based on the high-velocity ejection of positively charged ions or colloid particles accelerated in an electric field. The charge of the jet must be neutralized in order to maintain the electric neutrality on the spacecraft. This neutralization is done with an electron beam directed toward the jet. Thermionic emitters are typically used, but they consume significant power, reducing the efficiency of the thruster, and they require warm up, which slows the ignition time. A low-power (~100 W) electric thruster provides high thrust per propellant flow ratio and is particularly suited to low-thrust long-duration applications such as small-satellite orbit maintenance or small-spacecraft propulsion. Unfortunately, the poor efficiency of the neutralizing-beam source does not allow the design of a viable low-power thruster.

Field emitters can significantly improve the efficiency of the neutralizing source. The high-pressure environment (relative to vacuum expected in space) around the cathode is a major constraint for a field emitter. An investigation of the potential of a variety of field emitters (non-CNT) for electric propulsion neutralization is provided in Reference 16. CNT field emitters show a much improved resistance to operation in poor vacuum and hence provide a technological solution to the improvement of a neutralizing electron gun of an electric thruster. Investigation of the use of CNT cathodes for colloid thruster

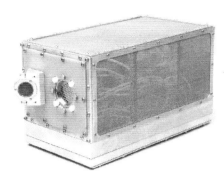 

FIGURE 8.8 Small colloid thruster prototype (left) featuring a CNT field emission cathode for beam neutralization (right), developed by Busek Co. Inc. (Courtesy of BUSEK Co.)

neutralization has been reported [17], which involves fabrication of MWNT cathodes and implementation on a micronewton thruster prototype. (Figure 8.8).

A survey of neutralizing source for FEEP (field emission electric propulsion) thrusters, a different type of electric thruster, is provided in Reference 18. Two types of CNT cathodes, among various electron sources, were studied and reported as interesting candidates for this application.

One step further in the miniaturization is the MEMS thrusters. These are a future generation of miniaturized thrusters (thruster-on-the-chip) that will enable missions with microspacecraft (a few kilograms) or orbit stabilization of microsatellites. One can expect field emission with CNT to play a key role in the development of this new microthruster technology.

8.3.4 Field Emission Displays

The largest potential market for CNT field emitters is for field emission display (FED), a technology that could replace most monitors used today in desktop and laptop computers and televisions. The major current technologies applied in the display industry are:

- The cathode ray tube (CRT), which produce images by scanning a high-velocity electron beam over a phosphor-coated screen where the phosphor emits light when bombarded by high-energy particles
- The liquid crystal display (LCD), which utilizes a two-dimensional array of liquid crystal elements sandwiched between crossed polarizers to electrically control the optical transmission factor of a backlit pixel (nonemissive technology)
- The plasma display panel (PDP), which utilizes plasma generated within the pixels to emit colored light

CRTs are superior to LCD in many aspects: wider field of view, higher brightness, better color rendering and contrast, and emissive technique. However, CRTs are bulky because of the volume required for the electron-beam scanning system. LCDs, on the other hand, are thin and flat. For these reasons, LCD technology has become common for small displays (handheld devices, instrumentation, automotive, etc.) and has gained an increasing share of the computer-monitor market for laptop and desktop applications. The PDP technology is far less common than CRT and LCD, and its application is currently limited to large and expensive monitors. The FED has a potential for replacing these three technologies. Like CRTs, an FED is based on the controlled bombardment of electrons on a phosphor to locally induce light emission. Whereas a CRT is composed of a single electron source producing a beam that is scanned over the surface of the display, an FED has an electron source for each individual pixel (or subpixel for a color display) as illustrated in Figure 8.9. The brightness of each pixel is controlled by adjusting the current on the phosphor element. FEDs, based on the same principle as CRTs, potentially have the same

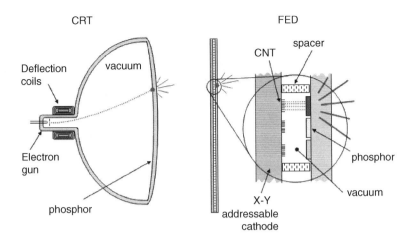

FIGURE 8.9 Schematic cross sections of a CRT and a FED.

advantages in terms of image quality. Their main advantage is that, because no beam scanning is required, the electron source can be placed very close to the phosphor element, leading to a very thin display. FEDs can be made for high resolution, and their size can vary from small for handheld devices to very large for wall-mounted TV monitors or commercial panels. A review of the materials challenges faced in the development of FEDs is provided in Reference 9.

An important characteristic of the FED is the nature of its phosphor. Indeed, like CRTs, FEDs rely on a phosphor material to convert the energy of the incident electron beam into light. High brightness and high efficiency are easier to achieve with higher-energy electrons. High-voltage (>10 KV) phosphors are readily available from CRT technology. Bright and efficient operation of low-voltage (a few 100 to a few 1000 V) phosphor is more challenging to obtain, and specific materials are being developed for such operating conditions.

High-voltage FEDs can be designed using phosphor similar to those applied in CRTs, and high brightness under low current densities is obtained with this readily available material. However, the cathode-to-anode gap must be large enough to avoid vacuum breakdown; hence, high-voltage FEDs are thick. High-voltage FEDs are also more complex and bulky because of the higher specification for the power supply and electric insulation, as well as the possible need for electrostatic optics within each pixel to focus the individual electron beams on their relative phosphor elements. Low-voltage FEDs are thinner and simpler, but they are currently limited by the poorer efficiency of the phosphor. FEDs also put a higher constraint on the field emitters because higher current densities are required to compensate for the lower energy of the electrons. During the 1990s, high expectations were placed on FEDs and major efforts were devoted to their development. Working FEDs were built, but their fabrication proved to be difficult to adapt to large-scale production, primarily because the field emitter technology in most cases was based on micromachined-tip arrays. Many companies that had invested a lot of time and money abandoned the technology in the early 2000s.

Progress in CNT field emission research has rekindled the interest in FEDs. The low turn-on voltage and simple fabrication processes of CNT emitters would allow fabrication of low-voltage displays at reduced cost compared with micromachined tips, enabling mass production of low-cost monitors, for applications ranging from small handheld devices to large wall-mounted systems.

At the time of this writing, no CNT-based FED is commercially available, but an intense competition is in place involving major companies such as Samsung, Motorola, Dupont, and several Japanese companies. Several CNT-based FEDs were demonstrated (up to 14-in. diagonal) and larger prototypes are said to be under development. Two approaches are being pursued in parallel: deposition of presynthesized CNTs into a patterned film (Samsung and Dupont) or direct growth of patterned CNT film on a cathode (Motorola and cDream).

 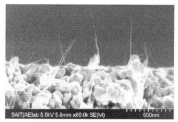

FIGURE 8.10 A fully sealed 4.5-in. FED prototyped in 1999 by Samsung Advanced Institute of Technology. Right: SEM image of the SWNT cathode. (Reprinted with permission from W.B. Choi et al. Appl. Phys. Lett., 75, 3129, (1999.) Copyright 1999, American Institute of Physics.)

The deposition approach has been pushed forward by Samsung, which demonstrated the first fully sealed FED prototype based on CNT field emitters in 1999, shown in Figure 8.10 [20]. The cathodes of this prototype were obtained by patterned deposition of a paste composed of SWNTs and an organic binder and subsequent heat treatment to remove the binder. Technologies more recently developed involve printing processes to deposit CNT-loaded ink in a patterned array, technology that would be easier to scale-up for production.

The *in-situ* CVD growth approach has been pursued by Motorola Labs and cDream, who have both developed low-temperature (<500°C) CVD processes to grow patterned MWNT films. The low temperature is important for industrial applications because it allows CNTs to grow on systems containing materials that cannot be heated to high temeprature, such as glass or transistors. Lower temperature also allows reducing the fabrication costs with lower power consumption of the CVD reaction and reduced constraints on their design.

A derivative application of CNT field emission exists with lighting elements. These elements can be used individually as an alternative to LEDs, can be assembled into arrays for large display, or could constitute a light source. CNT-based lighting elements have been demonstrated by ISE Electronics, a subsidiary of Noritake Corp (Figure 8.11). These elements are intended to be assembled into arrays to produce very large panel displays for advertisement and live events. Similarly, field emission lighting-tube prototypes were fabricated, where the phosphor coating is placed on the inner side of a glass tube along the middle of which a CNT-covered filament is placed [21]. Brightness comparable with commercial fluorescent tubes has been reported.

8.3.5 X-Ray Tubes

X-ray tubes are devices that allow controlled emission of x-rays. They are used in a variety of systems such as x-ray imaging in medical, industrial, and security applications; spectroscopic analysis (x-ray fluorescence, x-ray photoelectron spectroscopy); x-ray diffraction analysis; etc. The principle of x-ray generation in a tube involves bombardment of a metallic target with high-energy electrons (tens of keV). The interaction of these electrons with the target material is twofold as far as x-ray emission: their deceleration in the electric field of the target material results in the emission of a radiation with a continuous spectrum (bremstrahlung radiation), and electrons from deep electronic shells of the target atoms can be ejected or displaced, with the resulting relaxation leading to the emission of a discrete spectrum of x-rays characteristic of the target material (characteristic radiation). The intensity of x-ray emission is proportional to the flux of electrons on the target. The x-ray emitting area is determined by the zone bombarded by the electron beam. For applications where small emitting spots are required such as x-ray imaging of any application requiring x-ray optics, the tube is fitted with electron optics to concentrate the x-ray beam on a small area on the target. A typical structure of an x-ray tube is illustrated in Figure 8.12.

New applications of portable fluorescence analysis systems and space-deployed x-ray instruments demand miniature power-efficient x-ray tubes. Practically speaking, miniature x-ray tubes are limited by

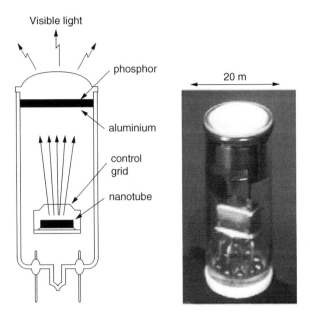

FIGURE 8.11 ISE Electronics high-voltage lighting element using CNTs, providing high brightness and efficiency comparable with LEDs. (Courtesy of ISE Corp.)

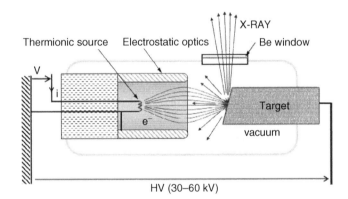

FIGURE 8.12 Schematic structure of an x-ray tube based on a thermionic source (grounded cathode configuration).

one of their most critical components: the electron source. Thermionic emitters do not allow appropriate efficiencies with low-power x-ray tubes because of the heat radiated from the filament. Field emitters could present many advantages over thermionic emission. Because field emitters do not require heat to generate electrons, they are more energy efficient and less likely to outgas species that would deteriorate the vacuum and contaminate the target. The cold nature of the emission would also prevent thermal drift of the cathode, allowing finer and more-stable electron focusing. Another major advantage of field emitters is that they do not have the inertia of thermionic sources. The emission of x-rays could be pulsed by simply switching the cathode, which would eliminate the need for a mechanical shutter in systems that require pulsed operation.

Until recently, field emission had not found practical applications in x-ray sources because no emitter structure could survive the rough conditions of operation in these high-voltage tubes. The potential benefits of field emission can be achieved only with field emitters that are mechanically, chemically, and thermally very robust. Carbon nanotubes are particularly interesting for this application. Indeed, not only are CNTs capable of producing high current densities at low extraction field, they are also mechanically,

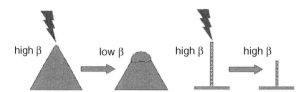

FIGURE 8.13 Comparison of the damage by arcing of a micromachined tip and a nanotube; the field enhancement at the nanotube tip is not as dramatically reduced if the emitter is physically damaged.

chemically, and electrically very robust. It has been shown that operating a CNT emitter under poor vacuum conditions or after electrical arcing between the emitter and the gate does not destroy the emitter but only lowers its performance [23]. Due to the very high aspect ratio of nanotubes, the field enhancement reduction consecutive to physical damage of the tip is much less likely to destroy the emitter than with micromachined tips, as illustrated in Figure 8.13.

Successful applications of CNT emitters in x-ray tubes have been recently reported [24,25], but these were limited to proof-of-concept systems. Driven by the need for lightweight and power-efficient tubes for a miniature x-ray diffraction/x-ray fluorescence (XRD/XRF) instruments for planetary exploration [26], a CNT field emission cathode for x-ray tubes has been recently reported [27]. Although planetary x-ray tubes are still under development at the time of writing, the technology presented herein has been extensively tested and is readily included in the production of commercial miniature x-ray tubes for handheld x-ray fluorescence spectrometers and OEM applications. A detailed account of cathode fabrication, characterization, and x-ray tube performance is given below.

8.3.5.1 Cathode Fabrication

An *in situ* growth technique is chosen for the fabrication of the CNT cathode rather than transplanting nanotubes grown elsewhere (postprocessing) to guarantee a strong attachment of the nanotubes to the substrate. With postprocessing techniques, the nanotubes are, in most cases, merely held on the substrate by van der Waals forces. Binders could be used to reinforce the CNT attachment, but the high-temperature bake-out (typically 450°C) of x-ray tubes before their sealing would prevent using organic compounds and complicate this approach. Although not necessarily a major drawback in low- to medium-voltage applications, the loose attachment of CNTs becomes a significant issue when the cathode is used in a high-voltage device (30 kV here). Indeed, CNTs can be removed from the cathode substrate by the electric field and subsequently deposited on polarized areas such as the electron focusing optics, thereby inducing field emission from supposedly nonemitting surfaces. The *in situ* growth technique providing good base attachment of the CNT is preferable in order to minimize this potential nanotube transfer.

Cathode fabrication is done using a thermal CVD process [28–31], which yields a randomly oriented MWNT film on a substrate on which metal catalysts have been deposited by ion-beam sputtering. For the results discussed here, molybdenum substrates were chosen for the x-ray tube application. The shape of the film is controlled during the catalyst deposition using a mask. Cathodes were grown in a circular film with diameters varying from 2 mm to 75 μm. Transmission electron microscopy (TEM) images revealed the internal structure of the nanotubes to be multiwalled with a good continuity of the graphene layers along the tube walls. Figure 8.14a and b show the typical microstructure of the films: randomly oriented nanotubes, several micrometers in length, and presenting various curvatures. The CNTs run along the surface of the substrate, forming hoops and in some cases pointing tips away from the surface. The nanotubes were found to be well attached to the substrate at their base through the catalyst particles. Figure 8.14c and d show two different sizes of CNT films controlled by adjusting the mask used during catalyst deposition.

8.3.5.2 Cathode Characterization

The cathode testing setup consists of a CNT cathode mounted in a holder, facing a flat anode with adjustable gap (Figure 8.15). The testing assembly is installed in a UHV chamber pumped by a dry turbo-molecular

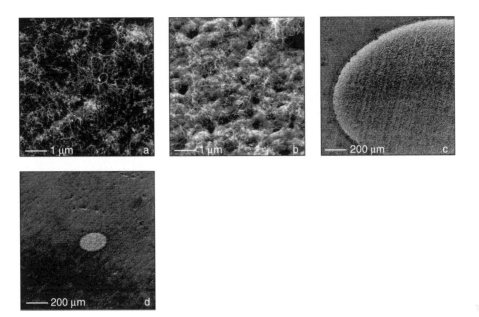

FIGURE 8.14 Scanning electron microscope images of CNT cathodes grown on Mo substrate. Film density is controlled by the catalyst formulation and growth conditions (a, high density; b, low density), cathode diameter is controlled by masking during catalyst deposition (c, 2-mm diameter; d 200-μm diameter).

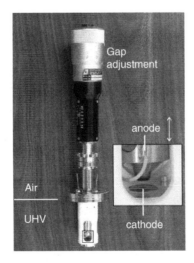

FIGURE 8.15 Instrumental setup for field emission characterization

pump backed with a dry diaphragm pump. All field emission measurements are performed in the 10^{-8}–mb range. A Keithley® 237 source-measure unit is used for application of the high voltage (up to 1100 V) and measurement of the emission current with pA sensitivity. Voltage cycles and data collection are controlled by a desktop computer using Testpoint® software. The cathode to anode gap is adjusted with micrometric precision and controlled optically through view ports. Typical values during field emission measurements were 200, 150, and 100 μm depending on the electric-field requirement.

Randomly oriented MWNT films with a wide range of density and structure were characterized. Dense films showed high turn-on voltages attributed to field screening from high densities of CNTs (high-emission site density but poor field enhancement), whereas very low-density films showed low turn-on

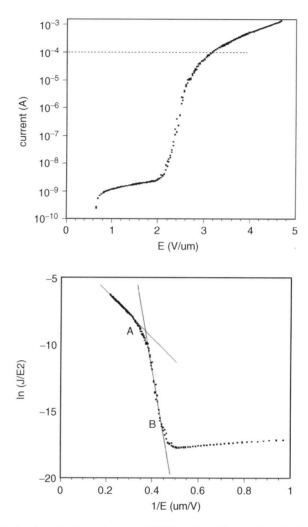

FIGURE 8.16 Field emission data of a 2-mm diameter MWNT cathode. Top: applied electric field vs. current density with typical operating current in x-ray tube. Bottom: FN plot showing two emission regimes.

voltage but poor current density (good field enhancement but very low-site density). A range of moderately low film densities was found to provide acceptable emission characteristics. Typical data recorded with such cathodes are shown in Figure 8.16. A rapid increase in the current is observed above 2 V/μm. When plotted in an FN representation, the data show typical field emission behavior. Two different regimes of emission are typically observed, with higher current showing a saturation mechanism. A 2-mm diameter cathode can reach currents of a few milliamperes before showing permanent degradation. When permanent degradation of the cathode does occur, the field emission properties are not completely lost, but rather the voltage required to achieve a given current merely increases. This suggests that either the emitting sites are not completely destroyed or they are replaced by other nanotubes in the film. The requirement for initial break-in of new cathodes and the hysteresis often observed in field emission data suggest a CNT reorganization within the film when an electric field is applied, giving the film a configuration more favorable for field emission. This reorganization is also interpreted as a cause of the robust behavior of the cathodes when run in severe conditions such as repetitive arcing. Smaller-diameter cathodes showed a significant improvement of the current density. Sustainable currents of over 1A/cm²

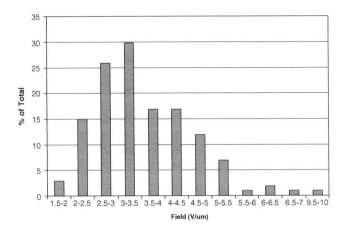

FIGURE 8.17 Variation of the electric field required to extract 100 µA from 2-mm diameter MWCNT films measured on 125 cathodes in diode geometry with 475-µm gap; cathodes were progressively ramped up to 500 µA before measurements. (Courtesy of Oxford Instruments Plr.)

were routinely measured with 100-µm diameter cathodes. This phenomenon requires further investigation but is believed to be the consequence of better field penetration in smaller-diameter films, which results in higher-emitting site density.

Reproducibility of the emission characteristics was measured using a specifically designed apparatus allowing characterization of several cathodes in parallel in a UHV chamber. These tests were run in a diode mode with a cathode to anode distance of 475 µm. Figure 8.17 shows the distribution of electric fields required to draw a current of 100 µA (after progressive run-in at 500 µA) measured on a series of 125 cathodes with 2-mm diameter films. Most cathodes require a low electric field to emit 100 µA, the typical operating current of a low-power tube. The dispersion of values is, however, large. Although this poor reproducibility has only a minor consequence on the fabrication of large-focus triode tubes (because the extraction field can be independently controlled by a gate voltage), it is a significant drawback for the serial fabrication of diode geometry tubes or microfocused tubes. It is expected that further refinement of substrate preparation and catalyst deposition procedures will improve the reproducibility of the cathode performances. This problem has, however, minimal consequences on the development of space-deployable tubes because thorough selection allows one to choose cathodes for optimum performance.

8.3.5.3 Miniature Field Emission X-Ray Tube

MWNT field emitters were installed in miniature low power x-ray tubes designed by Oxford Instruments X-ray Technologies Inc (Figure 8.18). The tube design is based on a small ceramic enclosure in which is installed a CNT emitter gated with a transmission grid for emission current regulation and electron optics to limit the size of the x-ray spot (Figure 8.19). Both reflection and transmission anode geometries were fabricated. The efficiency of these tubes was measured to be >80%, as compared with ~50% for filament tube. Lifetime tests are still in progress at the time of writing. A tube has been operated at 1.5 W for >100,000 10s pulses (66% duty cycle) without failure, illustrating the robust operation of a CNT source. This technology has been implemented in Oxford Instruments Eclipse II miniature x-ray sources that integrate a 3-W CNT field emission tube and power supply in a small package (160×38 mm, 300 g) that can be operated from batteries (Figure 8.20). These sources are commercially available in transmission and reflection anode configurations for x-ray spectroscopy applications.

This successful application of carbon nanotube field emitter in x-ray tubes demonstrates that substitution of the classic thermionic emitter by a CNT cold cathode can be technically beneficial and cost effective, even with devices as demanding as sealed high-voltage tubes.

FIGURE 8.18 Miniature field emission x-ray tube with gated MWNT 2-mm diameter Cathode. (Courtesy of Oxford Instruments Plr.)

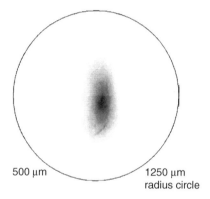

FIGURE 8.19 Image of the field emission x-ray tube emitting spot (1276 × 602 µm). (Courtesy of Oxford Instruments Plr.)

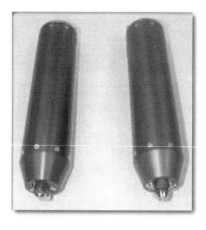

FIGURE 8.20 Oxford Instruments Eclipse II miniature x-ray sources that integrate a 3-W CNT field emission tube and power supply in a small package (160 × 38 mm, 300 g). (Courtesy of Oxford Instruments Plr.)

8.4 Conclusions

Application of CNTs for field emission has received much attention across the world. A tremendous progress has been made since the first reports of field emission with these materials. CNTs have been shown to be an excellent emitters, providing very low turn-on voltage and very high current densities while being very robust and promising low fabrication costs compared with micromachined field emitter arrays. Industrial applications of CNT field emitters are already reported: CNT cathodes are implemented in commercial miniature x-ray tubes, allowing a significant increase in the efficiency and reliability of portable x-ray systems. This application, however, remains limited to fairly small markets of high value-added products. The application of CNT field emitters in mass-produced systems still requires refinement of the cathode fabrication processes. Small startup and major corporate companies have dedicated increasing research and development efforts to solving these issues. The FED technology is regarded as the primary driver of the industrial application of CNT field emitters. Recent development of scalable processes for the fabrication of CNT cathodes is a step toward the deployment of this technology in the display industry.

References

1. R.H. Fowler and L. Nordhiem, Proc. Roy. Soc. Ser. A, 119, 173 (1928).
2. C. Adessi and M. Devel, Phys. Rev. B, 62(20), 13314 (2000).
3. C.A. Spindt, J. Appl. Phys., 39, 3504 (1968).
4. I. Brodie and P.R. Schwoebel, Proc. IEEE, 82, 1005 (1994).
5. W.A. De Heer, A. Châtelain, and D. Ugarte, Science, 270, 1179 (1995).
6. Y. Saito and S. Uemura, Carbon, 38, 169 (2000).
7. J-M. Bonard, H. Kind, T. Stöckli, and L-O. Nilsson, Solid-State Electron., 45, 893 (2001).
8. J.M. Bonard et al. Appl. Phys. A, 69, 245 (1999).
9. K. Dean, P. von Allmen, and B. Chalamala, J. Vac. Sci. Technol. B, 17, 1959 (1999).
10. C. Kim et al. J. Am. Chem. Soc., 124, 9906 (2002).
11. A.R. Wadhawan, K. Stallcup, and S.J. Perez, Appl. Phys. Lett., 79, 1867 (2001).
12. J.S. Suh and J.S Lee, Appl. Phys. Lett., 74, 2047 (1999).
13. V.V. Semet et al. Appl. Phys. Lett., 81, 343 (2002).
14. L. Nilsson et al. Appl. Phys. Lett., 76, 2071 (2000).
15. J. Suh, K. Jeong, and J. Lee, Appl. Phys. Lett., 80, 2392 (2002).
16. C.M. Marrese, Ph.D. thesis (Aerospace Engineering), University of Michigan (1999).
17. M. Camero-Castano et al. AIAA, 2000-3263 (2000).
18. M. Tajmar, AIAA, 2002-4243 (2002).
19. A.P. Burden, Int. Mat. Rev., 46, 213 (2001).
20. W.B. Choi et al. Appl. Phys. Lett., 75, 3129 (1999).
21. J-M. Bonard et al. Carbon, 40, 1715 (2002).
22. http://www.noritake-itron.jp/english/nano/
23. Q.H. Wang et al. Appl. Phys. Lett., 70, 3308 (1997).
24. Y. Saito, S. Uemera, and K. Hamagushi, Jpn. J. Appl. Phys., 37, L346 (1998).
25. G. Yue et al. Appl. Phys. Lett., 81, 355 (2002).
26. D. Vaniman et al. J. Geophys. Res., 103(E13), 31, 477 (1998).
27. P. Sarrazin et al. Advances in X-ray Analysis, Vol. 47 (2004).
28. L. Delzeit et al. J. Phys. Chem. B, 106, 5629 (2002).
29. A.M. Cassell, S. Verma, L. Delzeit, and M. Meyyappan, Langmuir, 17, 266 (2001).
30. L. Delzeit et al. Chem. Phys. Lett., 348 (2001).
31. C. Nguyen et al. Nanotechnology, 12, 363 (2001).

9

Carbon Nanotube Applications: Chemical and Physical Sensors

Jing Li
NASA Ames Research Center

9.1 Introduction

Nanotechnology provides the ability to work at the molecular level, atom by atom, to create large structures with fundamentally new molecular organization. It is essentially concerned with materials, devices, and systems whose structures and components exhibit novel and significantly improved physical, chemical, and biological properties; phenomena; and processes because of their nanoscale size. As one class of nanostructured materials, carbon nanotubes (CNTs) have been receiving much attention due to their remarkable mechanical properties and unique electronic properties as well as the high thermal and chemical stability and excellent heat conduction. CNTs have a large surface-to-volume ratio and aspect ratio with the diameter of a few nanometers and length up to 100 μm so that they form an extremely thin wire, a unique one in the carbon family, with the hardness of diamond and the conductivity of graphite. Diamond is the hardest substance found in nature and is an insulator, but graphite is one of the softest conducting materials (pencils use graphite, and graphite is also often used as a lubricant to allow two surfaces to slide freely). Because the electronic property of CNTs is a strong function of their atomic structure and mechanical deformations, such relationships make them useful when developing extremely small sensors that are sensitive to the chemical and mechanical or physical environment. In this chapter, CNTs for chemical and physical sensor applications are discussed. Sensor technology commands much attention because sensors have the ability to provide immediate feedback on the environment in places where our own five sensors of taste, sight, hearing, touch, and smell cannot go or are not sensitive enough. Advances in chemical and physical sensor technology are continuously needed for improved sensitivity and fast response to extract more-accurate and -precise information from changes in the environment.

0-8493-2111-5/05/$0.00+$1.50
© 2005 by CRC Press LLC

This chapter addresses CNT-based chemical sensors (e.g., nitrogen dioxide and ammonia sensors) and physical sensors such as AFM/STM probe tips, nanotweezers, nanobalance, infrared detectors, mechanical deformation sensors, and carbon nanothermometers. The working principle of these sensors and some preliminary data along with various sensing applications are discussed.

9.2   Carbon Nanotube Chemical Sensors

The chemical sensor market has been projected to be $40 billion dollars worldwide in less than 10 years. The potential uses of chemical sensors are monitoring and controlling environmental pollution; improving diagnostics for point care in medical applications; providing small, low-power, fast, and sensitive tools for process and quality control in industrial applications; and implementing or improving detection of warfare and security threats. In all these applications, there is a demand for improved sensitivity, selectivity, and stability beyond what is offered by commercially available sensors. To meet this demand, the CNT research community has considered many different approaches for chemical sensor applications, each based on changes in a specific property: chemiresistors and back gate field effect transistors relying on the conductivity change in the CNT, magnetic resonant sensors based on the frequency change caused by mass loading, and optical sensors due to scattering of light caused by the adsorption of chemical species.

9.2.1   Electrochemical Sensors

Electrochemistry implies the transfer of charge from one electrode to another electrode. This means that at least two electrodes constitute an electrochemical cell to form a closed electrical circuit. Another important general aspect of electrochemical sensors is that the charge transport within the transducer part of the whole circuit is always electronic. On the other hand, the charge transport in the sample can be electronic, ionic, or mixed. Due to the curvature of carbon graphene sheet in nanotubes, the electron clouds change from a uniform distribution around the C–C backbone in graphite to an asymmetric distribution inside and outside the cylindrical sheet of the nanotube (see Figure 9.1). Because the electron clouds are distorted, a rich π-electron conjugation forms outside the tube, therefore making the CNT electrochemically active. Electron donating and withdrawing molecules such as NO_2, NH_3, O_2, etc. will either transfer electrons to or withdraw electrons from single-walled nanotubes (SWNTs), thereby giving SWNTs more charge carriers or holes, respectively, which increases or decreases the SWNT conductance [1,2] Typical electrochemical interaction may be denoted as:

$$CNT + Gas \,\rule[0.4em]{8em}{0.4pt}\, CNT^{\delta e}\,Gas^{\delta+} \text{ or } CNT^{\delta+}\,Gas^{\delta e},$$

where δ is a number that indicates the amount of charge transferred during the interaction.

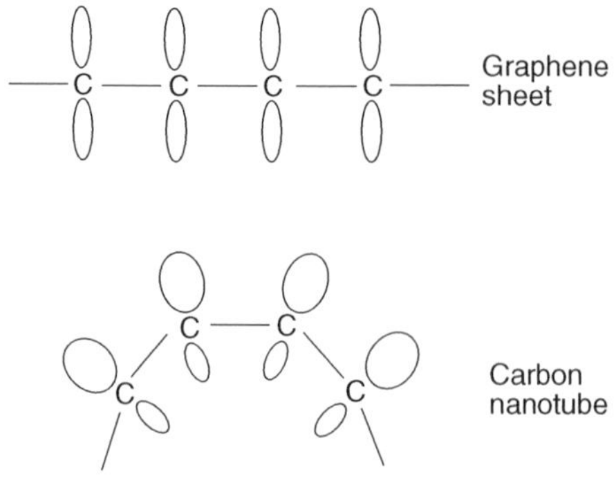

FIGURE 9.1   Schematic of electron distribution of plane graphene and CNT.

9.2.1.1   Back Gate Field Effect Transistor

The dependence of electronic properties on chemical environment was first reported by Kong et al. [3] based on the study of a single SWNT field-effect transistor (FET) for detecting nitrogen dioxide (NO_2) and ammonia (NH_3) [3]. This type of FET was constructed with a single tube as a channel to conduct the source-drain current; the back gate was made from polysilicon, and the source and drain electrodes were simply the large gold pads [4–6] as shown in Figure 9.2. The band gap of a semiconducting tube is proportional to the reciprocal of the tube diameter [7], $E_g \sim 1/d$ ($E_g \sim 0.5$ eV for $d \sim 1.4$ nm). The semiconducting tube (S-tube) connected by two metal electrodes forming a metal/S-tube/metal system exhibits p-type transistor characteristics.

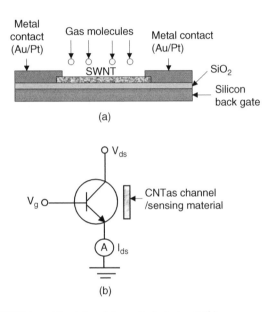

FIGURE 9.2   Schematic of CNT transistor (a) and its equivalent circuit (b).

Gas sensing experiments using the FET approach were carried out by placing an S-tube FET in a sealed 500-ml glass flask with electrical feed-through and passing diluted NO_2 or NH_3 in a diluent such as argon (Ar) or air with flow rate of 700 ml/min while monitoring the resistance of the SWNT. The conductance of SWNT substantially increased for NO_2 exposure and decreased for NH_3 exposure. The curves of conductance vs. time before and after the exposures of NO_2 and NH_3 from Reference 3 are shown in Figure 9.3. From the chemical nature of the molecules, NO_2 is known to be a strong oxidizer with an unpaired electron. Charge transfer takes place from the nanotube to NO_2 molecule, causing an enhancement of hole carriers in the SWNT and hence an increase in current. On the other hand, NH_3 is a Lewis base with a lone electron pair that can be donated to other species. Because theoretical calculations [3] find no binding affinity between NH_3 molecules and the (10,0) tube used in this experiment, the observed interaction between NH_3 molecules and the SWNT may be through some other species because it is known that NH_3 can interact strongly with adsorbed oxygen species on graphite [8].

Recently, a highly sensitive and selective CNT sensor was developed by coating the CNTs with polyethyleneimine (PEI) [9]. This device contains multiple SWNTs bridging the electrodes facilitated by chemical vapor deposition (CVD) from micropatterned catalyst (see Figure 9.4). The yield on multiple-tube devices appears to be higher than that of the single-tube device described above. These devices are also highly sensitive to the back-gate effect and, importantly, exhibit lower electrical noise. The CNTs on these devices are polymer functionalized with PEI to afford ultrahigh sensitivity for NO_2 detection (~100 ppt) while minimizing the sensitivity for NH_3 (~1%). The sensor response to NO_2 in the low concentration range is shown in Figure 9.4a. Unlike the as-grown p-type SWNT, the PEI-coated SWNT

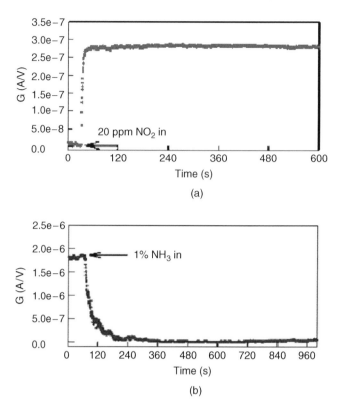

(a)

(b)

FIGURE 9.3 SWNT-FET sensors respond to NO_2 (a) and NH_3 (b) and corresponding chemical gating effect to semiconducting SWNT before gas exposure and after NO_2 and NH_3 exposures. Reprinted from J. Kong et al. Science, 287, 622 (2000). With permission from the American Association for the Advancement of Science.

becomes n-type due to electron transfer/donation by high-density amine groups on PEI. The PEI-coated SWNTs give a decreased conductivity change to NO_2 instead of increased conductivity change for the as-grown SWNTs discussed earlier. The conductance of the n-type devices decreased upon NO_2 binding due to electron transfer to NO_2 reducing the majority carriers in the nanotubes. The $I\text{-}V_g$ (current vs. gate voltage) curves recorded for the multitube device exposed to various NO_2 concentrations shifted progressively to the positive V_g side, accompanied by reduced n-channel conductance (see Figure 9.4a), consistent with the charge-transfer scenario. Similar threshold voltage shifts by NO_2 adsorption on individual PEI-functionalized n-type semiconducting SWNTs were also observed (see Figure 9.4b). The PEI-coated devices exhibit no change of electrical conductance when exposed to other molecules including CO, CO_2, CH_4, H_2, and O_2. The results showed that PEI-functionalized SWNTs are highly selective to strongly electron-withdrawing molecules. Another polymer, Nafion, was explored in this work to block certain types of molecules from reaching nanotubes, including NO_2. This allows for NH_3 detection in a more selective manner. Nafion is a polymer with a Teflon backbone and sulfonic acid side groups and is well known to be perm selective to –OH-containing molecules including NH_3 that tend to react with H_2O in the environment to form NH_4OH.

9.2.1.2 Chemiresistors

The chemiresistor is a type of sensor that measures the current change, which reflects the change in resistance or conductance, before and during the exposure of chemical species while a constant voltage is applied across the two ends of the sensing material. The room temperature sensitivity of the FET-based SWNT sensors discussed above is attributed to drastic changes in electrical conductivity of semiconducting SWNT induced through charge transfer of gas molecules. However, it is currently difficult to obtain

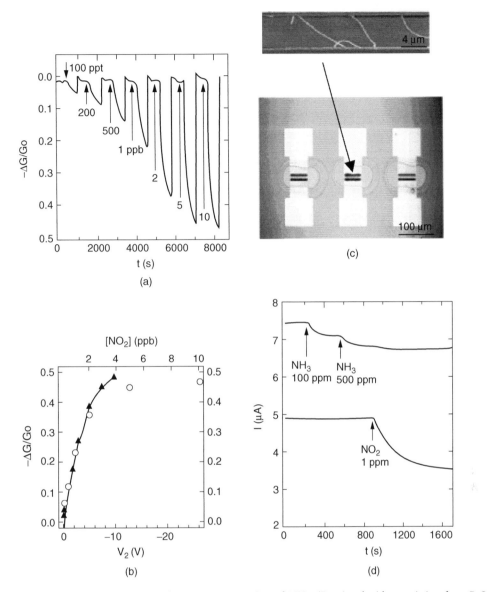

FIGURE 9.4 Multiple-tube transistor and its response to NO_2 and NH_3. (Reprinted with permission from P. Qi and H. Dai, Nano Lett., 3, 3, 347 (2003). Copyright 2003, American Institue of Physics.)

semiconducting SWNTs exclusively from as-grown samples, which typically tend to be a mixture of both metallic and semiconducting nanotubes. Even when nanotubes are directly grown on the platform by CVD, there is no control now for the growth of semiconducting tubes selectively. This often results in additional processing steps leading to fabrication complexity, low sensor yield, and poor reproducibility in sensor performance. In addition, the three-terminal system makes the FET-like device a costly platform for chemical sensor applications. The chemiresistor, in contrast, is a simple two-terminal device as discussed below.

A gas sensor (see Figure 9.5), fabricated by simple casting of SWNTs on an interdigitated electrode (IDE), has been demonstrated for gas and organic vapor detection at room temperature [10]. The sensor responses are linear from sub to hundreds of parts per million (ppm) concentrations with a detection limit of 44 ppb (parts per billion) for NO_2 and 262 ppb for nitrotoluene. The time is of the order of seconds for the detection response and minutes for the recovery. The variation of the sensitivity is less

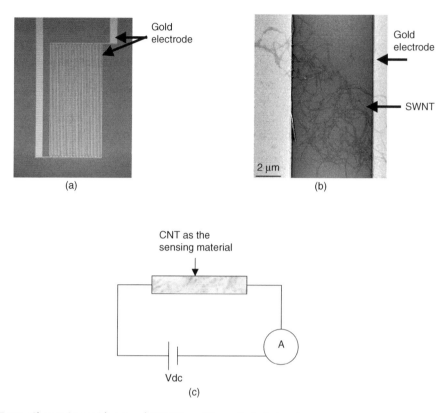

FIGURE 9.5   Chemresistor with network SWNTs and its equivalent circuit.

than 6% for all tested devices, comparable with commercial metal oxide or polymer microfilm sensors while retaining the room temperature high sensitivity of the SWNT transistor sensors and manufacturability of the commercial sensors. The extended detection capability from gas to organic vapor is attributed to direct charge transfer on individual semiconducting SWNT conductivity with additional electron hopping effects on intertube conductivity through physically adsorbed molecules between SWNTs. This sensor is described here in detail to give an example of the whole sensor development process.

The IDE was fabricated using the conventional photolithographic method (see Figure 9.5a) with a finger width of 10 μm and a gap size of 8 μm. The IDE fingers were made by thermally evaporating 20 nm of titanium (Ti) and 40 nm of gold (Au) on a layer of silicon dioxide (SiO_2) thermally grown on top of a silicon wafer. SWNTs form a network type of layer as sensing material across two gold electrodes (see Figure 9.5b). SWNTs produced by the high-pressure carbon monoxide disproportionation (HiPCo) process [11] were purified using a two-step method [12] with acid first to remove the residuals of metallic catalyst, followed by air oxidation at high temperature to remove the graphitic carbons. The final purified SWNT has purity over 99.6% and a surface area of 1587 m^2/g, both the highest values reported for SWNTs. Such high purity level is critical to ensure that the sensor response is from the nanotubes and not affected by the metal and amorphous carbon impurities. The purified SWNTs were then dispersed in dimethylformamide (DMF) to form a suspension. A 0.05-μl aliquot of 3 mg/l SWNT-DMF solution was drop deposited onto the interdigitated area of the electrodes. After the DMF evaporated, a network of nanotubes lay on the electrodes to bridge the fingers. The DMF was chosen to debundle the SWNT ropes because the amide group can easily attach to the surface of the nanotubes providing a uniformly suspended SWNT solution. This assures a good-quality SWNT deposition on top of the interdigitated area. The SWNTs on the IDE then were dried under vacuum to remove the DMF residue. The density of the SWNTs across the IDE fingers can be adjusted by varying the concentration of the SWNT-DMF solution. The total interdigitated area is 0.72 × 1.5 mm, which is big enough to hold about 0.05 μl solution

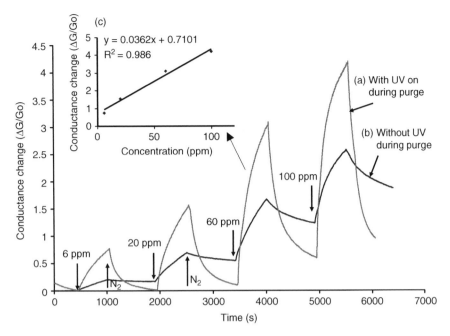

FIGURE 9.6 Representative sensor response for NO_2 (a) with UV light in the recovery, (b) without UV light for sensor recovery, (c) insertion: calibration curve from (b). The sample gas is NO_2 in a 400 cc/min nitrogen flow at room temperature. Ultrapure nitrogen is used for dilution and purging. (After J. Li et al. Nanoletters, 3, 7, 929 (2003).)

within the space. The large interdigitated area also ensures enough CNTs to give a reproducible sensor performance. The initial resistance of each sensor can be controlled by the density of SWNTs across the IDE fingers. The optimal resistance for sensor performance is about 10 kΩ from the electronic testing circuit point of view. Figure 9.6 shows that the SWNT sensor gives strong responses to NO_2 at four different concentrations. The recovery time is very long, of the order of 10 hours, due to the higher bonding energy between SWNT and NO_2. This long recovery time is the same as in previously reported CNT sensors [3]. The recovery time is accelerated to about 10 minutes by ultraviolet (UV) light illumination (Figure 9.6b). The UV exposure decreases the desorption energy barrier to ease the NO_2 desorption [13]. Indeed, UV illumination for 2 hours before the sensor testing helps reduce the baseline, initial conductivity of the nanotube. Figure 9.6c shows a good linear dependency between the response and the concentration of NO_2. The sensor response to nitrotoluene (Figure 9.7) shows faster and more reversible response than that for NO_2. However, the response is smaller due to weak bonds and less partial electron transfer between SWNTs and nitrotoluene molecules. The sensor response is linear with respect to concentration in the range investigated.

In the above investigations, the lowest detectable concentration was limited by the experimental setup [10]. The detection limit can be derived from the sensor's signal processing performance as described below. The sensor noise can be calculated using the variation of relative conductance change in the baseline using root mean square deviation (RMSD) [14]

$$V_{chisq} = \Sigma(y_i - y)^2, \tag{9.1}$$

where y_i is the measured data point and y is the corresponding value calculated from the curve-fitting equation. The RMS noise is calculated as

$$RMS_{noise} = \sqrt{\frac{V_{chisq}}{N}} \tag{9.2}$$

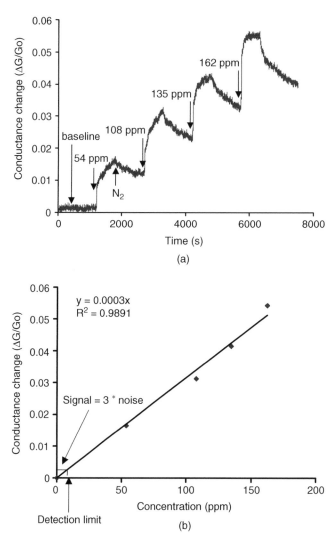

FIGURE 9.7 Representative sensor response for nitrotoluene. (a) Sensor response is a step function of concentration; (b) calibration curve from (a). The nitrotoluene vapor was evaporated using a bubbler with a 100 cc/min ultrapure nitrogen at room temperature, and this vapor stream was further diluted by nitrogen to a total flow of 400 cc/min. The purge gas is nitrogen as well. (After J. Li et al. Nanoletters, 3, 7, 929 (2003).)

where N is the number of data points used in the curve fitting. For example, in Figure 9.7a, 10 data points in the baseline before the nitrotoluene exposure were used. After plotting the data, a five-order polynomial fit was executed within the data point range, which gives not only the curve fitting equation but also the statistical parameters of the polynomial fit as well.

The sensor noise is 0.00026 for the nitrotoluene sensor in Figure 9.7a. The average noise level is 0.00050 ± 0.00028 for all the nitrogen oxide sensors. According to IUPAC definition [15], when the signal-to-noise ratio equals 3, the signal is considered to be a true signal. Therefore, the detection limit (DL) can be extrapolated from the linear calibration curve when the signal equals three times the noise:

$$DL\,(ppm) = \frac{3RMS}{slope}\,, \qquad\qquad (9.3)$$

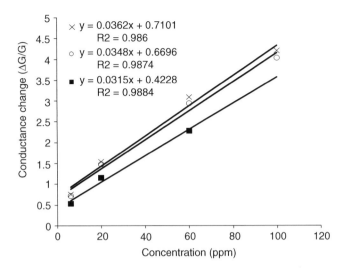

FIGURE 9.8 Calibration of three SWNT sensors response to NO_2 concentration. (After J. Li et al. Nano Lett., 3, 7, 929, 2003.)

Using the above equation, the NO_2 detection limit is calculated as 44 ppb from the average sensitivity of the three NO_2 sensors in Figure 9.8 and the average noise level. The nitrotoluene detection limit is similarly calculated as 262 ppb.

Reproducibility studies were carried out for the SWNT sensors. An example is shown in Figure 9.8 in which three sensors were tested for NO_2 at concentration levels of 6 ppm, 20 ppm, 60 ppm, and 100 ppm. All responses are linear, showing consistent sensitivity of 0.034 ± 0.002, defined by the slope or $(\Delta G/G)/\Delta(\text{concentration})$. Overall variation of the sensitivity for the fabricated devices is about 6%, comparable with and even better than that of metal oxide or polymer-based sensors [16,17], which shows excellent reproducibility of these SWNT sensors.

A striking feature emerges from Figure 9.7 for nitrotoluene and Figure 9.6 for NO_2 detection when the linear sensitivity curve is extrapolated to zero concentration. This leads to nonzero and zero response for NO_2 and nitrotoluene, respectively, suggesting two different sensing mechanisms for electrical response to molecular adsorption in SWNTs. One type of adsorption results in a direct charge transfer between donor or acceptor type of molecules (or groups) and an individual SWNT, leading to the modulation of the Fermi level in the semiconducting tubes (intratube modulation) causing a conductivity change [18]. This occurs for NO_2 and the nitrogen oxide group in nitrotoluene molecules. However, such intratube modulation from nitrotoluene molecules will be less significant than that for pure NO_2 because the stronger hydrophobic interaction between SWNTs and benzene groups will leave the nitrogen oxide groups farther away from the SWNTs. Another type of adsorption occurs in the interstitial space between SWNTs to form an SWNT-molecule-SWNT junction, leading to a hopping mechanism for intertube charge transfer between SWNTs and intertube modulation of the SWNT network in lieu of conductivity change. This occurs for all types of molecules and for both metallic and semiconducting SWNTs. The intratube modulation is similar to that of interaction between semiconductor metal oxides and donor or acceptor types of molecules, showing nonlinear (power law) response and large slope (sensitivity) feature over low concentrations [19]. The intertube modulation is similar to that of interaction between conductive polymers and physically adsorbed molecules, showing nearly linear and smaller slope (sensitivity) behavior over a broad range of concentrations [19].

The above phenomenological model explains the experimental observations. In nitrotoulene detection, the intertube modulation plays a key role.

In NO_2 detection, a nonlinear behavior should be observed if the response is extended to very low concentrations (in Henry's law regime), as it should reach zero at zero concentration. The electrical response is contributed from both inter- and intramodulation of electron states in the experimental conditions but should be dominated by intraeffect at low concentrations. Such a nonlinear feature has been observed previously using SWNT transistor array sensors [9] where each sensor was made of many metallic and semiconducting SNWTs between two electrodes. The behavior in Reference 9 can be characterized with two linear response regions with a slope of about 0.2 below 2 ppb and 0.03 for higher concentrations, indicating that the sensitivity from intratube modulation is at least one order of magnitude higher than that from intertube modulation. Clearly, these sensors show similar response behavior (sensitivity of 0.034 at higher concentrations) and would show higher sensitivity if they are applied to lower concentrations. Therefore, the detection limit can reach another order of magnitude of lower concentrations from the current 44 ppb derived from higher concentrations.

9.2.2   Resonator (Mass) Sensors

The resonator is a device that provides the frequency signal to reflect the chemical information around the device. Typical electrical characteristics of an unperturbed quartz resonator [20] can be modeled by a Butterworth-Van Dyke equivalent circuit [21], shown in Figure 9.9. The structure of a resonator sensor is usually a three-layer device with a layer of conductor, a layer of insulating material, and a layer of sensing material. The resonator electrical properties can be conveniently described in terms of the electrical admittance, $Y(f)$, defined as the ratio of current flow to applied voltage (Y is the reciprocal of impedance). In the equivalent circuit models of Figure 9.9, the admittance is given by

$$Y(f) = jwC_0^* + \frac{1}{Z_m}, \tag{9.4}$$

where Z_m is the electrical motional impedance; for the unperturbed resonator (Figure 9.9a),

$$Z_m = R_1 + jwL_1 + \frac{1}{jwC_0^*}. \tag{9.5}$$

When a quartz resonator is operated in contact with chemical species, the mechanical interaction between the surface of the resonator and the chemical species changes the motional impedance Z_m. This fact can be modeled by adding a complex electrical impedance element, $Z_e(w)$ (see Figure 9.9b).

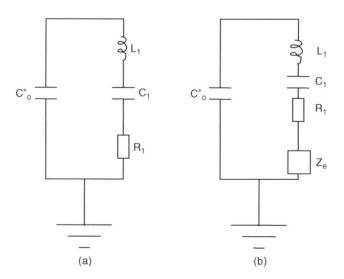

(a)   (b)

FIGURE 9.9 Equivalent-circuit models to describe the electrical characteristics of the resonator (for f near f_o) under various loading conditions: (a) unloaded (b) with complex impedance element Z_e to represent surface loading.

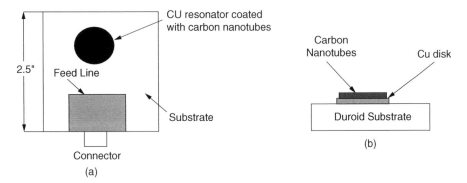

FIGURE 9.10 Configuration of a CNT resonant sensor. (a) Top view (b) side view. (Reprinted with permission from S. Chopra et al. Appl. Phys. Lett., 80, 24, 4632 (2002). Copyright 2002, American Institute of Physics.)

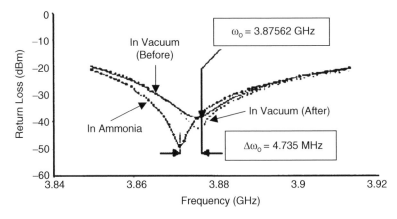

FIGURE 9.11 Frequency downshift due to the NH_3 exposure. (Reprinted with permission from S. Chopra et al. Appl. Phys. Lett., 80, 24, 4632 (2002). Copyright 2002, American Institute of Physics.)

Chopra et al. [22] have designed and developed a microwave resonant sensor [22] (see Figure 9.10), for monitoring ammonia. The sensor consists of a circular disk electromagnetic resonant circuit coated with either single- or multi-walled carbon nanotubes (MWNTs) that respond to ammonia by a change in its resonant frequency as a result of interaction of adsorbed ammonia molecules with the carbon nanotubes. The change in resonant frequency can be easily detected using a radio-frequency (RF) receiver that will enable the design of a remote sensing system. Chopra et al. used an 8753ES network analyzer that can measures the resonator's return loss to determine the resonant frequency. Their results show that the SWNTs give higher sensitivity to ammonia than the MWNTs. The frequency is downshifted 4.375 MHz upon exposure of the sensor to ammonia (see Figure 9.11). The resonant frequency shift can be attributed to the changes in the effective dielectric constant of the heterogeneous materials, the CNT in this case. The recovery and response times of this sensor are nominally 10 minutes. This technology is suitable for designing remote sensor systems to monitor gases inside sealed opaque packages and environmental conditions that do not allow physical wire connections.

Ong et al. [23] have developed a wireless, passive CNT-based gas sensor [23] for carbon dioxide (CO_2) and oxygen (O_2). This group used a MWNT-SiO_2 mixture as sensing material, and the sensor was fabricated by patterning a square spiral inductor and in interdigitated capacitor on a printed circuit board (PCB) (see Figure 9.12). As the sensor is exposed to various gases, the relative permittivity ε and the conductivity (proportional to ε'' [24]) of the MWNTs vary, changing the effective complex permittivity of the coating and hence the resonant frequency of the sensor. In this study, the frequency spectrum of

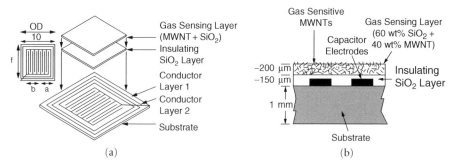

FIGURE 9.12 Scheme of a capacitive resonant sensor; (a) configuration, (b) sideview. (Reprinted from K.G. Ong et al. IEEE Sensors J., 2, 2, 82 (2002). © 2002 IEEE. With permission.)

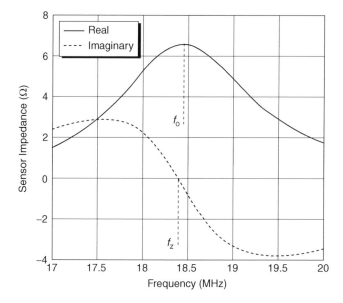

FIGURE 9.13 An illustrative measured impedance spectrum of the sensor-perturbed antenna, after subtracting the background antenna impedance. The resonant frequency f_o is defined as the maximum of the real portion of the impedance, and the zero-reactance frequency f_z is the zero crossing of the imaginary portion of the impedance spectrum. (Reprinted from K.G. Ong et al. IEEE Sensors J., 2, 2, 82 (2002). © 2002 IEEE. With permission.)

the sensor is obtained by directly measuring the impedance of spectrum of a sensor-monitoring loop antenna. The impedance of the loop antenna is removed from the measurement using a background subtraction, obtained by measuring the antenna impedance without the sensor present. A typical background-subtracted impedance spectrum is shown in Figure 9.13, where the resonant frequency f_o is defined as the frequency at the real impedance (resistance) maximum, and the zero-reactance frequency f_z is the frequency where the imaginary impedance (reactance) goes to zero. The complex permittivity, $\varepsilon' - j\varepsilon''$, of the coating material (both the MWNT-SiO$_2$ and the SiO$_2$ layers) are calculated from f_o and f_z.

$$\varepsilon'' = K \frac{\sqrt{f_o^2 - f_z^2}}{f_0^3}.$$

(9.6)

   The sensor response to CO$_2$ is shown in Figure 9.14. The frequency shifts are linear with the concentration of CO$_2$, 0.0004043/% CO$_2$ for ε' and -0.0003476/% CO$_2$ for ε''.

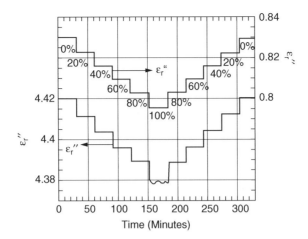

FIGURE 9.14 Measured ε_r' and ε_r'' values when the sensor is exposed to CO_2 concentrations varying from 0 (volume) to 100% and then back to 0%. The shifts are linear, with $\Delta\varepsilon_r' = -0.0004043/\%\ CO_2$ and $\Delta\varepsilon_r'' = -0.0003476/\%\ CO_2$. (Reprinted from K.G. Ong et al. IEEE Sensors J., 2, 2, 82 (2002). © 2002 IEEE. With permission.)

Due to the strong attraction of the MWNTs for water moisture, the CO_2 sensor suffers a dramatic increase of response time in high-humidity environments. The high operation temperature improves the response time of the sensor. In order to correct for the humidity effect, a SiO_2-coated sensor and an uncoated sensor were tested parallel to the MWNTs sensor. Neither of these sensors responded to CO_2 and both responded to temperature and humidity uniquely. Using a calibration algorithm enables CO_2 concentrations to be measured to within $\pm 3\ CO_2\%$ in a changing humidity and temperature environment. This sensor is reversible for O_2 and CO_2 but irreversible for NH_3. Sensor response time is approximately 45 sec, 4 minutes, and 2 minutes for CO_2, O_2, and NH_3, respectively.

9.2.3 Thermal Sensors

It follows from the first law of thermodynamics that any process in which the internal energy of the system changes is accompanied by absorption or evolution of heat. The thermal sensor for chemical species detection is based on the heat generated by a specific reaction as the source of chemical information. There are two properties of heat that are unique with respect to any other physical parameter: heat is totally nonspecific and cannot be contained, i.e., it flows spontaneously from the warmer (T_1) to colder (T_2) part of the system. From the sensing point of view, the strategy is to place the chemically selective layer on top of a thermal probe and measure either the change in temperature of the sensing element or the heat flux through the sensing element (see Figure 9.15).

As a thermally sensitive layer, the carbon nanotubes have been used to measure the change of the thermalelectric power during gas exposure [25]. The CNT samples studied in this work were made as lightly compacted mats of tangled SWNT bundles. The material consisted of ~50 to 70% carbon as SWNTs with a mean bundle diameter ~15 nm. The thermoelectric response of this mat to 1 atm overpressure of hellium gas at T = 500 K is shown in Figure 9.16a. The thermopower rises exponentially

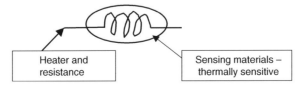

FIGURE 9.15 A typical catalytical bead as a thermosensor for gas detection.

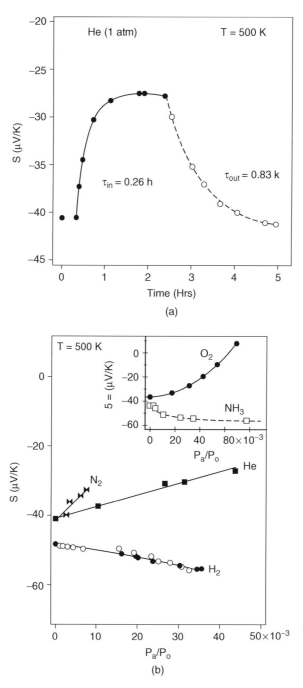

FIGURE 9.16 Sensor performance of a SWNT thermoelectric *nano-nose*. (a) The thermoelectric power response characteristics to 1 atm overpressure of He gas (filled circles) and during subsequent application of a vacuum over the sample (open circles); (b) calibration of thermopower vs. partial pressure of gas H_2, He, N_2, O_2, and NH_3. Reprinted from C.K.W. Adu et al. Chem. Phys. Lett., 337, 31–35 (2001). With permission from Elsevier.

with time, saturating at ~12 mV/K. Removing the He induces an exponential decay of thermopower with time. The thermoelectric response (*s,p*) of a bundle of SWNTs to a variety of gases (He, N_2, H_2, O_2, and NH_3) have been tested (see Figure 9.16b). The mechanism can be understood in terms of the change in the thermoelectric power of the metallic tubes due to either a charge-transfer-induced change of Fermi energy [3] or to the creation of an additional scattering channel for conduction electrons in the tube wall. The scattering channel is identified with impurity sites associated with the adsorbed gas molecules. Mathematical models have been developed to understand the mechanisms. The SWNT mat thermopower ($S \times T$) has been argued as a consequence of the percolating pathways through the metallic tube components in the mat [26]. According to the Mott relation, the thermoelectric power associated with the diffusion of free carriers in a metal can be written compactly as a logarithmic energy derivative of the electrical resistivity (p) [27]

$$S = CT \frac{d}{dE} \left[\ln \rho(E) \right]_{E_F} , \tag{9.7}$$

where

$$C = \left(\frac{\pi^2 k_b^2}{3|e|} \right) ;$$

k_b is Boltzmann's constant; T is the temperature; $\rho(E)$ is the energy-dependent resistivity, i.e., $|e|$ is the charge on the electron; and the derivative is evaluated at the Fermi energy E_f. The total bundle resistivity can be separated into an intrinsic part and an extrinsic part caused by the adsorption of the gas molecules. If the molecules under study are physisorbed, i.e., van der Waals bonding to the tube walls, they would induce only a small perturbation on the SWNT band structure, and an almost linear relationship between S and should be obtained. If the molecules are strongly adsorbed on the SWNTs, then the S vs. ρ plot would show a curvature that indicates chemisorption. It has been claimed that a significant difference in slope of S vs. ρ curves for He and N_2 exists at a fixed temperature, which can be the basis for the utility of a SWNT thermoelectric *nano-nose*.

9.2.4 Optical Sensors

The interaction between the sensing material and the chemical species causing a change in optical properties is the basis of an optical sensor. In optical sensors, the optical beam is guided out of the spectrophotometer, allowed to interact with the sample, and then reintroduced into the spectrophotometer in either its primary or secondary form for further processing. So far, there has been no work reported on the use of CNTs as optical sensors for chemical detection.

9.3 Carbon Nanotube Physical Sensors and Actuators

9.3.1 Flow Sensors

Ghosh et al. [28] have reported a CNT-based flow sensor that is based on the generation of a current/voltage in a bundle of SWNTs [28], when the bundle is kept in contact with flowing liquid. Generation of such current/voltage was predicted by Kral and Shapiro [29] who theoretically investigated the behavior of nanotubes in a flowing liquid [29]. Generally, an electric current in a material is produced when flow of free-charge carriers is induced in the material. According to Kral and Shapiro, the generation of an electric current in a nanotube is essentially due to transfer of momentum from the flowing liquid molecules so as to have a dragging effect on the free-charge carriers in the nanotube. The predicted relationship between electric current and the flow velocity is linear. In sharp contrast, Sood and coworkers [28] found that the induced voltage fits logarithmic velocity dependence over nearly six decades (see Figure 9.17). The magnitude of the voltage/current induced along the nanotube by flow depends significantly on the ionic strength of the flowing liquid. The ionic liquid in flow causes an asymmetry by the velocity gradient (shear) at the liquid-solid interface. Thus, one can imagine that charging imbalance is

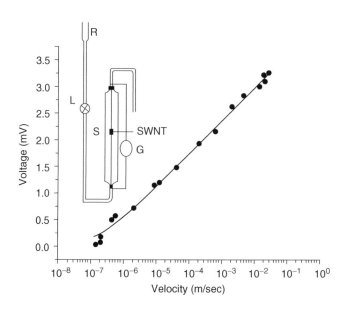

FIGURE 9.17 Variation of voltage (*V*) developed as function of velocity (*u*) of water. Solid line is a fit to functional form as $V = \alpha \log (u + 1)$, where α and are constants. Inset shows experimental setup where R is the reservoir, L is the valve controlling the liquid flow, S is the cylindrical glass flow chamber, and G is the voltmeter. (Reprinted from S. Ghosh, A.K. Sood, and N. Kumar, *Science*, 299, 1042–1044 (2003). With permission from American Advancement of Science.)

shear-deformed by the velocity gradient, resulting in an asymmetric fluctuating potential. The experimental setup is shown in Figure 9.17A. For a very low flow velocity of 5×10^{-4} m/s, a voltage of 0.65 mV is generated and for a flow velocity of the order of 10^{-5} m/s, saturation was observed. The experimental data points fitted to an empirical equation $V = \alpha \log (\mu\beta +1)$ showing the relation of generated voltage to the flow velocity.

9.3.2 Force Sensors — Pressure, Strain, Stress

A typical example for force sensors using CNT is the scanning microscopic probe, such as atomic force microscopy (AFM) and scanning tunneling microscopy (STM) tips. This has been described in Chapter 6. In this section, the focus will be on the CNT pressure, stress, and strain sensors.

Wood and colleagues have demonstrated the potential of SWNTs as molecular and macroscopic pressure sensors [30]. High hydrostatic pressures were applied to SWNTs by means of a diamond anvil cell (DAC), and micro-Raman spectroscopy was simultaneously used to monitor the pressure-induced shift of various nanotube bands. The use of direct macroscopic pressure in the DAC allows the simple manipulation of the nanostructures for comparison with peak shifts derived from internal pressure of immersion media. The data confirm recent results independently obtained from internal pressure experiments with various liquids, where the peak shifts were considered to arise from compressive forces imposed by the liquids on the nanotubes. It is also shown that the nanotube peak at 1580 cm^{-1} (the G band) shifts linearly with pressure up to 2000 atm and deviates from linearity at higher pressure. This deviation is found to be coincident with a drop in Raman intensity for the disorder-induced peak at 2610 cm^{-1} (the overtone of the D* band), possibly corresponding to the occurrence of reversible flattening of the nanotubes.

The cohesive energy density and the internal pressure are interchangeable parameters, similar to the energies or forces at the surface of liquids, which describe the powerful cohesive forces that hold liquids together. A balance between attractive and repulsive molecular forces exists in the interior of a liquid, which, if disturbed by a second phase (in this case the nanotube aggregates), induces the transfer of a large hydrostatic stress from the surrounding liquid to the suspended particles. The magnitude of this

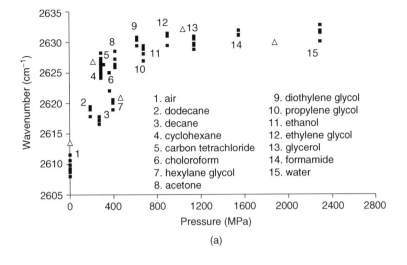

(a)

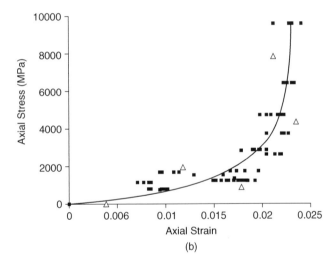

(b)

FIGURE 9.18 Correlation between the pressure and its induced wavenumber shift in the Raman spectrum. (a) Peak position of D* band on immersion of SWNT aggregates in liquids with various cohesive energy densities (the triangles show the pressure data from the DAC experiments; (b) compressive stress-strain curve for SWNT determined from the two calibration graphs in (a): triangles show the data from the DAC experiments, squares are from the experiments on liquids. (Reprinted with permission from J.R. Wood et al. J. Phys. Chem. B, 103, 10388–10392 (1999). Copyright 2003, American Chemical Society.)

stress can be determined from the energy of vaporization ΔEv (J/mol) and the molar volume of the liquid V (cm³/mol), the thermodynamic quantities that describe the attractive strength between molecules. Values of the cohesive energy density (CED) can be calculated from the enthalpy of vaporization ΔHv (J/mol) or from the solubility parameter δ ((J/cm³)$^{1/2}$ or MP$_a$$^{1/2}$) of the liquid, using the relation [31]

$$CED = \delta^2 = \frac{\Delta E_V}{V} \approx \frac{\Delta H_V - RT}{V},$$ (9.8)

where T (K) is the temperature and R (J/mol K) is the universal gas constant. The cohesive energy density has the units of pressure or stress. Figure 9.18a shows the plot of peak position for the D* peak from the

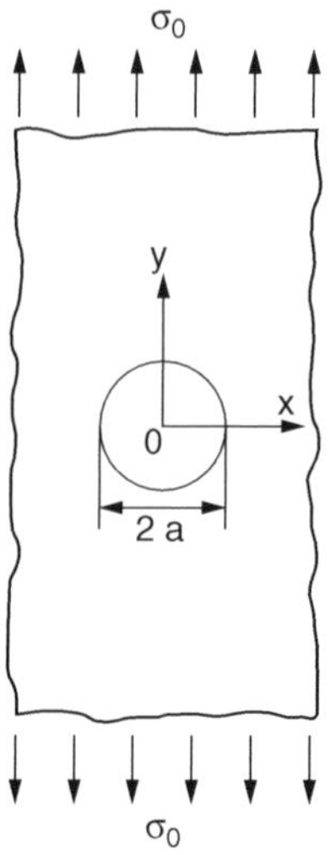

FIGURE 9.19 The specimen configuration considered in the present study is a circular hole of radius, a, in a thin polymer plate under unidirectional tensile stress, σ_0. y is the axis of applied stress and x is perpendicular to y in the plane of the plate. (According to Q. Zhao, M.D. Frogley, and H.D. Wagner, Comp. Sci. Technol., 62, 147 (2002).)

DAC experiments and for all the liquids tested against their respective cohesive energy density. It can be seen that there is a strong concurrence between the two data sets.

A trend similar to that in Figure 9.18a can be obtained for the surface tension of the liquid γ_l instead of the CED by plotting the wavenumber against $\gamma_l (1/V)^{1/3}$, in agreement with the semiempirical relation between these two parameters [32]. Although it is anticipated that the argon medium also has internal pressure acting on the nanotubes, this is expected to be negligible because the intermolecular interactions in liquid argon are very small. Figure 9.18b shows the compressive stress-strain curve using the liquid and pressure data from Figure 9.18a. Both data sets are strikingly similar, even though the methods of inducing pressure (and thus the spectral shifts from which Figure 9.18b is derived) are very different. It is apparent that the modulus is not a unique value but increases with increasing strain. This is similar to the behavior of elastomers in compression [33], which are also network structures. The shape of the curve is not unexpected when it is considered that SWNT have high moduli despite being highly flexible. The compressive modulus ranges between 100 GP_a at low strains and about 5 TPa at high strains. This could explain the scatter in the data of Krishnan et al. [34], slightly different applied strains can yield large differences in Young's modulus.

The carbon nanotubes were also used as embedded strain sensors for mapping the strain profile in a polymer matrix [35]. The strain is transferred from polymer to carbon nanotubes and can be measured when the carbon nanotubes are distanced 1 μm from each other. When mapping with the Raman technique, the measured wavenumber shifts represent the mean response of all the nanotubes at the focal region of the laser [36]. To measure the individual components of the strain (or stress), the Raman signal from nanotubes in a particular direction must be selected out. This can be achieved by physically orienting the nanotubes in the polymer or by using polarized Raman spectroscopy. The experiment was done by both methods to map the stress around a circular hole in a polymer plate under uniaxial tension (see Figure 9.19). The in-plane stress field for this situation has distinct components in the directions parallel and perpendicular to the applied stress. The Raman techniques can be used to quantify the individual stress components: oriented SWNT/PUA with unpolarized Raman spectroscopy and SWNT/Epoxy with polarized Raman spectroscopy.

9.3.3 Electromechanical Actuator

The direct conversion of electrical energy to mechanical energy through a material response is critically important for such diverse needs as robotics, optical fiber switches, optical displays, prosthetic devices, sonar projectors, and microscopic pumps. SWNT-based electromechanical actuators were shown to generate higher stresses than natural muscle and higher strains than high-modulus ferroelectrics [37].

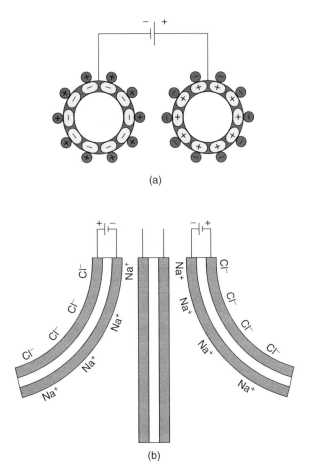

(a)

(b)

FIGURE 9.20 (a) Schematic illustration of charge injection in a nanotube-based electromechanical actuator. (b) Scheme of the actuator deflected by charging. According to Ray Baughman et al. Science, 284, 1340 (1999).

Unlike conventional ferroelectric actuators, low-operating voltages of a few volts generate large actuator strains. Predictions based on measurements suggest that actuators using optimized nanotube sheets may eventually provide substantially higher work densities per cycle than any previously known technology.

Baughman and coworkers [37] have constructed an actuator by stocking SWNT sheets on opposite sides of a tape. The tape with nanotubes on both sides was suspended in 1 *M* NaCl, and no dopant intercalation was required for device operation. Changing the applied voltage injects electronic charge into a SWNT electrode, which is compensated at the nanotube-electrolyte interface by electrolyte ions (forming the so-called double layer). Both calculations [38] and experimental results [39] for charge transfer complexes of graphite and conducting polymers show that the strain due to quantum mechanical effects (changes in orbital occupation and band structure) changes sign from an expansion for electron injection to a contraction for hole injection for low-charge densities but with either sign for high-density charge injection. Like natural muscles [40], the nanotube sheet actuators are arrays of nanofiber actuators, providing a novel type of actuation because the SWNTs add high surface area to the mechanical properties, electrical conductivity, and charge transfer properties of graphite. Figure 9.20a is a schematic illustration of charge injection in a nanotube-based electromechanical actuator. Figure 9.20b shows the schematic edge view of a cantilever-based actuator operated in aqueous NaCl, which consists of two strips of SWNTs (shaded) that are laminated with an intermediate layer of double-sided tape (white). When a charge is injected (a DC potential of a few volts or less) with different polarities, the sandwich type of actuators are deflected toward the left or right and the tip is tilted up about a centimeter. When a square wave potential was applied, oscillation was visually observed up to at least 15 Hz as the actuator pushed the electrolyte back and forth.

The ideal actuator material would operate at low voltage and at least match the performance of skeletal muscle: 10% strain; 0.3 MP$_a$ stress generated, and strain rate: 10% per second. Although nanotube electromechanical actuators function at a few volts, compared with electrostrictive elastomer actuators at KV [41], but to meet the strain rate is a challenge. A study on increasing the actuation rate of electromechanical CNT has been carried out to meet the performance of skeletal muscle [42]. The results of this study demonstrate that resistance compensation can provide significant improvement in the charging rate, and consequent actuation strain rate, for carbon nanotube sheets operated in an organic electrolyte. The strain rate increased with increasing potential pulse amplitude and a more negative potential limit. The amount of strain produced also increased with longer pulse times. The improvements in strain rate are somewhat offset when large negative potential limits are used due to the introduction of faradic reactions in the electrolyte medium that do not contribute to actuation.

9.3.4   Temperature Sensors

Bulk MWNTs were manipulated by AC electrophoresis to form resistive elements between gold micro-electrodes and were demonstrated to potentially serve as temperature and anemometry sensors [43]. The bundled MWNT as sensing elements for microthermal sensors can be driven in constant current mode configuration and used as anemometers. The MWNT sensor chip was packaged on a PCB for data acquisition and was put inside an oven. The oven temperature was monitored by a Fluke type-K thermocouple attached on the surface of the PCB. Then, the resistance change of the MWNT sensors was measured as the temperature inside oven was varied. The temperature-resistance relationship for the MWNT sensors was measured, and a representative data set shows a linear relationship between the resistance and the temperature in the range of 20 to 60°C [43]. The temperature coefficient of resistance (TCR) for the MWNT sensors can be calculated based on following equation:

$$R(T) = R_0\left[1 + \alpha(T - T_0)\right],\tag{9.9}$$

where R_0 is the resistance at room temperature T_0 and α is the temperature coefficient of resistance. The *I-V* measurements of the resulting devices revealed that their power consumption was in the μW range. Besides, the frequency response of the testing devices was generally over 100 kHz in constant current mode operation.

9.3.5   Vision Sensors

As described earlier, the band gap of semiconducting carbon nanotube correlates to the diameter and chirality. This makes it a potential candidate for IR sensors. Devices for IR detection require large-area, densely packed detectors in a well-defined array structure with efficient but also easy light coupling, high responsivity, good pixel-to-pixel uniformity, low dead-pixel count, and excellent stability [44]. Therefore, large-area, ordered, and well-aligned high-density arrays of MWNTs are desirable for this application. The basic cell of the MWNT IR detector array [45] is illustrated in Figure 9.21. The incoming radiation is incident on a group of MWNTs that comprise a single pixel of the array. Radiation greater than the band gap is absorbed, which increases the conductivity of the cell. Such change in conductivity serves to detect IR radiation in the wavelengths from far-IR to mid-IR. The uniformity of the arrays gives an improved signal-to-noise ratio. These arrays can be formed on appropriate substrates, including curved surfaces that qualify them for function as a "smart skin."

Thus far, the MWNTs for IR detection are still at the conceptual stage. More development work needs to be carried out and some questions need to be considered: how to fine tune the CNTs and the dielectric template matrix in which they are embedded for minimizing the dark current and maximizing the

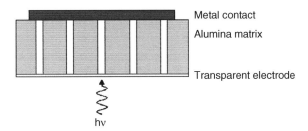

FIGURE 9.21 Basic element of MWNT IR detector array. (According to J. Li et al. Appl. Phys. Lett., 75, 367 (1999).)

absorption; how to optimally trade-off the fill factor and spatial resolution in order to get peak quantum efficiency, which in turn relates to one of absorption and surface reflections; etc.

9.3.6 Acoustic Sensors

CNTs have been used to mimic nature's methods to detect sound in the same manner as an ear [46]. The arrays of MWNTs bend in response to acoustic energy, and the motion is transformed to an electrical signal that can be transmitted using standard electrical techniques. The resulting electrical signal is much like the signal from a traditional microphone except that it contains information about much fainter sounds. MWNTs are naturally directional — they always bend away from the source of the sound. Directionality in normal hearing is derived from the use of two ears, but nanotube technology can potentially provide directional information from a single sensor. This is a unique technology that will enable a new class of innovative microsensors and microactuators for real-world applications in gas or liquid environments. The primary product of this research will be the demonstration of acoustic sensing with *artificial stereocilia arrays*. The miniaturization and directional sensitivity intrinsic to this approach could ultimately lead to revolutionary advances in acoustic detection and signal processing.

9.4 Summary and Outlook

Sensor development using CNTs is at its early stages. A summary of early results along with the basics is provided in this chapter. The potential in various sectors is outlined below. The challenges are common to all these sectors: inexpensive fabrication steps, sensor robustness, reliability, reproducibility, and system integration.

9.4.1 Industry

Carbon nanotube-based chemical sensors possess high sensitivity, small size, and low power consumption, which can be used to quickly verify incoming raw materials at the delivery point. The technology can significantly reduce the amount of time and money spent analyzing those materials in a laboratory, as well as reduce the amount of materials handling. Most changes in chemical processes can be reflected in the changing composition of the vapor phase surrounding or contained within the process. Thus, the vapor-phase sensors enable the quick assessment of the chemical status of most industrial processes. Examples are found across many sectors, including food processing (coffee roasting and fermentation), petrochemical (plastics manufacture and gasoline blending), and consumer products (detergents and deodorants). Much like vision inspection is used to assess the visual integrity (color, shape, and size) of products, olfactory inspection assesses the chemical integrity (consistency and presence of contaminants).

Sensors and control systems in manufacturing provide the means of integrating different, properly defined processes as input to create the expected output. In this integrated system, physical sensors are key components for achieving the precision and accuracy of the operation. Because the carbon nanotube sensors can be embedded and amended for remote control, they can be implanted in the materials to give an instantaneous feedback for *in situ* strain-stress measurement, temperature measurement, mechanical deformation test, and pressure test. They can be used remotely for imaging and acoustic sound wave detection using IR detectors and acoustic sensors respectively.

9.4.2 Environment

Increasing awareness and new regulations for safety and emission control make environmental monitoring one of the most desired among the numerous industrial and civil applications for which the development of reliable solid-state gas sensors is demanded. Current methods for air-quality control approved by the standards consist of analytical techniques, which need the use of very costly and bulky equipment. For applications in this arena, sensors that are able to selectively detect various gases at a concentration level of a few points per billion and in the form of low-cost portable handheld devices for continuous *in-situ* monitoring are needed. With unique advantages of high sensitivity, small size, low power consumption, and strong mechanical and thermal stability, CNT-based chemical sensors are best fit for this type of application. Combined with CNT-based physical sensors for temperature, anemometry, pressure, and humidity monitoring, a CNT-based integrated sensing/monitoring system can be built to get a comprehensive information simultaneously.

9.4.3 Defense

Chemical sensors are very focused for security and defense applications due to their portability and low power consumption. Carbon nanotube sensors potentially can offer higher sensitivity and lower power consumption than the state-of-the-art systems, which make them more attractive for defense applications. Some examples include monitoring filter breakthrough, personnel badge detectors, embedded suit hermiticity sensors, and other applications. Additionally, a wireless capability with the sensor chip can be used for networked mobile and fixed-site detection and warning systems for military bases, facilities, and battlefield areas. Physical sensors, such as carbon nanotube IR detectors and acoustic sensors, potentially show very high sensitivity, which can be directly used in monitoring the troop movement remotely with the wireless capability.

9.4.4 Medical and Biological

It is believed that chemical sensors would provide physicians with a quicker and more-accurate diagnostic tool. Applications could include obtaining objective information on the identity of certain chemical compounds in exhaled air and excreted urine or body fluids related to specific metabolic conditions, certain skin diseases, or bacterial infections, such as those common to leg or burn wounds. Additionally, the chemical sensors may provide more-accurate, real-time patient monitoring during anesthesia administration.

Although the chemical sensors are used for point-of-care diagnostics, the present bio-inspired technology helps the CNT-based physical sensors in medical and bioengineering applications as well. For example, a CNT-based acoustic sensor with higher sensitivity to faint sound can act as a "nanostethoscope" to probe nano-microscale biological activity by "listening to the music of life." For instance, the nanostethoscope may one day detect, monitor, and characterize the "sounds" of cancerous cells, which are known to have a more intense intracellular activity than healthy cells. The combination of IR detector and acoustic sensor based on MWNTs grown on curved surface of substrate can make "smart skin" that responds to light, temperature, and sound or fluid flows. As described before, the CNT actuators can be operated at low voltages of a few volts and generate large actuator strains; therefore, they have a potential of being used as "artificial muscles" using optimized nanotube sheets that may eventually provide substantially higher work densities per cycle than any previously known technology.

References

1. B.K. Pradhan et al. Mat. Res. Spc. Symp. Proc., 633 (2001).

2. G.U. Sumanasekera et al. Phys. Rev. Lett., 85, 5 (2000).

3. J. Kong et al. Science, 287, 622 (2000).

4. S.J. Tans, A.R. M. Verschueren, and C. Dekker, Nature, 393, 49 (1998).

5. T. Soh et al. Appl. Phys. Lett., 75, 627 (1999)

6. J. Kong et al. Nature, 395, 878 (1998).

7. M.S. Dresselhaus, G. Dresselhaus, and P.D. Eklund, *Science of Fullerenes and Carbon Nanotubes*, Academic Press, San Diego (1996).

8. A. Cheng and W.A. Steele, J. Chem. Phys., 92, 3867 (1990).

9. P. Qi and H. Dai, Nano Lett., 3, 3, 347 (2003).

10. J. Li et al. Nano Lett., 3, 7, 929 (2003).

11. P. Nikolaev et al. Chem. Phys. Lett., 313, 91 (1999).

12. M. Cinke et al. Chem. Phys. Lett., 365, 69 (2002).

13. R. Chen et al. Appl. Phys. Lett., 79, 6951 (2001).

14. H. Martins and T. Naes, *Multivariate Calibration*, Wiley, New York (1998).

15. L.A. Currie, Pure Appl. Chem., 67, 1699 (1995).

16. T.K.H. Starke and G.S.V. Coles, IEEE Sensor J., 2, 14 (2002).

17. B. Matthews et al. IEEE Sensor J., 2, 160 (2002).

18. J. Zhao et al. Nanotechnology, 13, 195 (2002).

19. J. Gardner and P.N. Bartlett, *Electronic Noses: Principles and Applications*, Oxford (1999).

20. S.J. Martin, G.C. Frye, and A.J. Ricco, Anal. Chem., 65, 2910 (1993).

21. J.F. Rosenbaum, *Bulk Acoustic Wave Theory and Devices*, Artech, Boston (1988).

22. S. Chopra et al. Appl. Phys. Lett., 80, 24, 4632 (2002).

23. K.G. Ong et al. IEEE Sensors J., 2, 2, 82 (2002).

24. S. Ramo, J.R. Whinnery, and T. Van Duzer, *Fields and Waves in Communication Electronics*, Wiley, New York (1984).

25. C.K.W. Adu et al. Chem. Phys. Lett., 337, 31–35 (2001).

26. G.U. Gollins et al. Meas. Sci. Technol., 11, 237 (2000).

27. R.D. Barnard, *Thermoelectricity in Metal and Alloys*, Wiley, New York (1972).

28. S. Ghosh, A.K. Sood, and N. Kumar, Science, 299, 1042–1044 (2003).

29. P. Kral and M. Shapiro, Phys. Rev. Lett., 86, 131 (2001).

30. J.R. Wood et al. J. Phys. Chem., B, 103, 10388–10392 (1999).

31. E.A. Grulke, in *Polymer Handbook*, 3rd ed., J. Brandrup and E.H. Immergut, Eds., Wiley, New York (1989).

32. J.H. Hildebrand et al. *The Solubility of Nonelectrolytes*, 3rd ed., Reinhold, New York (1948).

33. L.R.G. Treloar, *The Physics of Rubber Elasticity*, 3rd ed., Oxford University Press, Oxford (1975).

34. A. Krishnan et al. J. Phys. Rev. B, 58, 14013 (1998).

35. Q. Zhao, M.D. Frogley, and H.D. Wagner, Comp. Sci. Technol., 62, 147 (2002).

36. J.R. Wood, Q. Zhao, and H.D. Wagner, Composites, A32, 391 (2001).

37. R. Baughman, et al. Science, 284, 1340 (1999).

38. H. Gao, J. Mech. Phys. Solids, 1, 457 (1993).

39. P.D. Washabaugh and W.G. Knauss, Int. J. Fract., 65, 97 (1994).

40. J.A. Spudich, Nature, 372, 515 (1994).

41. R. Pelrine et al. Science, 287, 836 (2000).

42. J. Barisci and R. Baughman, Smart Mater. Struct., 12, 549 (2003).

43. V.T.S. Wong and W.J. Li, Micro Electro Mechanical Systems, 2003. MEMS-03 Kyoto. IEEE The Sixteenth Annual International Conference, 2003.

44. D.A. Scribner et al. Proc. IEEE, 79, 66 (1991).

45. J. Li et al. Appl. Phys. Lett., 75, 367 (1999).

46. Flavio Noca et al. Nanoscale Ears based on Artificial Stereocilia, The 140th Meeting of the Acoustical Society of America/NOISE-CON 2000 in Newport Beach, California.

10

Applications: Biosensors

Jun Li
NASA Ames Research Center

10.1 Introduction

Biomolecules such as nucleic acids and proteins carry important information of biological processes. The ability to measure extremely small amounts of specific biomarkers at molecular levels is highly desirable in biomedical research and healthcare. However, it is very challenging to find practical solutions to meet these needs. Current technologies rely on well-equipped central laboratories for molecular diagnosis, which is expensive and time consuming, often causing delay in medical treatments. There is a strong need for smaller, faster, cheaper, and simpler biosensors for molecular analysis [1]. The recent advancement in carbon nanotube (CNT) nanotechnologies has shown great potential in providing viable solutions. CNTs with well-defined nanoscale dimension and unique molecular structure can be used as bridges linking biomolecules to macro/micro- solid-state devices so that the information of bioevents can be transduced into measurable signals. Exciting new biosensing concepts and devices with extremely high sensitivities have been demonstrated using CNTs [2].

As the size of the materials reach the nanometer regime, approaching the size of biomolecules, they directly interact with individual biomolecules, in contrast to conventional macro- and micro-devices, which deal with assembly of relatively large amount of samples. Nanomaterials exhibit novel electronic, optical, and mechanical properties inherent with the nanoscale dimension. Such properties are more sensitive to the environment and target molecules in the samples. Although a big portion of nanomaterials are isotropic nanoparticles or thin films, high-aspect ratio one-dimensional nanomaterials such as CNTs and various inorganic nanowires (NWs) are more attractive as building blocks for device fabrication. The potential of CNTs and NWs as sensing elements and tools for biomolecular analysis as well as sensors for gases and small molecules have been recently recognized [3]. Promising results in improving sensitivity, lowering detection limit, reducing sample amount, and increasing detection speed have been reported using such nanosensors [4–6]. CNTs integrated with biological functionalities are expected to have great potential in future biomedical applications. This chapter summarizes the recent progress in the development of biological sensors using CNTs and highlights the potential future directions.

As described in previous chapters, CNTs are unique one-dimensional quantum wires with extremely high surface-to-volume ratio. As a result, their electronic properties are very sensitive to molecular adsorption. Particularly, in a semiconducting single-walled CNT (SWNT), all carbon atoms are exposed at the surface so that a small partial charge induced by chemisorption of gas molecules is enough to deplete the local charge carrier and cause dramatic conductance change [4,7–10]. Because biomolecules typically carry many ions, they are expected to affect CNT sensing elements and transducers more dramatically than are simple gases and small molecules [11,12]. Sensing devices have been fabricated for various applications using single CNTs [13–15], single semiconducting SWNT field-effect-transistors (FETs) [4,11,12,16], vertically aligned nanoelectrode arrays [6,17,18], and random networks or arrays [9,19–21]. Many studies have also reported complex nanostructure based on the hybrid bio-/nano-systems. These materials are heterogeneous assembly of biological molecules with solid-state nanomaterial, many of which use CNTs as templates for biomolecule assembly [22,23] or as conducting wires connecting biomolecules [24]. Such hybrid approaches have demonstrated the potential to combine the biorecognition functionalities with desired solid-state electronic properties in self-assembly of heterogeneous nano- and biosystems for biological sensing.

10.2   Fabrication of Carbon Nanotube Biosensors

10.2.1   CNT Growth

Even though much progress on CNT growth has been made in the past decade, it is still challenging to produce CNTs with desired properties for specific applications. Particularly, new methods are desired that can be directly integrated into device fabrication. High-temperature techniques such as electrical arc-discharge and laser ablation produce CNTs with the highest quality in terms of the graphitic structure. These techniques yield nanotubes along with other carbonaceous materials and metal catalysts. Individual CNTs were picked out of this mixture and placed on solid surfaces to fabricate devices such as FETs [25–27]. However, these techniques are slow and expensive, not applicable for mass-production of devices. Chemical vapor deposition (CVD) has provided a solution for the direct growth of CNTs in device fabrication at much lower temperature (about 850 to 1000°C). Wafer-scale SWNT FET arrays have been fabricated using CVD [28,29]. Incorporation of an electrical field was found to provide additional control of CNT growth along desired horizontal directions [30].

Plasma-enhanced CVD (PECVD), with the advantage of compatibility with semiconductor processing, has also attracted extensive attention in CNT growth for device fabrication [31–34]. A single CNT or an array of CNTs can be grown at sites defined by lithographic techniques down to tens of nanometers [35,36]. The alignment can be precisely controlled by an electrical field normal to the substrate surface. However, CNTs grown by PECVD are defective multiwalled nanotubes (MWNTs), or more appropriately referred to as multiwalled nanofibers (MWNFs) [37,38], (see Chapter 4) with the graphitic layers not perfectly parallel to the tube axis. Many bamboo-like closed shells are formed along the tube axis instead of presenting well-defined hollow channels running from one end all the way to the other end [32–34,36–38]. Such nanostructures, although not ideal, were found to be sufficient for many sensor applications [6,17,18]. With continued effort in the development of growth techniques, it is expected that CNTs with desired properties, quality, and quantity could be obtained for various applications.

10.2.2   Device Integration

An essential task for device fabrication is to integrate nanoscale CNTs into functional devices that can pass information to the macroscopic world. Lithography-based micro- and nanofabrication techniques provide methods to fill this gap. Figure 10.1 summarizes four types of most commonly used biological sensing devices based on CNTs. Depending on the sensing applications, different device architectures and fabrication routes are required to successfully achieve the desired functions. Common to all, CNTs are the critical components of the sensing devices, which are integrated either directly or indirectly during the fabrication route. A variety of methods, ranging from advanced micro- or nanolithographic techniques to manual placing, has been used to date. Generally, CNTs are either the sensing elements whose

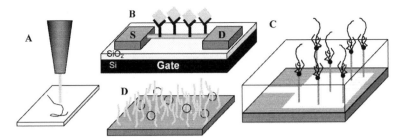

FIGURE 10.1 Schematic of four types of CNT devices for biological sensing. (A) A single CNT attached to a microtip to probe a single cell or single molecule, (B) a filed-effect-transistor device using a single semiconducting CNT, (C) a random CNT network or three-dimensional porous film as an electrochemical biosensor, and (D) a nanoelectrode array based electrochemical biosensor.

properties are changed upon specific biological events, or the transducers reduce that transfer the signal to the measuring units. The sensing device could use either a single CNT or an ensemble of CNTs.

The most straightforward biological sensing employs a single CNT to probe the biochemical environment in a single living cell or interrogate a single biomolecule (as shown in Figure 10.1A). The CNT probe can be attached to the pulled tip of an electrode [39] for electrical, electrochemical, and electrophysiology measurements. Normally, for such applications, the sidewalls of the CNT probe have to be shielded or insulated to reduce the background so that the small signal from the very end of the probe can be detected. The CNT probes provide the best spatial resolution as well as ultrahigh sensitivity and faster measuring speed than do conventional probes. The small size, high aspect ratio, and mechanical robustness of CNTs are also employed as the physical probe for high-resolution scanning-probe microscopy (SPM) as discussed in Chapter 6. This is a powerful technique that illustrates the structure of single molecules such as DNA and proteins with the resolution down to a few nanometers.

A single semiconducting CNT can be used to construct an FET as shown in Figure 10.1B using lithographic techniques. Such a device consists of a semiconducting CNT connected to two contact electrodes (source and drain) on an oxide-covered Si substrate, which could serve as the gate electrode [4,25–27]. Conventional FETs fabricated with semiconductors such as Si have been configured as sensors by modifying the gate oxide (without the gate electrode) with molecular receptors or ion-selective membranes for the analytes of interest [40]. The binding or adsorption of charged species could produce an electric field, which depletes or accumulates carriers within the semiconducting material similar to that by the gate potential. As a result, the conductance between the source and the drain electrodes is dramatically changed. Chemical sensors based on such a mechanism are referred to as chemical field-effect-transistors (ChemFET), which have been widely used in many applications [40].

The drawback of FET is the high cost of fabrication. Recent studies have demonstrated that random networks of SWNTs between two microelectrodes can behave as a thin film transistor and may be used as sensors [21]. Similar random CNT networks [34,41–43] and vertical CNT arrays [19,20] on metal electrodes can also be used as biosensors as shown in Figure 10.1C. The large surface-to-volume ratio and good electrical conductance along the tube axis make CNTs attractive for enzyme-based EC sensors. The highly porous CNT networks and arrays serve both as large immobilization matrices and as mediators to improve the electron transfer between the active enzyme site and the EC transducer. Improved electrochemical behavior of NADH [43], neurotransmitters [9], and enzymes [42] has been reported.

The fourth type of CNT device is also an EC sensor based on an array of vertically aligned CNTs embedded in SiO_2 matrices as schematically shown in Figure 10.1D. The EC signal is characteristic of the reduction/oxidation (redox) reaction of the analytes instead of nonspecific charges sensed by FETs, resulting in high specificity comparable with fluorescence-based optical techniques that are commonly used in today's biology research. In addition, a high degree of miniaturization and multiplex detection can be realized for molecular diagnosis using an individually addressed microelectrode array integrated with microelectronics and microfluidics systems [1,44–46], which has advantages over optical techniques, particularly for field applications requiring quick and simple measurements. However, the sensitivity of EC techniques using traditional macro- and microelectrodes is orders of magnitude lower than laser-based fluorescence techniques, limiting their applications. Nanoscale-sensing elements such as CNT nanoelectrode arrays have been actively pursued to seek solutions for improving sensitivity of EC techniques.

The performance of electrodes with respect to temporal and spatial resolution is known to scale inversely with the electrode radius [47–49]. It is of interest in biosensing to reduce the radius of electrodes to 10 to 100 nm, approaching the size of biomolecules. It has been demonstrated that an MWNT array, with an average diameter of ~30 to 100 nm, can be integrated into a nanoelectrode array for ultrasensitive chemical and DNA detection [6,17,18]. The nanoelectrode array is fabricated with a bottom-up scheme resulting in a precisely positioned and well-aligned MWNT array embedded in a planarized SiO_2 matrix [36,50]. MWNT arrays are first grown on metal films deposited on a Si surface using PECVD and then subjected to tetraethyloxysilicate (TEOS) CVD for gap-filling with SiO_2. The excess SiO_2 can be subsequently removed allowing exposure of the MWNT tips via a mechanical polishing step. The open ends of MWNTs exposed at the dielectric surface act as nanoelectrodes. Figure 10.2 shows scanning electron microscopy (SEM) images of a CNT array grown on 3×3 multiplex microelectrodes. Each microelectrode is about 200×200 µm^2 and individually addressable. The location of CNTs can be controlled by the catalyst spots defined by UV or e-beam lithography. If the catalyst spot is less than 100 nm, a single CNT can be grown from each spot as shown in Figure 10.2D. Otherwise, a bundle of CNTs are formed as shown in Figure 10.2C. The dielectric encapsulation and polishing procedures ensure that only the very end of CNTs is exposed whereas the sidewall is insulated as shown in Figure 10.2E and F. Such nanoelectrode array has shown characteristic electrochemical behavior for redox species both in bulk solution and immobilized at the CNT ends. Dramatic improvements in sensitivity and time constant were reported in References 6, 17, and 18.

10.2.3 Biofunctionalization

A common feature for biological sensing is the requirement of immobilization of biomolecules with specific functionalities on the sensing device. These biomolecules serve as probes to either specifically bind particular species in the testing sample or catalyze the reaction of a specific analyte. Such an event produces a change in chemical or physical properties that can be converted into a measurable signal by the transducer. The specific recognition of the target molecules is the essential feature for biological sensing. The common probe and target (analyte) recognition mechanisms include (a) antibody/antigen interactions, (b) nucleic acid hybridizations, (c) enzymatic reactions, and (d) cellular interactions. Depending on the device and its sensing mechanism, different functionalization methods have to be adapted. Current functionalization methods can be divided into two categories: (a) covalent binding to the open ends of CNTs and (b) covalent and noncovalent binding to the sidewall of CNTs.

CNTs, from a structural point of view, are very similar to a roll of graphitic sheets. The sidewalls have very inert chemical properties similar to graphite basal planes. The open ends of CNTs, on the other hand, are similar to graphite edge planes, which are much more reactive due to the dangling sp^2 bonds [51,52]. For measuring chemical force of single molecules with CNT SPM tips [14] or using CNT nanoelectrodes for biosensing, the open end of the CNTs needs to be functionalized. Wong et al. [53] demonstrated that the open end of a SWNT is rich with –COOH group, which could be used for selective covalent bonding of primary amine molecules through amide bonds facilitated by the coupling reagents N-hydroxysuccinimide (NHS) sometimes aided with dicyclohexylcarbodiimide (DCC) [14,53]. Williams et al. [24] used similar methods to functionalize the open end of a SWNT with a peptide nucleic acid (PNA) with the sequence NH_2-Glu-GTGCTCATGGTG-$CONH_2$, where Glu is a glutamate amino-acid residue and the central block represents nucleic-acid bases. The primary amine terminated DNA oligo-probes can also be covalently functionalized to the open ends of a MWNT array embedded in SiO_2 matrix by similar carbodiimide chemistry using water-soluble coupling reagents 1-ethyl-3(3-dimethyl amino-propyl carbodiimide hydrochloride (EDC) and N-hydroxysulfo-succinimide (sulfo-NHS) [54].

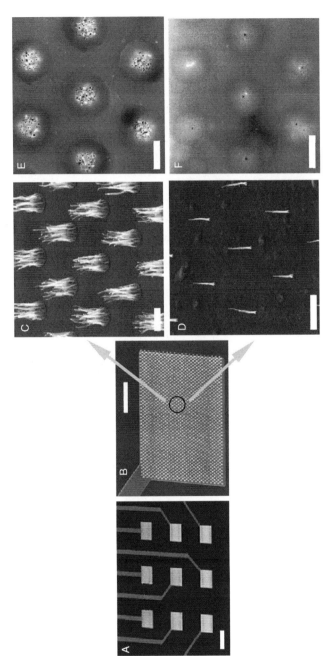

FIGURE 10.2 SEM images of (A) a 3 × 3 electrode array; (B) array of MWNT bundles on one of the microelectrode pads; (C) and (D) array of MWNTs at UV-lithography and e-beam patterned Ni catalyst spots, respectively; (E) and (F) the surface of polished MWNT nanoelectrode array grown on 2 µm and 100 nm spots, respectively [6]. (A) to (D) are 45° perspective views and (E) to (F) are top views. The scale bars are 200, 50, 2, 5, 2, and 2 µm, respectively.

For sensors using the FET configuration, functionalization of the sidewall of CNTs is required [11,12,55,56]. Because semiconducting SWNTs are used as the conducting channels whose electronic properties are monitored upon the binding of charged target molecules at the surface, the graphitic sp^2 sidewall structure has to be preserved to maintain its inherent properties. Such a sidewall structure is strongly hydrophobic and chemically inert, which raises problems in biocompatibility and biofunctionalization for specific recognition. It has been reported that proteins such as streptavidin and HupR can adsorb strongly onto the MWNT surface presumably via hydrophobic interactions between the aromatic CNT surface and the hydrophobic domains of the proteins [22]. A designed amphiphilic α-helical peptide has been found to spontaneously assemble onto the SWNT surface in aqueous solution with the hydrophobic face of the helix noncovalently interacting with the CNT surface and the hydrophilic amino acid side chains extending outwards from the exterior surface [23]. Chen et al. [55] reported a noncovalent sidewall functionalization scheme whereby a variety of proteins were immobilized on SWNTs functionalized by π-stacking of the conjugate pyrenyl group of 1-pyrenebutanoic acid succinimidyl ester. The succinimidyl ester group reacts with amine groups on lysine residues of proteins to form covalent amide linkages. However, all such noncovalent interaction-based immobilization methods lack specificity, particularly the direct nonspecific binding of proteins to CNTs needs to be suppressed for biological sensing. Extensive washing using conventional protocols was not sufficient to remove such nonspecifically bound proteins. For this purpose, a surfactant (Triton-X 100) was coadsorbed on the CNT surface with poly (ethylene glycol) (PEG) [56]. Such a coating was found to be effective in resisting nonspecific adsorption of streptavidin. Amine-terminated PEG can be used so that the biotin moiety can be added to the PEG chains through covalently linking with an amine-reactive biotin reagent, biotinamidocaproic acid 3-sulfo-*N*-hydroxysuccinimide ester [56]. The functionalized CNTs have demonstrated specific recognition to streptavidin. A similar method using mixed polyethylene imine (PEI) and PEG coating was reported for functionalizing biotin to the sidewall of the SWNT in FET devices [11]. These methods can be extended to the recognition of other biomolecules based on specific interactions of antibody-antigen and complimentary DNA strands.

Besides the application in FET-sensing devices, the biofunctionalized CNTs also show much better solubility so that further chemistry or biochemistry can be applied. In addition, the biorecognition can be employed to regulate the assembly of supramolecular structures, which may lead to new sensing devices. Because the integrity of the CNT structure is not as critical in such applications as in FETs, covalent sidewall functionalizations are also actively pursued by directly attaching functional groups to the graphitic surface such as fluorination and hydrogenation or using carboxylic acid groups at the defect sites to form amide or ester linkages [57]. Such sidewall functionalizations may particularly be applicable to MWNT sensors in which CNTs serve only as the probe materials to transduce signals from the measured molecules, and their own electronic properties are insensitive to the environment. For example, a MWNT electrode functionalized with antibody or enzyme can be used as a single EC probe to study biochemistry in a single cell [39]. Sidewall functionalization of the MWNT array can also greatly increase the enzyme (glucose oxidase) loading in electrochemical glucose sensors [20]. Some other studies on enzyme-based sensors using CNT-casted glassy carbon electrodes have demonstrated that even spontaneous adsorption or polymer wrapping can improve the enzyme loading and improve the electron transfer [41–43]. From the biological side, peptides with selective affinity for carbon nanotubes were recently reported [58], which could lead to new methods for bio-/nanointegration.

10.3   Biosensing Applications

10.3.1   Single-Cell and Single-Molecule Sensors

The most direct application of CNTs for biological sensing is to use them as single probes to gain great spatial resolution. With the small size, such probes can be inserted into a single cell for *in situ* measurements with minimum disturbance and ultrahigh sensitivity. Vo-Dinh et al. [59] demonstrated an antibody-based nanobiosensor for the detection of benzo[*a*] pyrene tetrol (BPT), a biomarker for human

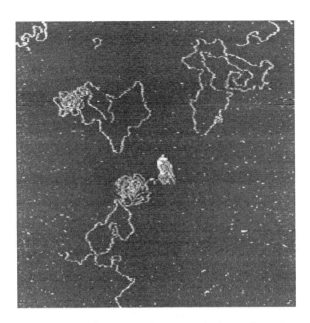

FIGURE 10.3 An SPM image of DNA molecules on $2.3 \times 2.3\ \mu m^2$ mica surface submerged under 1 mM MgCl$_2$ buffer solution obtained with a single MWNT tip [15].

exposure to the known carcinogen benzo[a] pyrene (BaP) by simply pulling an optical fiber to nanometer size at the tip. The distal end was covalently coated with anti-BPT antibodies through silane linkers. The nanobiosensors were inserted into individual cells, incubated 5 minutes to allow antigen-antibody binding, and then removed for the detection. Such a nanobiosensor, based on fluorescence spectroscopy, shows a sensitivity down to $\sim 1.0 \times 10^{-10}\ M$ for BPT [59,60] and an absolute limit of detection for BPT of ~ 300 zeptomoles (10^{-21} mole) [61]. A single MWNT nanoelectrode probe [39] can be adapted using similar techniques to study the electrophysiology phenomenon in a single cell or the reactivity of a single molecule. The electrical signal has the potential to reach single molecular sensitivity [49].

Single CNTs attached to an SPM tip have also attracted intensive interests, as discussed in Chapter 6, due to their small diameter, high aspect ratio, large Young's modulus, and mechanical resilience [13]. They can be used as physical probes to obtain high-resolution image of macromolecules or cell surfaces. Li et al. [15] demonstrated the method for implementing and characterizing CNTs as SPM tips for *in situ* imaging of DNA molecules on mica surface within a buffer solution. A magnetically driven oscillating probe in an atomic force microscope (AFM) with a silicon nitride cantilever (spring constant of $k = 0.1$ N/m) was used at a frequency of ~ 30 kHz. A bundle of MWNTs was attached to the pyramidal tip with an acrylic adhesive. The diameter of the bundles ranged from tens to hundreds of nanometers. Typically, a single MWNT extended out and was used as a probe for AFM imaging. Lambda DNA molecules were spontaneously bound to the surface from 2 μg/ml solutions with the presence of ~ 1 mM MgCl$_2$ for enhancing the DNA/mica interaction. Figure 10.3 shows an AFM image of DNA molecules in a 2.3×2.3 μm^2 area. Single DNA molecules were clearly resolved. The resolution of DNA images is very uniform and consistent with the diameter of the CNT tip (~ 5 nm in diameter). SWNT tips could provide even higher resolution [14,53], and the double-helix structure of DNA may be resolved using SWNT tips. The tip can also be functionalized to provide additional chemical force information [14,53].

Woolley et al. [62] reported the direct haplotyping of kilobase-size DNA using SWNT AFM probes. The haplotype of a subject — the specific alleles associated with each chromosome homolog — is a critical element in single nucleotide polymorphisms (SNPs) mapping that leads to a greater comprehension of the genetic contribution to risk for common diseases such as cancer and heart disease. However, the current methods for determining haplotypes have significant limitations that have prevented their use in large-scale genetic screening. For example, molecular techniques for determining haplotypes, such as allele-specific or single-molecule polymerase chain reaction (PCR) amplification, are hampered by the need to optimize stringent reaction conditions and the potential for significant error rates. Using AFM with high-resolution SWNT probes, multiple polymorphic sites can be directly visualized by hybridizing specifically labeled oligonucleotides with the template DNA fragments of ~100 to 10,000 bases. The positions of streptavidin and IRD800 labels at two sequences in M13mp18 were demonstrated. The SWNT tips, with tip radii less than 3 nm (~10 base resolution), made it possible for high-resolution multiplex detection to differentiate different labels such as streptavidin and the fluorophore IRD800 based on their size. This concept has been further applied for the determination of haplotypes on the UGT1A7 gene [62], which is studied for its role in cancer epidemiology. The direct haplotyping using SWNT AFM probes represents a significant advance over conventional approaches and could facilitate the use of SNPs for association and linkage studies of inherited diseases and genetic risk [63].

The electronic properties of biomolecules such as DNAs have been extensively studied in the past few years due to their potential for single molecular sensing. However, DNA molecules were found not to be very conductive even in the duplex form. There are also difficulties in assembling them into reliable devices. Williams et al. [24] reported a study to couple SWNTs covalently with PNA, an uncharged DNA analogue. The hybridization of DNA fragments to the PNA sequence was measured with an AFM under ambient conditions and indicated that the functionalization is specific to the open ends of SWNTs with rare attachment to the sidewall. PNA was chosen due to its chemical and biological stability. The uncharged PNA backbone gives rise to PNA–DNA duplexes that are more thermally stable than their DNA–DNA counterparts because there is no electrostatic repulsion. This method unites the unique properties of SWNTs with the specific molecular recognition features of DNA, which may provide new means to incorporate SWNTs into larger electronic devices by recognition-based assembly as well as electronic biological sensing.

10.3.2 FET-Based Biosensors

The extreme high sensitivity and potential for fabricating high-density sensor array make nanoscale FETs very attractive for biosensing, particularly because biomolecules such as DNA and proteins are heavily charged under normal conditions. SWNT FETs are expected to be more sensitive to the binding of such charged species than chemisorbed gas molecules. However, the wet chemical environment with the presence of various ions and presence of other biomolecules makes it much more complicated than gas sensors. Extensive efforts have been made to understand the fundamental issues of SWNT FET in wet environments.

First, the behavior of SWNT FETs has to be well understood in ambient environment. Several studies indicated that SWNT FETs fabricated on SiO_2/Si substrates with the structure similar to that in Figure 10.2B exhibit hysteresis in current versus gate voltage. This is not desired for electronics applications but is the nature of nanoscale chemical sensors relying on the interaction between SWNT and molecular species in the environment. The hysteresis was attributed to charge traps either in SiO_2 film or SWNTs [64–66]. However, Kim et al. [67] found that the major cause can be attributed to water molecules, especially the strong SiO_2 surface-bound species, which are difficult to remove even by pumping in vacuum. Passivating the SWNT FET with poly(methyl methacrylate) (PMMA) can nearly remove the hysteresis. The choice of passivation polymers and the humidity in the environment can strongly affect electronics properties of SWNT FETs. Bradley et al. [68] coated the FET device with a charged polymer and found that the hysteresis can be used as a humidity sensor. The mobile ions in the hydrated polymer film were suspected to be responsible for the hysteresis.

If water adsorption and mobile ions can significantly affect the performance of SWNT FETs, the measurement of the binding of charged biomolecules on SWNT FETs in liquid environments is even more challenging. Star et al. [11] explored the device response in a buffer and found that it is essentially obscured by the abundant ions in the enviroment. So far, not much success has been reported in using SWNT FETs for biosensing in practical biological environments. To get around this problem, the SWNT FET after incubation was washed and dried before characterization in an ambient environment [11].

The device was indeed effective for specific biotin-streptavidin binding. A mixed PEI and PEG polymer coating has been used to avoid nonspecific binding. Biotin molecules were attached to the polymer layer for specific molecular recognition. The SWNT FET device coated with PEI/PEG polymer alone changes from p-type to n-type device characteristics probably due to the electron-donating properties of of the NH_2 groups of the polymer. The attachment of biotin molecules through covalent binding to the primary NH_2 group reduced the overall electron-donating function of PEI and convert the device back to p-type characteristics. Finally, the specific binding of streptavidin with biotin was found to almost totally remove the gating effect and cause dramatic decrease in source-drain current at negative gate voltages. Nonspecific binding was observed in devices without the polymer coating, whereas no binding was found for polymer-coated but not biotinylated devices. Streptavidin, in which the biotin-binding sites were blocked by reaction with excess biotin, produced essentially no change in device characteristic of the biotinylated polymer-coated devices.

Despite the complexity, Rosenblatt et al. [16] successfully demonstrated a high-performance electrolyte-gated SWNT FET. The uncoated device was submerged in a drop of water solution containing 10 mM NaCl. A water gate voltage V_{wg} was applied to the droplet through a silver wire. Such electrolyte-gated SWNT FET showed p-type characteristics with high device mobilities and transconductances, very promising for biosensing applications. Interestingly, the electrolyte gating approaches the ultimate limit where the capacitance is governed by quantum effects and not electrostatics.

Recently, Besteman et al. [12] realized that proteins carry pH-dependent charged groups that could electrostatistically gate a semiconducting SWNT, creating the possibility of constructing a nanosize protein or pH FET sensors. Even more interesting, redox enzymes go through a catalytic reaction cycle where groups in the enzyme temporarily change their charge state resulting in conformational changes in the enzyme. They successfully demonstrated that the enzyme-coated SWNT FETs can be used as single-molecule biosensors. As shown in Figure 10.4A, the redox enzyme glucose oxidase (GOx) is immobilized on SWNT using a linking molecule, which on one side binds to the SWNT through van der Waals coupling with a pyrene group and on the other side covalently binds the enzyme through an amide bond as developed by Chen et al. [55]. The FET preserves the p-type characteristic but shows much lower conductance upon GOx immobilization, which is likely the result of the decrease in the capacitance of the tube due to GOx immobilization because GOx blocks the liquid from access to the SWNT surface.

The GOx-coated SWNT did show a strong pH dependence as well as high sensitivity to glucose. Figure 10.4B shows the real-time measurements where the conductance of a GOx-coated SWNT FET has been recorded as a function of time in milli-Q water. No significant change in conductance is observed when more milli-Q water is added as indicated by the first arrow. However, when 0.1 M glucose is added, the conductance increases by 10% as indicated by the second arrow. The inset (a) of Figure 10.4B shows the similar behavior repeated with another GOx-coated device in contrast to no change with a bare device shown in inset (b). Clearly, GOx activity is responsible for the measured increase in conductance upon the addition of glucose.

Both studies on liquid gated FETs use very low ionic strength with 10 mM NaCl [16] and 0.1 mM KCl [12], respectively. The salt concentrations are more than 10 times lower than physiology buffers. Experiments are needed to explore the measurements under more practical physiological buffers to confirm whether the high ionic concentration will obscure the response. This is critical for further biosensor applications.

10.3.3 Nanoelectrode Array–Based Electronic Chips

An embedded CNT array minimizes the background from the sidewalls whereas the well-defined graphitic chemistry at the exposed open ends allows the selective functionalization of –COOH groups with primary amine terminated oligonucleotide probes through amide bonds. The wide electropotential window of carbon makes it possible to directly measure the oxidation signal of guanine bases immobilized at the electrode surface. Such nanoelectrode array can be used as ultrasensitive DNA sensors based on an electrochemical platform [6,17,18]. As shown in Figure 10.1D and Figure 10.5A (see color insert following

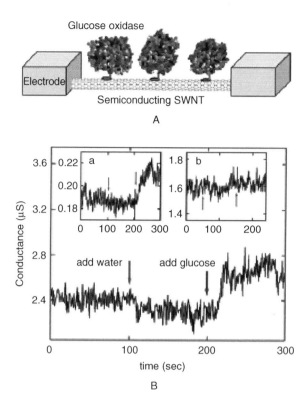

FIGURE 10.4 (A) Schematic of an SWNT FET biosensor with GOx immobilized on the SWNT surface. (B) Real-time electronic response of the SWNT sensor to glucose, the substrate of GOx. The conductance is measured in 5 μl-milli-Q water. The source-drain and liquid-gate voltage is kept constant at 9.1 mV and –500 mV, respectively. At the time indicated by the first arrow, 2-μl milli-Q water is added to the solution. After a while (at the second arraow), 2 μl 0.1 mM glucose in milli-Q water is added. Inset (a) shows experiment repeated with a second device and inset (b) is the same experiment on a device without GOx. (Reprinted from Besteman et al. Nano Lett., 3 (6), 727–730 (2003). With permission.)

page 146), oligonucleotide probes of 18 bases with a sequence of [Cy3]5-CTIIATTTCICAIITCCT-3[AmC7-Q] are covalently attached to the open end of MWNTs exposed at the SiO$_2$ surface. This sequence is related to the wild-type allele (Arg1443stop) of BRCA1 gene [69]. The guanine bases in the probe molecules are replaced with nonelectroactive inosine bases, which have the same base-pairing properties as guanine bases. The oligonucleotide target molecule has a complimentary sequence [Cy5]5-AGGAC-CTGCGAAATCCAGGGGGGGGGGG-3 including a 10 mer polyG as the signal moieties. Hybridization was carried out at 40°C for about 1 hour in ~100 nM target solution in 3 × SSC buffer. Rigorous washing in three steps using 3 × SSC, 2 × SSC with 0.1% SDS, and 1 × SSC respectively at 40°C for 15 minutes after each probe functionalization and target hybridization process was applied in order to get rid of nonspecifically bound DNA molecules, which is critical for getting reliable electrochemical data.

Such solid-state nanoelectrode arrays have great advantages in stability and processing reliability over other electrochemical DNA sensors based on mixed self-assembled monolayers of small organic molecules. The density of nanoelectrodes can be controlled precisely using lithographic techniques, which in turn define the number of probe molecules. The detection limit can be optimized by lowering the nanoelectrode density. However, the electrochemical signal is defined by the number of electrons that can be transferred between the electrode and the analytes. Particularly, the guanine oxidation occurs at rather high potential (~1.05 V vs. saturated calomel electrode [SCE]) at which a high background is produced by carbon oxidation and water electrolysis. This problem can be solved by introducing Ru(bpy)$_3^{2+}$ mediators to amplify the signal based on an electrocatalytic mechanism [70]. Combining the

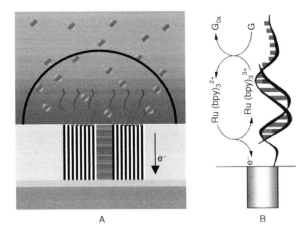

FIGURE 10.5 Schematic of the MWNT nanoelectrode array combined with Ru(bpy)$_3^{2+}$ mediated guanine oxidation for ultrasensitive DNA detection [18]. [**See color insert following page 146.**]

MWNT nanoelectrode array with Ru(bpy)$_3^{2+}$ mediated guanine oxidation (as schematically shown in Figure 10.5B), the hybridization of less than ~1000 oligonucleotide targets can be detected with a 20 × 20 µm² electrode, with orders of magnitude improvement in sensitivity compared with previous EC-based DNA detections [6,17].

Figure 10.6A shows three consecutive AC voltammetry (ACV) scans in 5.0 mM Ru(bpy)$_3^{2+}$ in 0.20 M NaOAc buffer solutions after hybridizing the polyG tagged BRCA1 targets on a MWNT array electrode (with average tube-tube spacing of ~1.5 µm). The AC current is measured by applying a sinusoidal wave of 10 Hz frequency and 25 mV amplitude on a staircase potential ramp. Well-defined peaks are observed around 1.04 V with the first scan clearly higher than the almost superimposed subsequent scans. The background is almost a flat line at zero. The peak current at ~1.04 V consisting of two parts with:

$$I_p = I_{mediator} + I_{amplifiedG}, \qquad (10.1)$$

where $I_{mediator}$ is the oxidation current of bulk Ru(bpy)$_3^{2+}$ mediators coincidently superimposed with the mediator amplified oxidation current of guanine bases $I_{amplifiedG}$ at almost the same potential. Due to the electrocatalytical mechanism, the signal of guanine is much larger than that from guanine bases alone (I_G), i.e.,

$$I_{amplifiedG} = N \times I_G \gg I_G, \qquad (10.2)$$

where N is the amplification factor (normally >5). However, guanine oxidation is irreversible and contributes only in the first scan, and the current in subsequent scans mainly corresponds to $I_{mediator}$. As a result, the net signal associated with guanine bases can be derived as

$$I_{p1} - I_{p2} = I_{amplifiedG} \propto [G]. \qquad (10.3)$$

As shown in Figure 10.6B, subtracting the second scan from the first one gives a well-defined positive peak (continuous line) whereas subtracting the third scan from the second gives a much smaller negative peak (dotted line). The difference between further scans is almost a flat line at zero. The high quality of the data indicates that there is still plenty of room to lower the detection limit of target DNAs.

In practical applications, the target DNA typically consists of over hundreds of bases, with a large quantity of guanine bases in each target molecule. The Ru(bpy)$_3^{2+}$ mediators can efficiently transport electrons between guanine bases and the CNT electrode. As a result, all inherent guanine bases in the DNA target that dangle in the hemispherical diffusion layer of the Ru(bpy)$_3^{2+}$ mediators can efficiently

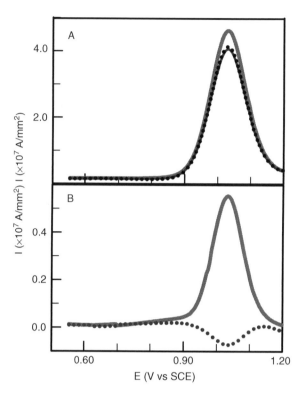

FIGURE 10.6 (A) Three consecutive AC voltammetry measurements of the low-density MWCNT array electrode functionalized with oligonucleotide probes with the sequence [Cy3]5-CTIIATTTCICAIITCCT-3[AmC7-Q] and hybridized with oligonucleotide targets with the sequence [Cy5]5-AGGACCTGCGAAATCCAGGGGGGGGGGGG-3 [6]. The thick, thin, and dotted lines correspond to the first, second, and third scan, respectively. The measurements were carried out in 5 mM Ru(bpy)$_3^{2+}$ in 0.20 M NaOAC supporting electrolyte (at pH=4.8) with an AC sinusoidal wave of 10 Hz and 25 mV amplitude on top of a staircase DC ramp. (B) The difference between the first and the second scans (solid line) and between the second and the third scans (dotted line) [6]. The positive peak corresponds to the increase in Ru(bpy)$_3^{2+}$ oxidation signal due to the guanine bases on the surface. The negative peak serves as a control representing the behavior of a bare electrode.

serve as signal moieties, resulting in a large signal compared with the techniques using redox tags [45]. Thus the detection limit can be lowered to less than 1000 target molecules, approaching the limit of laser-based fluorescence techniques [17]. The use of inherent guanine bases as signal moieties makes it possible to skip the expensive and time-consuming labeling procedure. Other advantages of EC detection such as the ability to apply extra stringent control using local electrical field could be realized with this system. It is also applicable in enzymatic biosensors for pathogen detection by immobilizing proteins such as enzymes and antibodies at the electrodes.

10.3.4 Nanonetworks and Thin Films

Besides being used as building elements in well-defined devices, both SWNTs and MWNTs can be cast as a random network or thin film on conventional electrodes [41–43] or used as a three-dimensional porous film [9,19,20]. CNTs serve both as large immobilization matrices and as mediators to improve the electron transfer between the active enzyme site and the electrochemical transducer. CNT-modified glassy carbon electrodes exhibit a substantial (~490 mV) decrease in the overpotential for -Nicotinamide adenine dinucleotide (NADH) oxidation. Various enzymes such as GOx and flavin adenine dinucleotide (FAD) can spontaneously adsorb onto CNT surface and maintain their substrate-specific enzyme activity over prolonged time [41]. Biosensors based on enzymes that catalyze important biological redox reactions (such as glucose oxidation) can be developed.

10.3.5 CNT Templated Bioassembly

Besides being used as building blocks in device fabrication, CNTs have also attracted extensive interest as nanoscale templates for self-assembly of CNT-biomolecular complex. Such bio-/nanomolecular assembly could incorporate the specific recognition properties of biomolecules with desired electronic properties of CNTs to develop novel biosensors and bioelectronic materials. SWNT-PNA hybride molecular wires by covalently functionalizing PNA fragments to the open end of SWNTs was described in Section 10.3.1. Several studies have also been carried out to investigate the self-assembly processes driven by the hydrophobic forces both at the inner and outer surfaces of CNTs.

The phenomenon of molecules inserting into the confined space in the inner channel of CNTs, particularly in liquid environments, is of great fundamental interest. Even though it is a very challenging problem for experimental exploration due to the lack of proper characterization techniques, computer simulation studies did show interesting results. Recent molecular dynamics simulations demonstrated that SWNTs can be used as molecular channels for water transport [71]. Gao et al. [72] further demonstrated that a DNA molecule could be spontaneously inserted into a SWNT in a water solute environment. The van der Waals and hydrophobic forces were found to be important for the insertion process, with the former playing a more dominant role in the DNA-CNT interaction. It suggests that the encapsulated CNT-DNA molecular complex can be exploited for applications such as DNA modulated molecular electronics, molecular sensors, electronic DNA sequencing, and nanotechnology of gene delivery systems.

On the other hand, the external CNT surface has hydrophobic properties similar to graphite surface, which is known to be a model substrate to study molecular self-assembly. The decoration of a CNT surface with self-assembly of biological molecules has been demonstrated in various systems including lipids [73], oligonucleotides [74], proteins [22,55], and peptides [23], etc. Richard et al. [73] found that both sodium dodecyl sulfate (SDS) and synthetic lipids form supramolecular structures made of rolled-up half-cylinders on the nanotube surface. Depending on the symmetry and the diameter of the CNT, it can form rings, helices, or double helices. Permanent assemblies can be produced from the mixed micelles of SDS and different water-insoluble double-chain lipids after dialysis of the surfactant. Biomolecules such as histidine-tagged proteins can be immobilized on such lipid-decorated CNTs for the development of new biosensors and bioelectronic materials.

For sensor applications, biomolecules with specific recognition functions such as proteins can also be directly packed on the external surface of CNTs, but they have to remain functional. A good criterion for the conservation of the functional properties of the protein is its ability to form ordered arrays. Balavoine et al. [22] reported a study of streptavidin assembly on CNTs to form helical crystalization. Streptavidin is very useful in many biochemical assays, such as labeling and affinity chromatography due to its high affinity for (+) biotin (Ka ~ 10^{15}). The assembly was carried out in solutions by spontaneous adsorption. MWNTs were prepared by the arc-discharge method and stored as a suspension in methanol (~2 mg/ml). A 100-μl aliquot of MWNT suspension was dried under an ethane gas flow and resuspended in 20 ml of a 40% aqueous solution of methanol. This suspension was sonicated to disperse the MWNTs before the addition of 20 μl streptavidin solution (~10 μg/ml) in a buffer containing 10 mM Tris (pH = 8) and 50 mM NaCl, and allowed to stand at room temperature for 45 minutes. Such a sample was deposited on carbon film covered grid and was negatively stained with a 2% uranyl acetate solution for transmission electron microscope imaging. In appropriate conditions, the MWNT surface was found almost completely covered with streptavidin, presumably due to the interaction with its hydrophobic domains. Even though most assemblies are disordered, some regular-spaced helical structures were observed at proper conditions. Another water-soluble protein, HupR, was also studied and showed ordered arrays on a wider range of MWNT diameters than streptavidin.

A 29-residue amphiphilic α-helical peptide was also specifically designed to coat and solubilize CNTs as well as control the assembly of the peptide-coated CNTs into macromolecular structures through

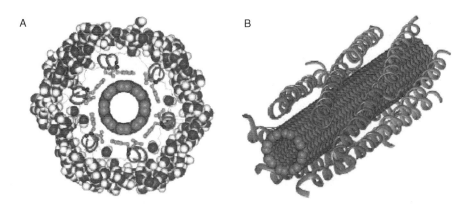

FIGURE 10.7 Model of the amphiphilic peptide helices assembled on a SWNT surface. (A) Cross-section view showing six peptide helices wrapped around a SWNT. The backbone of each peptide is denoted by a helical ribbon. The hydrophobic Val and Phe side chains are packed against the SWCNT surface. A 5-angstrom thick water shell was used in the energy refinement. (B) View of peptide-wrapped SWCNT with 12 peptide helices. The peptide formed two layers with head-to-tail alignment. [**See color insert following page 146.**] (Reprinted with permission from G.R. Dieckmann et al. J. Am. Chem. Soc. 125 (7), 1770–1777, 2003. Copyright 2003, American Chemical Society.)

peptide-peptide interactions [23]. Figure 10.7 shows the cross-sectional view of the molecular structure and the perspective view of the helical backbones of the model illustrating the assembly of such molecules on a SWNT surface. Six such α-helices are sufficient to surround the circumference of an individual SWNT while maintaining typical interhelical interactions. The hydrophobic face of the helix with apolar amino acid side chains (Val and Phe) presumably interacts noncovalently with the aromatic surface of CNTs, and the hydrophilic face extends outward to promote self-assembly through charged peptide-peptide interactions. Electron microscopy and polarized Raman studies reveal that the peptide-coated CNTs assemble into fibers with CNTs aligned along the fiber axis. The size and morphology of the fibers can be controlled by manipulating solution conditions that affect peptide-peptide interactions.

Whereas the above-mentioned studies are based on nonspecific adsorptions, Wang et al. [58] have used phage display to identify peptides with selective affinity for CNTs. Binding specificity has been confirmed by demonstrating direct attachment of nanotubes to phage and free peptides immobilized on microspheres. Consensus binding sequences show a motif rich in histidine and tryptophan at specific locations. The analysis of peptide conformations shows that the binding sequence is flexible and folds into a structure matching the geometry of carbon nanotubes. The hydrophobic structure of the peptide chains suggests that they act as symmetric detergents. An IgG monoclonal antibody against the fullerene C_{60} [75] was also studied to show binding to CNTs with some selectivity [76].

10.4 Summary and Future Directions

The potential of CNT-based nanodevices for ultrasensitive biological sensing has been recognized, and several demonstrations confirmed the exciting possibilities ahead. The reduction of the size of sensing and transducing elements to the size of biomolecues, i.e., 1 to 100 nm, makes it possible to detect down to single molecules. The development in this direction may revolutionize the filed of biotechnology. However, though the sensitivity improves, the reliability may pose a problem. Particularly at the level to detect a handful of molecules in the sample, the signal will fluctuate in a large range due to statistical reasons. Using a high-density array of such devices is desired to increase the statistics and improve the reliability. In the meantime, the dynamic range can also be increased. Extensive efforts have to be made in both device fabrication and assay development to solve these issues before the great potential for practical applications can be realized.

References

1. W.G. Kuhr, Nat. Biotechnol., 18, 1042 (2000).
2. J. Li, H.T. Ng, and H. Chen, Carbon nanotubes and nanowires for biological sensing, in *Protein Nanotechnology: Protocols, Instrumentation and Applications*, T. Vo-Dinh, Ed., Humana Press, Totowa, NJ (2003).
3. J. Li and H.T. Ng, Carbon nanotube sensors, in *Encyclopedia of Nanoscience and Nanotechnology*, H.S. Nalwa, Ed., American Scientific Publishers, Santa Barbara, CA (2003).
4. J. Kong et al. Science, 287, 622 (2000).
5. Y. Cui et al. Science, 293, 1289 (2001).
6. J. Li et al. Nano Lett., 3 (5), 597 (2003).
7. P.G. Collins et al. Science, 287, 1801 (2000).
8. G.U. Sumanasekera et al. Phys. Rev. Lett., 85 (5), 1096 (2000).
9. H.T. Ng et al. J. Nanosci. Nanotech., 1 (4), 375 (2001).
10. J. Li et al. Nano Lett., 3 (7), 929 (2003).
11. A. Star et al. Nano Lett., 3 (4), 459 (2003).
12. K. Besteman et al. Nano Lett., 3 (6), 727 (2003).
13. H. Dai et al. Nature, 384, 147 (1996).
14. S. Wong et al. Nature, 394, 52 (1998).
15. J. Li, A. Cassell, and H. Dai, Surf. Interface Anal., 28, 8 (1999).
16. S. Rosenblatt et al. Nano Lett., 2 (8), 869 (2002).
17. J. Koehne et al. Nanotechnology, 14, 1239 (2003).
18. J. Koehne, et al. J. Matr. Chem., 14 (4), 676 (2004).
19. J. Li et al. J. Phys. Chem. B, 106, 9299 (2002).
20. S. Sotiropoulou and N.A. Chaniotakis, Anal. Bioanal. Chem., 375, 103 (2003).
21. E.S. Snow et al. Appl. Phys. Lett., 82 (13), 2145 (2003).
22. F. Balavoine et al. Angew. Chem. Int. Ed., 38 (13/14), 1912 (1999).
23. G.R. Dieckmann et al. J. Am. Chem. Soc., 125 (7), 1770 (2003).
24. K.A. Williams et al. Nature, 429, 761(2002).
25. P.G. Collins, M.S. Arnold, and P. Avouris, Science, 292, 706 (2001).
26. S.J. Tans, A.R.M. Verschueren, and C. Dekker, Nature, 393, 49 (1998).
27. M.S. Fuhrer et al. Science, 288, 494 (2000).
28. J. Kong et al. Nature, 395, 878 (1998).
29. N.R. Franklin et al. Appl. Phys. Lett., 81 (5), 913 (2002).
30. A. Ural, Y. Li, and H. Dai, Appl. Phys. Lett., 81 (18), 3464 (2002).
31. Z.F. Ren et al. Science, 282, 1105 (1998).
32. L. Delzeit et al. J. Appl. Phys., 91, 6027 (2002).
33. M. Meyyappan et al. Plasma Sources Sci. Technol., 12, 205 (2003).
34. B.A. Cruden et al. J. Appl. Phys., 94 (6), 4070 (2003).
35. Z.F. Ren et al. Appl. Phys. Lett., 75(8), 1086 (1999).
36. J. Li et al. Appl. Phys. Lett., 82 (15), 2491 (2003).
37. V.I. Merkulov et al. Appl. Phys. Lett., 80(3), 476 (2002).
38. M.A. Guillorn et al. J. Appl. Phys., 91(6), 3824 (2002).
39. J.K. Campbell, L. Sun, and R.M. Crooks, J. Am. Chem. Soc., 121, 3779 (1999).
40. A.J. Bard, *Intregrated Chemical Systems: A Chemical Approach to Nanotechnology*, Wiley, New York 27–33 (1994).
41. A. Guiseppi-Elie, C. Lei, and R.H. Baughman, Nanotechnology, 13, 559 (2002).
42. B.R. Azamian et al. J. Am. Chem. Soc., 124, 12664 (2002).
43. M. Musameh et al. Electrochem. Commun. 4, 743 (2002).
44. R.G. Sosnowski et al. Proc. Natl. Acad. Sci. U.S.A., 94, 1119 (1997).
45. R.M. Umek et al. J. Molec. Diagn. 3 (2), 74 (2001).

46. N.D. Popovich and H.H. Thorp, Interface, 11 (4), 30 (2002).
47. R.M. Wightman, Anal. Chem., 53, 1125A (1981).
48. R.M. Penner et al. Science, 250, 1118 (1990).
49. F. –R. F. Fan and A.J. Bard, Science, 267, 871 (1995).
50. J. Li et al. Appl. Phys. Lett., 81 (5), 910 (2002).
51. R.L. McCreery, Carbon electrodes: structural effects on electron transfer kinetics, in *Electroanalytical Chemistry* 17 Bard, A.J., Ed. Marcel Dekker, New York 221–374 (1991).
52. J.M. Nugent et al. Nano Lett., 1(2), 87 (2001).
53. S.S. Wong et al. J. Am. Chem. Soc., 120, 8557 (1998).
54. C.V. Nguyen et al. Nano Lett., 2(10), 1079 (2002).
55. R.J. Chen et al. J. Am. Chem. Soc., 123, 3838 (2001).
56. M. Shim et al. Nano Lett., 2(4), 285 (2002).
57. Y.-P. Sun et al. Acc. Chem. Res., 35, 1096 (2002).
58. S. Wang et al. Nat. Mater., 2(3), 196 (2003).
59. T. Vo-Dinh et al. Nat. Biotechnol., 18, 764 (2000).
60. T. Vo-Dinh, J. Cell. Biochem. Suppl., 39, 154 (2002).
61. T. Vo-Dinh, B.M. Cullum, and D.L. Stokes, Sensors Actuators B, 74, 2 (2001).
62. A.T. Woolley et al. Nat. Biotechnol., 18, 760 (2000).
63. T. Andrew and C.A. Mirkin, Nat. Biotechnol., 18, 713 (2000).
64. M.S. Fuhrer et al. Nano Lett., 2, 755 (2002).
65. M. Radosavljevi et al. Nano Lett., 2, 761 (2002).
66. J.B. Cui et al. Appl. Phys. Lett., 81, 3260 (2002).
67. W. Kim et al. Nano Lett., 3 (2), 193 (2003).
68. K. Bradley et al. Nano Lett., 3 (5), 639 (2003).
69. Y. Miki et al. Science, 266, 66 (1994).
70. M.F. Sistare, R.C. Holmberg, and H.H. Thorp, J. Phys. Chem. B, 103, 10718 (1999).
71. G. Hummer, J.C. Rasalah, and J.P. Noworyta, Nature, 414, 188 (2001).
72. H. Gao et al. Nano Lett., 3 (4), 471 (2003).
73. C. Richard et al. Science, 300, 775 (2003).
74. Z. Guo, P.J. Sadler, and S.C. Tsang, Adv. Mater., 10(9), 701 (2002).
75. B.C. Braden, Proc. Natl. Acad. Sci. U.S.A., 97, 12193 (2000).
76. B.F. Erlanger et al. Nano Lett., 1, 465 (2001).

11

Applications: Composites

E. V. Barrera
Rice University

M. L. Shofner
Rensselaer Polytechnic Institute

E. L. Corral
Rice University

11.1 Introduction

With continued developments in the synthesis and production of carbon nanotubes, composite materials containing nanotubes are a near-term application and will see innovations that take advantage of their special properties. New conducting polymers, multifunctional polymer composites, conducting metal matrix composites, and higher fracture-strength ceramics are just a few of the new materials being processed that make them near-term opportunities. Moreover, high conductors that are multifunctional (electrical and structural), highly anisotropic insulators and high-strength, porous ceramics are more examples of new materials that can come from nanotubes. Figure 11.1 shows some of the opportunities that will emerge from new nanotube-reinforced composites.

Most investigators who are developing new composite materials with nanotubes work with nanotube concentrations below 10 wt.% due to the limited availability of nanotubes. Collective studies show broader promise for composite materials with concentrations as high as 40 and 50 wt.% that may be limited only by our ability to create a complete matrix and fiber registry with high surface-area nanofibers. Within this chapter, numerous discussions will be presented that identify the potential and application range of nanotube-reinforced composites. The similarities and differences compared with conventional composites will be apparent, and currently understood behavior, future directions, and near-term applications will be discussed.

Prediction of the future is not attempted here, but an effort is made to present the current and near-term properties of nanocomposites, showing where these materials will have useful roles in engineering applications. The classes of materials can be categorized along similar classes of conventional composites and will be identified by the matrix materials. Although many of the nano enthusiasts are working in

0-8493-2111-5/05/$0.00+$1.50
© 2005 by CRC Press LLC

Radiation protection
Heat dissipation coatings
Static discharge
High strength/lightweight parts
Heat engine components

Deicing coatings
Lightning protection
Stress sensors
High strength/light weight parts
Heat engine components

Organic LEDs
High strength/light weight housings
Electrically conductive ceramics

Paintable polymers
High strength/light weight parts
Heat engine components

Anti-fouling paints
UV protective coatings
Corrosion protection

FIGURE 11.1 Applications for polymer nanotube composites.

the area of polymer composites, efforts in metal and ceramic matrix composites are also of interest. Polymer nanocomposite studies have, in may cases, focused on nanotube dispersion, untangling, alignment, bonding, molecular distribution, and retention of nanotube properties. Studies of metal and ceramic systems have had additional issues of high-temperature stability and reactivity. Identifying modes of processing and new nanotube chemistries that provide for stabilization and bonding to the high-temperature matrices have been the central goals (high-temperature matrices herein are considered materials that melt or soften at temperatures higher than the nanotube degradation temperatures). Most investigators processing nanotubes into other materials treat nanotube addition in the same way they would any micron-size additive. In some cases the mixing is treated from a highly chemical route and the more micro- to macroscopic issues are not readily addressed. Certainly, it would be useful and more commercially economical if the more conventional processing routes would work for producing new nanotube composites, and in those cases the only limiting issue would be the cost of the nanotubes themselves.

The origin of nanotube-reinforced composites is the nanotubes themselves, but processing them into various matrices has been a spin-off from earlier work on incorporating fullerenes into metals and other matrices, polymer compounding, producing filled polymers (considerations to carbon black), assembling laminate composites, and polymerizing molecular rigid rod polymers. Metal systems are derived from powder metallurgy, methods for producing nanocrystalline materials, electrolytic approaches, and some ventures into melt processing. Ceramics have their origin in sol-gel processing, colloidal and slurry processing, *in situ* processing, and other tape and film casting methods. Special approaches have been identified that decrease either the processing time or the temperature in order to enable ceramic processing with stable nanotubes. All these methods are considered to take full advantage of the nanotube properties so that advanced materials for a range of applications can be produced. Nanotube composites can be replacement material for existing materials where properties not as superior to nanotube composites or to create nanocomposites for applications where composites have traditionally not been used before. Although most investigators are not after new nanocomposites with slightly improved properties compared with existing polymers, conventional alloys, or typical ceramics, these enhancements and replacement materials will provide the near-term imprudence to foster the "nanoengineered" composites achieving even greater superiority to existing systems. Therefore, some of the excitement generated from nanotubes will be described by showing the general promise that can come in electrical, mechanical, and thermal enhancements, thus enabling properties that will lead to new engineering designs and the quest for new systems not yet fully realized.

The section entitled "Nanotube Superiority" will connect the general properties of nanotubes to composite design. The evolution of nanotube processing, along with continued advances in purification, functionalization, and separation, have produced numerous starting conditions for composite development. The continued interest in developing longer nanotubes, similar to continuous reinforcements, will further enable the materials that can be produced with nanotubes. The current interest in processing "neat" nanotube continuous fibers also sparks opportunities as the properties of these fibers continue to improve [1,1a]. Therefore, the applications identified in this chapter are perhaps the tip of the iceberg as far as nanocomposites are concerned because they describe new materials that are based on the currently attainable nanotube properties. The composites currently investigated are materials on the learning curve for what is still to come. Therefore, one of the most important near-term outcomes from nanocomposites today will be the knowledge that is gained about preparing them for the development of the nanocomposites of the future.

Subsequent sections describe nanocomposites for electrical, mechanical, and thermal service. The sections will be separated into applications for polymer-, metal-, and ceramic-based systems, respectively.

11.2   Nanotube Superiority

The outstanding properties of nanotubes have given cause for the development of composite materials. Modeling has further shown that nanotubes will produce materials with superior properties and has aided in identifying the issues that must be addressed. Modeling of electrical conduction has shown that because of the small size of nanotubes, a network will form at very low concentrations (~1 wt.%) and may affect multifunctionality [2,3]. The network that forms, particularly at high concentrations, may limit homogeneous dispersion. Models that describe the mechanical properties of nanotube-reinforced composites show the promise of composite systems while demonstrating that alignment and bonding will be issues to be reckoned with [4–6]. The models for composite systems for thermal enhancement have shown that the thermal conductivity can vary (even decrease) as a function of alignment and dispersion of nanotubes in various matrices. In fact, thermal properties can depend on whether the nanotubes act as a thermal transport source or as a thermal scattering center. Their high aspect ratio and ease of network formation contribute to thermal transport while their size and hollow structure contribute to their use as scattering centers [7,8]. Clearly, modeling has shown the promise and pitfalls that must be overcome for producing nanotube-reinforced materials.

11.2.1 The Role of Nanotubes (A View from the Recent Modeling of Composites)

The rule of mixtures [9], whether it is for electrical, mechanical, or thermal conditions, is a general sign for the usefulness of a reinforcement in a matrix. There are issues as to whether or not a continuum model can adequately model a nanospecies in a matrix where separation of the various nanoparticles from each other is also on the nanoscale. Nevertheless, the rule of mixtures provides an adequate "back of the envelope calculation" for the potential of nanotubes in a matrix, and therefore it is of interest to demonstrate the promise of nanotube-filled composites. The thermal model presented in Figure 11.2 and Figure 11.3 for the thermal enhancement of nanotubes in a zirconia matrix shows the usefulness of single-walled nanotubes (SWNTs) for enhancing and reducing the thermal conductivity of the composite. The model assumes an infinite matrix, steady-state condition where K is independent of temperature. Fibers are considered to be surrounded by the matrix where the volume fraction f is less than 0.5. Both nanotube ropes and unroped nanotubes are considered for one-, two-, and three-dimensional conditions as shown in Figure 11.2. Because SWNTs are insulating in the transverse direction and a high thermal conductor along the tube length, a broad range of material properties is expected from their distribution and degree of alignment as seen in Figure 11.3. The equivalent inclusion method used here does not consider a transition zone between the nanotubes and the matrix, nor does it consider other scattering conditions from matrix alterations generated from the presence of the nanotubes. Still, the inclusion of nanotubes is shown to produce a broad range of thermal materials. When specifically coupled with zirconia, a low thermal conductivity potentially tough ceramic, a new multifunctional ceramic system may well result.

11.2.2 Nanotube Chemistry

Nanotubes are considered to have varying degrees of defects depending on whether they are multiwalled nanotubes (MWNTs) or SWNTs. Computational models show that unroped SWNTs are the nanotubes of choice for mechanical enhancement of composites, but the variety of conditions of various nanotubes can play a role with and without chemical functionalization. Initially, nanotube functionalization was thought to occur at the various defect sites (MWNTs being highly defective compared with SWNTs), but functionalizations both at nanotube ends and along the sidewalls without disruption or degradation of the tubes have been demonstrated. A variety of functionalized nanotubes are being developed for composite applications, including fluoronanotubes (f-SWNTs); carboxyl-nanotubes with various end functionalization; and numerous covalently bonded SWNTs such as amino-SWNTs, vinyl-SWNTs, epoxy-SWNTs, and many others to provide for matrix bonding, cross-linking, and initiation of polymerization. Wrapped nanotubes (w-SWNTs) with noncovalent bonding are also another variety of nanotubes and find particular use when the electrical properties of the nanotubes need to be preserved. Figure 11.4A shows the end of a nanotube with initial carboxyl groups that are replaced by an initiator to the styrene

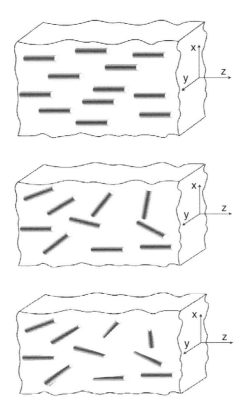

FIGURE 11.2 The one-, two-, and three-dimensional configurations for nanotube ropes and unroped nanotubes in a matrix section. The configurations are for an equivalent inclusion method of modeling the thermal conduction properties of SWNTs in a zirconia matrix [8].

polymerization process. The insert, Figure 11.4B, shows the eventual polymeric structure that could result with nanotubes end-functionalized to the polymer system. These polymeric structures lead to some degree of cross-linking in a typical thermoplastic polymer system or may occur as a chain mill system. Sidewall functionalization certainly brings in a high degree of cross-linking as seen in Figure 11.5A. Whether the functionalized nanotubes are added as a reinforcement or at a level to impart hybrid polymer formation, the chain mobility may be highly restricted while enabling a high degree of stiffness and strength. Figure 11.5B to D shows three functionalizations used in composite applications. In Figure 11.5B, f-SWNTs are produced and used as an intermediate step to exfoliate the nanotubes from the tangles and ropes so that further functionalizations as seen in Figure 11.5C can occur to unroped nanotubes. Figure 11.5D shows a case where the SWNT is an integral part of the epoxy curing agent and, in turn, yields a new hybrid polymer system. The functionalization will enable a wide variety of composite systems that include low-concentration reinforced systems, traditional fiber/nanotube systems, hybrid polymers, copolymer systems of various combinations of nanotube-filled polymers, and situations with full integration that create fully integrated nanotube composites (FINCs).

Nanotube chemistry not only is for polymeric systems but will be evolved for metals (particularly for Al, Cu, potentially Ti) and ceramics that are carbide, oxide, nitride based and many other varieties. Nanotube functionalization in metals and ceramics will likely play a role in nanotube stabilization, defect refinement, and overcoating methodologies. If ends are considered defects, then methodologies for welding nanotubes may well include other species that would be delivered to the nanotubes from a metal or ceramic matrix. These opportunities coupled to a variety of processing modes may well produce next-generation metal hybrids and porous, high-conducting, high-toughness ceramic structures [4,10,11].

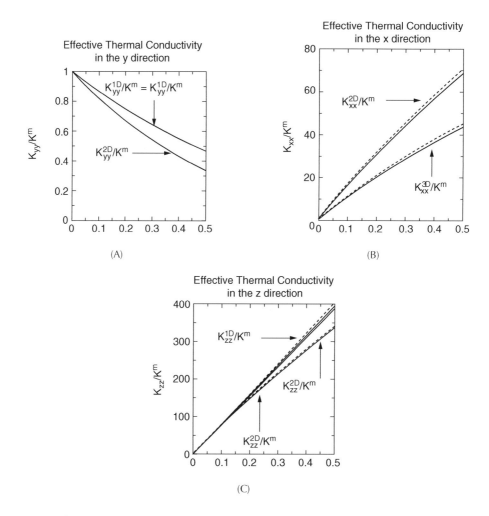

FIGURE 11.3 The ratio of the thermal conductivity of the K_{ii} direction compared with that for the zirconia matrix where ii is (A) yy, (B) xx, and (C) zz [8].

Fiber Dimension		Thermal conductivity
⬭ 1 nm diameter		
-------- 3 μm		Along the fiber direction 2000 W/mK
⬭ 30 nm diameter 10 μm length		Cross the fiber diameter 0.4 W/mK
Zirconia		Isotropic 2.5 W/mK

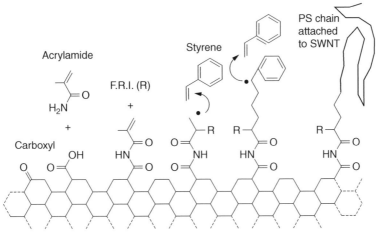

Single-Walled Nanotube open (armchair) edge

(A)

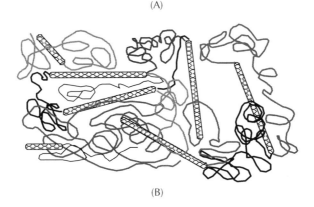

(B)

FIGURE 11.4 (A) SWNT end functionalization with bonding to a styrene polymer linear chain. A P-SWNT with COOH is altered with an amide initiator for bonding to the styrene monomer. (B) End-functionalized SWNTs linked in a linear polymer system. (From E.V. Barrera, J. Mater. 52(38), 2000. With permission.)

11.3 Polymer Nanocomposites

11.3.1 Electrical Properties

The most accessible near-term application for nanotube-reinforced polymer composites involves their electrical properties. The intrinsic high conductivity of nanotubes makes them a logical choice for creating conductive polymers. Because they possess high aspect ratios, nanotubes have the added advantage of achieving percolation at lower compositions than spherical fillers. Percolation describes the range of compositions where a three-dimensional network of the filler is formed. When percolation is reached, the conductivity increases dramatically because the network has created a conductive path. The percolation behavior can be expressed by a power law equation [12]:

$$\sigma_v \approx (v\text{-}v_c)^t, \tag{11.1}$$

where σ_v is the conductivity as a function of the filler concentration, v is the volume fraction of filler, v_c is the volume fraction where percolation occurs, and t is the critical exponent. Percolation composition ranges from as low as 0.0225 to 0.04 wt.% have been reported without having achieved ideal dispersion [13]. Many researchers have reported percolation thresholds less than 1 wt.% nanotubes [4,14–17] where

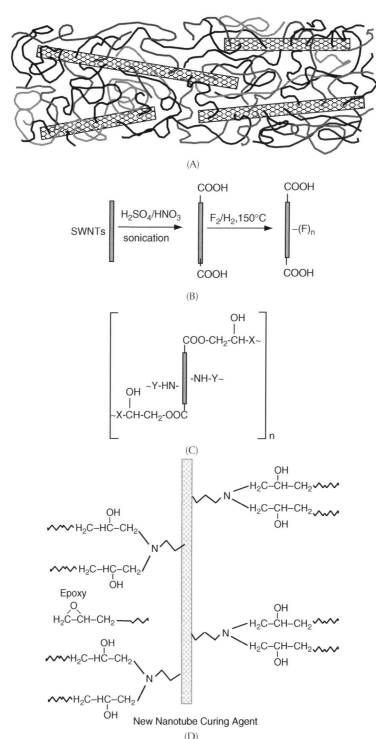

FIGURE 11.5 (A) Nanotubes sidewall cross-linked in a polymer matrix. (From E.V. Barrera, J. Mater., 52(38), 2000. With permission.) (B) One route to producing fluoronanotubes for use in composite applications. This two-step process (acid wash/fluoronation) assures nanotube exfoliation for better dispersion in both epoxies and thermoplastics [10]. (C) Functionalization of f-SWNTs where fluorine is removed and replaced with an amide functionalization. (D) Nanotube sidewall functionalization in an epoxy curing agent leading to a new hybrid curing agent to be linked to the epoxy resin. (From Jiang Zhu et al., Nano Lett., 3(8), 1107, 2003. With permission.)

in some cases good dispersion is achieved. Low nanotube concentrations with useful electrical enhancements are of interest, and this offsets their current high cost ($500/g for SWNTs in 2003) with respect to other fillers. Furthermore, for a number of composites the neat polymer properties such as elongation to failure and optical transparency are not significantly decreased [4,13]. Research has indicated that the interfacial region between the polymer matrix and the nanotubes influences the conductivity of the composite and the percolation behavior. Conductivity measurements on composites, where the matrix and the nanotubes are in intimate contact, can show higher percolation compositions. MWNTs in a poly(vinyl alcohol) (PVOH) matrix have shown that percolation can occur between compositions of 5 to 10 wt.% [18]. In this case the nanotubes were coated with a layer of adsorbed polymer and found to interact strongly with the matrix. The broader range of compositions of 5 to 10 wt.% in this system allows for greater control of the conductivity, whereas composites with lower percolation thresholds are more sensitive to small composition changes and may be less stable for electrostatic dissipative (ESD) applications. Still, discharge times for ESD have been very low and the ESD is maintained in different relative humidities and is therefore superior to many ESD materials currently available [4]. Another method of creating conductive polymers that may reduce percolation thresholds even further is by creating immiscible polymer blends with one polymer containing nanotubes [19]. This approach has already lowered the carbon black concentration required for percolation in poly(ethylene-co-vinyl acetate)/high density polyethylene blends.

Depending on the conductivity level achieved, conducting polymers have applications as ESD materials, electromagnetic interference (EMI) materials, and high-conducting materials. ESD materials are characterized as possessing a surface resistivity between 10^{12} and 10^5 Ω/square. These materials could be used for carpeting, floor mats, wrist straps, and electronics packaging. For EMI applications, the resistivity should be less than 10^5 Ω/square. EMI applications include cellular phone parts and frequency shielding coatings for military aircraft and electronics. High-conducting materials would serve as weight-saving replacements for metallic materials when produced as multifunctional composites (mechanical/electrical).

The potential for highly conductive polymers exists through the combination of intrinsically conducting polymers and nanotubes, particularly after separation of the metallic nanotubes is more readily achieved. MWNTs in polyaniline create a composite material with a greater conductivity than either of the starting components [20]. *In situ* polymerization creates an interaction between the polymer and the nanotubes that promotes charge transfer. Composites made by deposition of conducting polymers onto MWNTs can be used as supercapacitor devices [21]. Furthermore, opportunities can be gained by mixing metallic nanotubes with conjugated polymers to produce high-conductive polymers with metallic-like conduction. It is likely that the nanotubes will contribute to better stability of the conjugated polymers while taking advantage of aligned nanotube configurations. Table 11.1 shows a number of composite systems where electrical enhancements have been observed. Conducting plastics have applications in the automotive industry as body panels and bumpers because they can be painted without a conducting primer. High conductors will serve as interconnects and may well replace wiring harnesses in transportation systems where weight saving is an issue.

TABLE 11.1

Polymer	Nanotube Type	Nanotube Concentration (wt. %)	Percolation Threshold (wt. %)
ABS [4]	SWNT	0.5–10	
Epoxy [13]	MWNT	0.0225–0.15	0.0225–0.04
Polyimide [14]	SWNT	0.01–1.0	0.02–0.1
PMMA [15]	SWNT	0.1–8.0	<1
Polystytrene [16]	MWNT	1–5	<1
PmPV and PVOH [17]	MWNT	0.037–4.3	0.055 for both composites
PVOH [18]	MWNT	5–50	5–10

11.3.2   Mechanical Properties

With a tensile modulus on the order of 1 TPa and a tensile strength 100 times greater than steel at one sixth the weight, SWNTs are ideal reinforcements for polymer matrix composites, but to fully exploit their exceptional properties, nanotubes should be homogeneously dispersed and aligned within a polymer matrix. Two main approaches to combining the nanotubes and the polymers are *in situ* polymerization and compounding of the nanotubes and polymer. *In situ* polymerization has been shown to produce direct bonding between the nanotubes and the polymer [22], and methods of polymerization involving sonication have been shown to produce high dispersion [14]. Incipient wetting combined with high shear mixing has also been shown to disperse nanotubes in thermoplastic polymer matrices [4,23,24]. Incipient wetting involves dispersing the nanotubes in a solvent, overcoating the polymer, and removing the solvent to create an initial dispersion of nanotubes on the polymer particles. The overcoated polymer is mixed with high shear mixing using Banbury-type mixing or dual screw extrusion.

Composites containing isotropically oriented nanotubes generally possess lower mechanical properties than do composites with aligned nanotubes [25,26]. Methods such as fiber spinning and solid freeform fabrication have been used to create anisotropic orientations. Variations of fiber spinning such as melt spinning, wet spinning, and gel spinning have been used to process nanotube/polymer composites. In these processes a polymer melt, polymer solution, or gel is extruded through a nozzle with a postextrusion drawing step to further orient the nanotubes and polymer molecules. By these methods, nanotube composite fibers with isotropic pitch, poly(methyl methacrylate) (PMMA), polyethylene, PVOH, and poly(*p*-phenylene benzobisoxazole) (PBO) matrices have been produced [24,27–31]. Fibers produced by fiber spinning showed good nanotube alignment when measured by x-ray diffraction, electrical conductivity, and Raman spectroscopy tests. The degree of alignment has been shown to decrease with increasing nanotube concentration and to increase with increasing draw ratio [24]. Static and dynamic mechanical property improvements have been observed in nanotube-reinforced polymers compared with the neat polymers, and anisotropic nanotube orientation has shown greater improvements over isotropic nanotube orientations in polymers. Enhancements of ~450% in tensile modulus have been observed at nanotube loadings of 20 wt.% even though functionalization was not used [24]. In composite fibers containing highly aligned nanotubes, simultaneous increases in tensile strength, tensile modulus, and strain to failure have been realized [30].

Because some processing approaches, such as fiber spinning and extrusion, more easily promote alignment conditions, other approaches that are extrusion based may well be brought to the forefront because aligned nanotube conditions can be achieved in this way from them. Solid freeform fabrication (SFF) techniques are these types of manufacturing approaches where aligned nanotubes could occur from their extrusion-based approaches. SFF describes a family of additive manufacturing processes where parts are built in layers using geometrical information from a three-dimensional computer model [31]. Extrusion freeform fabrication (EFF) is a type of SFF where a cylindrical billet of material is deposited from an extrusion head that can move in the *x-y* plane, and the material is deposited onto a stage that moves in the *z* direction allowing successive layers to be built. Where fiber spinning produces composite fibers to be incorporated into tapes, woven into fabric, or used as reinforcing agents in polymers, SFF processing creates usable shapes in basically one step containing nanotubes of a prescribed orientation.

Extrusion-based SFF techniques with nanotube feedstocks have potential applications for space-based manufacturing and remote manufacturing. Nanotube additions to the polymers used by these processes improve the functionality of SFF processes. Nanotubes have been dispersed in poly(acrylonitrile-co-butadiene-co-styrene) (ABS) and processed by EFF to create test samples containing oriented nanotubes. The degree of alignment seen in initial samples is less than that for drawn fibers, but a postextrusion drawing step should improve the alignment [26]. Figure 11.6 compares the tensile modulus measurements made on composite fibers containing 5 wt.% nanotubes produced from fiber spinning and SFF.

The large surface area of nanotubes has also been used to improve the load transfer capabilities of micron-size fiber-reinforced epoxy composites [32,33]. In this application, nanotubes were used as a sizing on the larger fiber. MWNTs have been grown by chemical vapor deposition on high-modulus

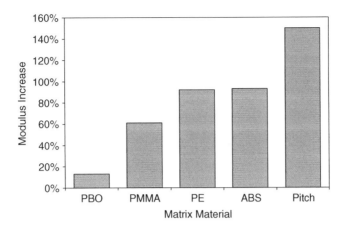

FIGURE 11.6 Comparison of modulus improvement with 5 wt.% nanotubes in polymer matrices. Intrinsically stronger matrices tended to show less improvement at this nanotube loading.

pitch-based carbon fibers creating a multiscale composite. Testing on single fiber composites has shown improved interfacial bonding with the addition of nanotubes to the carbon fiber's surface. An improvement of 15% was observed with respect to an uncoated fiber through the measurement of critical aspect ratio. The improvement was found to be a result of the nanotube coating and not related to the catalyst or chemical vapor deposition process. The improvement in load transfer was attributed to stiffening of the epoxy at the interface. A similar procedure has been used to improve the mechanical properties of micron-size glass fiber/epoxy composite. A solvent containing SWNTs and epoxy were sprayed onto a glass fiber weave (another form of incipient wetting), and then the fiber weave was infiltrated with epoxy. The test results showed that the composite sizing improved the flexural strength 13% with respect to the uncoated glass fiber/epoxy composite even when functionalization was not used. This approach provides for matrix infiltration while avoiding the problem of increased resin viscosity resulting from the introduction of nanotubes. Even when mixing a small concentration of nanotubes (1wt.%) directly into epoxies, significant increases in viscosity occur particularly when the nanotubes become well dispersed.

Table 11.2 shows a number of composites where their mechanical enhancements are described. Structural applications for nanotube-reinforced polymers include aviation, automotive, military, and space-related parts. The combination of high strength and light weight makes these materials candidates for aircraft parts and car bodies. Bulletproof vests could adapt to bulletproof garments made of woven nanotube composite fibers that would cover the entire torso instead of just the front and back. Lightweight radiation/impact shields and space suits could also be fabricated from these composites. With consideration to small-concentration composite systems, the z-axis properties of composites could be improved, thereby enabling composite systems for a broader application base.

11.3.3 Smart Materials

The Raman spectrum of carbon nanotubes, particularly the D* band at 2610 cm^{-1} in air, is sensitive to its strain state [34]. By monitoring this band and linear band shift when the strain is in the elastic range, nanotubes can be used as strain sensors in polymers. With SWNTs dispersed in the matrix, the stress field was mapped around an E-glass fiber embedded in epoxy and around a hole in a thin polymer film [34,35]. In addition to shifts in Raman signals, changes in the electrical behavior can be used to monitor stress and strain. A film of isotropically oriented SWNTs embedded in epoxy was attached to a brass sample in a universal testing machine. The voltage measured by a four-point probe shifted linearly in tension and compression with applied strain and stress as shown in Figure 11.7 [36,37]. These results indicate that nanotubes could be embedded in high-performance composites as reinforcing agents and strain sensors allowing for nondestructive monitoring and distributed sensing of large structures.

TABLE 11.2

Polymer	Nanotube Type	Nanotube Concentration (wt.%)	Advantages	Disadvantages
PMMA [22]	MWNT	1–20	Direct bonding +30% tensile strength	Tensile strength drop above 7 wt. %
Polyimide [14]	SWNT	0.01–1	High dispersion +60% storage modulus	Decreased transparency at 500 nm
ABS [4,26]	SWNT	0.5–10	High dispersion +93% modulus	Decreased strain to failure Melt fracture
PMMA [23]	MWNT and SWNT	Up to 4	High dispersion +300% impact resistance	No change in modulus
PE [24]	SWNT	1–20	+450% modulus High alignment High dispersion	Decreasing alignment and dispersion with increasing nanotube content Melt fracture
Pitch [25]	MWNT	1–10	+90% strength +150% modulus	Not able to spin fibers above 5%
PMMA [28]	SWNT	1–8	+94% modulus High alignment	Melt fracture Decreased strain to failure
PVOH [29]	SWNT	60	30% strain to failure High strength and modulus High alignment	
PBO [32]	SWNT	5–10	+20% modulus +60% tensile strength +40% strain to failure High dispersion	No change in electrical conductivity
Epoxy/Carbon Fiber [32]	MWNT		+15% interfacial strength	
Epoxy/Glass Fiber [33]	SWNT		+13% flexural strength	

Polymer/nanotube composites spun into fibers could be used in the fabrics with multifunctional capabilities. SWNT/PVOH fibers containing 60 wt.% SWNTs have a tensile strength similar to spider silk and elongation to failure similar to the toughest silk [29]. In addition to high strength, these fibers could be woven into textiles to create garments with sensing and EMI shielding capabilities. Aligned nanotubes in fibers allow for directional measurements based on the macroscopic fiber's orientation in the fabric weave.

11.3.4 Polymer Coatings

Incorporating nanotubes into polymers for use as protective coatings and paints is another large application range for nanotubes. Thermally conducting coatings with nanotube composites have uses in the aviation and space industries. These coatings could serve as aids in deicing aircraft wings in cold weather through the application of current to the coating. Another application for nanotube/polymer coatings is heat dissipation on space vehicles. The coating would act as a space radiator to replace liquid heat rejection systems. Protective coatings containing nanotubes could be used to mitigate damage to aircraft caused by lightning strikes. Composites possessing sufficient conductivity and a percolated structure could dissipate the charge over the entire fuselage. Biocatalytic composites created using SWNTs have potential applications as antifouling paints and biosensor coatings [38]. The nanotubes were found to have a stabilizing effect on the leaching of enzyme from the film, and the composites have shown increased bioactivity over the materials without nanotubes. Composites and coatings may both find their way into shielding systems for radiation and impact protection. The addition of MWNTs to an epoxy that cures via ultraviolet radiation could serve as a coating to resist radiation [39]. Because nanotubes can be

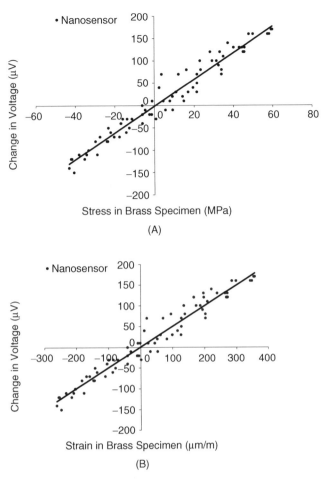

FIGURE 11.7 Specimen subjected to tension and compression cycles. (A) Change in voltage in the carbon nanotube film as a function of stress in brass specimen. (B) Change in voltage in the carbon nanotube film as a of function strain in brass specimen. (From P. Dharap et al., J. Nanotechnol., 15, 379, 2004. With permission.)

functionalized with additional hydrogen species, the composite systems processed with them could serve as radiation protection from exposure to secondary radiation events. Imparting nanotubes into the midplane of composite panels or on the surface as a coating or in the other laminates could well serve as radiation protection, as protection against lightning strikes, or as structural layers.

11.3.5 Polymeric Devices

The unique electronic structure of nanotubes makes them ideal candidate materials for composite devices. A composite with SWNTs and the conjugated polymer poly(3-octylthiophene) possesses properties suitable for use in photovoltaic cells [40]. Photoinduced electron transfer occurring between the nanotubes and the polymer causes increased short-circuit current and fill factor in diodes containing the composite material as compared with diodes with the neat polymer. Composites with poly(*m*-phenylenevinylene-*co*-2,5-dioctyloxy-p-phenylenevinylene) (PmPV) have been shown to be effective in organic light-emitting diodes as electron transport layers [41]. Actuators based on SWNT/Nafion composites have potential applications as microswitches and artificial muscles [42]. Unlike actuators made from doping Nafion with metal, the SWNT composite actuators showed no relaxation with time, and the conductivity required to achieve actuation was much lower than that of metal (1 wt.% SWNTs vs. 30 wt.% metal). Deflection amounts were found to increase with increasing nanotube concentration.

Nanotube/polyimide composites have been investigated for use as ultrafast optical switches [43]. The materials displayed optical delay times less than 1 ps at a wavelength of 1.55 µm and showed potential as all-optical switches. MWNTs dispersed in a photoresist epoxy have applications in the production of microelectromechanical systems (MEMS) such as electroplating molds, sensors, and actuators [39]. Dimensionally, their application size is small, but they are processed as nanotubes in polymers and are considered composite systems on that basis.

11.4   Nanotube-Metal Composites

Whether or not nanotubes in metals would overcome the age-old problem of carbon reinforcement in metals is yet to be fully understood. The "no defect" description of SWNTs is probably what sparked an interest in developing nanotube-metal composites (NMCs). Carbon fibers have been plagued with bonding and reaction issues in aluminum and other metals, and the absence of defects suggests these issues may be evaded (Al_4C_3 formation between Al and carbon fibers). Three main processing routes to producing NMCs are being explored and are directed toward producing nanotube nanostructures, advanced alloys, nanocrystalline materials, coatings, and bulk composites. The applications for NMCs are nanowires, lightweight structures, electronic materials for avionics, wear coatings, novel magnetic and superconducting systems, and new multifunctional metals.

To promote the development of NMCs of various types and sizes, wetting and adhesion of metals to the nanotubes are very important. Dujardin et al. and others [44–46] showed that the determining factor for wetting metals to nanotubes is surface tension and resolved that a cutoff occurs between 100 and 200 mN/m. They also indicated that this limit implies that aluminum, copper, and iron (865 mN/m, 1270 mN/m, and 1700 mN/m, respectively) will not easily wet on the nanotube surfaces (Dujardin et al. studied MWNTs). Therefore, various routes for improving the wetting are necessary in order to achieve strong interactions between the nanotubes and the matrix.

Even with identified ways to promote interactions, many of the processing modes used to date involve processing below the melting temperature of the metals when in contact with the nanotubes to provide for nanotube stability. These processes include thin-film evaporation, electroless plating, other chemical routes, and hot pressing and extrusion of powders. Few methods have ventured into the realm of processing nanotubes in molten metals because reactivity and potential nanotube degradation would be extremely high at this stage.

11.4.1   Nanowires

Nanotubes have been used as templates for producing nanowires. Although these may not be thought of as composites, the combination of the metal overcoat or metal inside the nanotube with the nanotube constitutes a nanocomposite system [46–51]. To form the nanowires with the metals on the outside of the nanotubes, processing methods of evaporation, electroless plating, and electroplating have been used. The applications for these nanowires vary from use at the nanoscale for electronics miniaturization to fillers for bulk systems as a way to preserve the nanotubes in the processing of large-scale composite systems. In these cases, the coating is designed to bond to the nanotube, and the coating can be one that easily interacts with the matrix system. An example of this system would be the activated nanotube bonded with nickel, and the nickel-coated nanotube to be processed in copper [51]. Some of the investigators have shown that oxidized nanotubes are essential for achieving a metal bond, and others have shown that metals like titanium, nickel, and niobium are also good for direct bonding to the nanotubes. There are cases where nanotubes are coated with metals but further processed to alter or remove the nanotube, resulting in metal nanotubes. In these cases the NMC is the precursor and provides for a way to make nanosized metal hollow fibers.

Nanotubes have also been processed with metals inside the nanotubes. The following three approaches have been used to accomplish this: (a) encapsulation of metals in nanotubes during the formation of the nanotubes using an arc-discharge, (b) encapsulation of metals in the tubes by opening the tubes and depositing the filling material using wet chemistry techniques, and (c) encapsulation of metals by capillary action [51]. These approaches have produced metals encapsulated in nanotubes for potential use as nanowires, composite fibers, and microcrystals. Even these systems have been studied as precursor systems where the nanotubes have been converted into carbide nanorods by reactions with oxides and halides [51]. Therefore, a variety of composite nanowire conditions can be created where the limiting factors are the bonding, the lengths of the nanowires, the lengths that metals can be put inside the nanotubes, and the practical use of these composites for real applications. Electronic and superconducting applications may require only small amounts of these nanowires, but development of coatings and bulk composite systems from these nanowires will require large amounts of the composite fibers. Some investigators have considered this in recent developments of coatings and bulk composite systems.

11.4.2 NMC Coatings

Understanding that bonding to the nanotubes is a clear direction for the development of coating systems, systems of nickel, nickel-phosphorous, and copper have been studied because they can be processed by either electroless plating or electroplating [47,51,52]. These coatings have been generally produced as wear coatings for electrode surfaces for sliding electrical contacts. Encouraging improvements in wear properties have been shown to occur from these coatings, and they appear to have better properties than do metal-graphite systems. One of the questions to consider in implementing these new sliding contacts is whether the thermal enhancement of the material can be used in providing longer life to the contacts by their receiving better cooling and less wear. Therefore, this multifunctional use of the nanotubes may create an opportunity for these types of contacts to see near-term use.

11.4.3 Bulk NMCs

Bulk NMCs have been processed using conventional powder metallurgy where matrices of aluminum, silver, and copper have been used [53–55]. This approach has been used because the processing is accomplished below the melting temperature of the metals (melting temperature for aluminum is 660°C). Powder metallurgy routes using powders and in some cases extrusion have produced an Al metal matrix composite, copper electrodes, and macroscopic composite wires (where extrusion was used). Some of these materials have taken advantage of precoating the nanotubes using electroless plating. In some cases, nickel was plated on the nanotubes before they were powder processed into a bulk composite. These composites were used for electrodes, wear contacts, and produced systems where mechanical properties were enhanced by grain-size refinement. The properties were stable after annealing because the nanotubes prevented grain growth from occurring. In some cases, nanotubes could be used to produce ceramic-coated nanotubes for use in metal matrix composites, thereby creating nanosize ceramic reinforcements for metal matrix composites. This type of application for nanotubes is further discussed in the upcoming sections on ceramic matrix composites.

11.5 Ceramic Matrix Composites

11.5.1 Processing

Ceramic matrix composites (CMCs) can be made either by conventional powder-processing techniques, used for making polycrystalline ceramics, or by some new and rather unconventional techniques, which sometimes are no more than variants on the processing of monolithic ceramics [56]. The processing of CMCs is a critical part of the overall design of a CMC composite because its final properties vary with its density, composition, phases that form, and defect structure. Nanotube-reinforced ceramic composites (NCCs) are especially dependent on composite processing to ensure the survivability of the reinforcements during sintering. In most cases it is essential that the nanotube integrity be maintained throughout processing. In the following sections, some of the most relevant processing techniques for NCCs are described. It is important to note that ceramics can benefit from nanotubes, as reinforcements in the areas of strength and toughness and thermal and electrical enhancement. Thermal enhancements may be related to decreases in thermal conductivity to produce more insulating ceramics or may become highly anisotropic to produce heat pipes and high thermal conductors. A key to producing advanced NCCs with nanotubes lies in the ability to stabilize the nanotubes to temperatures that are much higher than their degradation temperature in air (~600°C). In this way, more conventional processing routes can be used to produce more economic NCC structures.

11.5.2   Conventional Ceramic Processing Techniques

Direct mixing of nanotubes within the ceramic matrix powder followed by cold pressing and sintering is considered to be equivalent to conventional processing of ceramics. In faster fabrication methods, the ceramic composite material requires the incorporation of various organics to process the materials by extrusion or injection molding and tape casting. These methods require complete removal of the organic binder to occur before a fully sintered body with a near-theoretical density can be obtained. It is quite useful for the binder to be removed below the degradation temperature of the nanotubes to ensure their stability to higher temperatures. During sintering, the matrix material shrinks considerably and results in a composite with numerous cracks, and these ever-present microcracks are the source of microstructural defects that decrease the ceramic strength. More importantly, for ceramic composites processed with nanotubes, these reinforcements present physical limitations during sintering. It has been found that the presence of nanotube reinforcements in a metal-oxide ceramic matrix hinders the grain growth during hot pressing and can limit the densification to approximately 90% of the theoretical density [57]. Some form of hot pressing is frequently used in the consolidation stage of processing CMCs [56]. The simultaneous application of pressure and high temperature can accelerate the rate of densification leading to pore-free, fine-grained compacts being formed.

11.5.3   Novel Techniques

11.5.3.1   Colloidal Processing

During the fabrication of ceramic materials, problems with mechanical property reproducibility stem from the presence of microscopic flaws. Significant improvements in the formulation of high-strength ceramics have emerged within the past 20 years using colloidal processing of submicron ceramic powders. A dispersion of submicron ceramic powders is produced using colloidal processing techniques. The structure of the green body depends on the processing technique and conditions, as well as the underlying colloid science. For example, dispersions of nanotubes can be stabilized using surfactant additives, and the matrix ceramic particles can be stabilized using either polymeric additives or surfactants. These additives are adsorbed on the surfaces of the different materials minimizing the van der Waals attractive forces between them. Control over the surface forces between the different phase materials results in a highly dispersed composite suspension. Additionally, colloidal processing methods can be employed to coat nanotube surfaces with submicron ceramic powders [58].

11.5.3.2   Spark Plasma Sintering

In order to obtain fully dense nanocrystalline ceramics and avoid damaging nanotube reinforcements during sintering, spark-plasma sintering (SPS) is considered the leading technique for composite consolidation to achieve full density because it is a rapid sintering process [59,60]. This technique is a pressure-assisted fast sintering method based on a high-temperature plasma (spark plasma) that is momentarily generated in the gaps between powder materials by electrical discharge during on-off DC pulsing. It has been suggested that the DC pulse could generate several effects: spark plasma, spark impact pressure, Joule heating, and an electrical field diffusion effect [61]. Through these effects, SPS can rapidly consolidate powders to near-theoretical density through the combined effects of rapid heating, pressure, and powder surface cleaning. Fully dense SWNT/Al_2O_3 nanocomposites were fabricated using SPS. Figure 11.8

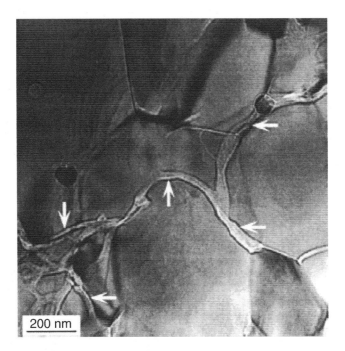

FIGURE 11.8 Transmission electron micrograph of a 5.7 vol% SWNT/Al$_2$O$_3$ nanocomposite fabricated using SPS. The white arrows indicate SWNTs within Al$_2$O$_3$ grains. (From G.-D. Zhan et al. Appl. Phys. Lett., 83(6), 1228, 2003. With permission.)

shows a transmission electron micrograph of the nanocomposite, where the SWNTs are found between grain boundaries.

11.5.4 Applications: Carbon Nanotube-Ceramic Composites

11.5.4.1 Ceramic Nanotube Composite Systems

There is a great deal of interest in developing NCCs in order to enhance the mechanical properties of brittle ceramics. Conventional ceramic processing techniques were first employed to develop MWNT-reinforced SiC ceramic composites where a 20% and 10% increase in strength and fracture toughness, respectively, were measured on bulk composite samples over that of the monolith ceramic [62]. These increases are believed to be due to the introduction of high-modulus MWNTs into the SiC matrix, which contribute to crack deflection and nanotube debonding. Furthermore, good interfacial bonding is required to achieve adequate load transfer across the MWNT-matrix interface, which is a condition necessary for further improving the mechanical properties in all NCC systems. Therefore, nonconventional processing techniques such as colloidal processing and *in situ* chemical methods are important processing methods for NCC systems. These novel techniques allow for the control of the interface developed after sintering through manipulation of the surface properties of the composite materials during green processing.

A number of material systems have been explored during the development of NCCs. Figure 11.9 shows a diagram of the number of ceramic systems that have been used to develop NCCs from 1998 to present. Most of the work has focused on using oxide ceramic materials as the matrix material with either SWNT or MWNT reinforcements. Carbide and nitride ceramic materials have not been as widely used for the matrix material due to the extreme processing conditions required to consolidate and sinter the matrix material. Average sintering temperatures for most nitride and carbide ceramic materials are well above 1800°C. The following sections deal with specific novel processing routes for NCCs and their related applications.

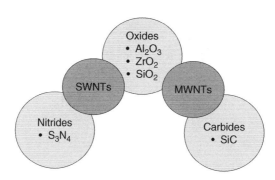

FIGURE 11.9 Ceramic materials being used in the development of SWNT and MWNT ceramic matrix composites.

11.5.4.2 Ceramic Nanotubes

Zirconia-coated MWNTs and partially stabilized zirconia (PSZ)-coated MWNTs were processed for the creation of hollow zirconia and PSZ nanotubes. The preparation of hollow ceramic metal oxide nanotubes is an important and challenging problem that has easily been addressed through the direct use of carbon nanotubes [63]. The procedure that produced zirconia and PSZ nanotubes involves chemical reactions on acid-treated MWNTs surface sites that result in coated MWNTS. Subsequent heat treatments oxidize the carbon and leave an intact ceramic nanotube.

The zirconia-coated MWNTs contained a 50/50 mixture of monoclinic and tetragonal phases. After heating to high temperatures to obtain monophasic nanotubes, the nanotubes collapsed because of the first-order nature of the monoclinic-tetragonal transformation. In order to avoid this problem, the PSZ-coated MWNTs were used instead of pure-zirconia–coated MWNTs. The zirconia nanotube diameters are 40 nm, and the wall thickness is 6 nm. The PSZ nanotubes are 40 to 50 nm in diameter and 6 nm in wall thickness. Hollow ceramic nanotubes may have several applications related to the thermal management of ceramics and as template structures for other advanced composite applications.

11.5.4.3 Thermal Management in Ceramics

Reduced thermal conductivity of ceramics used for thermal barrier coatings (TBCs) are of high interest to the gas turbine industry because they were used to protect critical components in the hot sections of gas turbine engines. Even small improvements in the thermal properties of TBCs are beneficial because turbine blades operate at extreme thermal and mechanical conditions and are expensive components. Reduction in thermal conductivity of a 200-μm PSZ TBC from 1.9 W/m*K to 1.1 W/m*K (42% reduction) will reduce the temperature of the superalloy blade surface by 32°C [64]. These turbine blades usually operate at temperatures approaching their melting points, so a small reduction in blade temperature will significantly extend their lifetime. For example, a 13°C reduction in the operating blade temperature will double the lifetime of the turbine blade [65]. SWNT-PSZ composites have already been developed as a means to stimulate the porous nature of plasma-sprayed TBCs through the inclusion of small amounts (1 to 3 wt.%) of dispersed SWNT bundles [66]. An isotropic dispersion of SWNT bundles between small (~100 nm) ceramic ZrO_2 particles reduced thermal conductivity by 40 to 50% by altering the nature of the porosity, contributing to an extended grain boundary region, and by serving as the basis for polycrystalline templates, similar to the synthesis of ceramic nanotubes. Conventional processing methods were used to process PSZ ceramic slurries that were subsequently tape cast into 100- to 200-μm tapes, stacked together, and pressed to form a thick (250 to 450 μm) tape. A combination of commercial plasticizers, binders, and dispersants was used to form the slip within a solvent and ball milled for 5 hours. The laminated samples were subjected to a two-stage heat treatment, first in air for binder burnout and second in partial vacuum for sintering at 1100°C, from 1 to 12 hours depending on the desired sample density, 50 to 57% theoretical density and 50 to 43% porosity, respectively. The vacuum condition for the sintering stage helps inhibit the oxidation of the SWNTs in the PSZ matrix.

11.5.4.4 Ceramic-Coated MWNT Microrods

Ceramic-coated MWNT microrods have shown great promise to be used as mechanical reinforcements for brittle ceramics. MWNTs coated with SiO_2 were used as microrod reinforcements in brittle inorganic ceramics [67]. Nonconventional processing techniques involve using MWNT-surfactant structures as templates for the synthesis of SiO_2 glass rods *in situ* creating SiO_2-MWNT microrods. In order to avoid formation of aggregates, MWNTs were mixed with CTAB surfactant in aqueous solution. In the presence of surfactants, MWNTs were believed to form co-micelle structures with surfactant molecules via strong van der Waals interactions and can be well dispersed in the aqueous solution. These cationic surfactant-MWNT co-micelles were used as templates for the synthesis of SiO_2 glass rods by the addition of sodium silicate to the MWNT-surfactant aqueous solution. Because of electrostatic interactions, silicate anions will adsorb onto the surface of the cationic surfactant-MWNT co-micelles. Upon slowly lowering the pH of the solution from 11 to 9.5, cross-linking of silicates occurs. The concentration of CTAB surfactant used to create the SiO_2-MWNT microds was slightly above its critical micelle concentration of 0.8 mM, so liquid crystal micelle nanorods do not form. By forming micrometer-size glass rods, MWNTs can be dispersed. The glass microrods are then mixed with SiO_2 powders and pressed into tablet discs. The hardness of composite discs increased proportionally to the percentage of the SiO_2-MWNT glass rods. The hardness of composite discs containing 60 wt.% microrods increases by ~100%, as compared with a pure SiO_2 disc. The wt.% of MWNTs in the composite discs is only 6 wt.%. The increase in composite disc hardness is due to both the SiO_2 outer coating and the MWNTs.

Mechanisms of hardening due to silica-MWNT microrods are believed to occur at the glass rod–matrix interface. In order to increase the hardness of ceramics, the fibers have to bridge between two sides of cracks. At higher percentages of added fibers, the probability is higher that cracks will encounter the added fibers, and hardness is increased. Furthermore, silica-MWNT pull-out would suggest that the mechanical strength of the silica-MWNT glass rods is stronger than the interactions between the glass rods and the matrix. As a final application note, these glass rods can also be used to reinforce other inorganic ceramic composites and metal matrix composite systems.

11.5.4.5 Ceramic-SWNT Aqueous Dispersions Fabricated Using Rapid Prototyping

Colloidal processing techniques can be applied to a number of ceramic material systems providing good dispersion levels of the composite materials in aqueous suspension. As an example, SWNT-Si_3N_4 composite slurries based on colloidal processing are fabricated via a rapid prototyping technique called robocasting [68]. CTAB, a cationic surfactant, is used to develop moderate levels of dispersed SWNT bundles (~50-nm diameter) within Si_3N_4 slurries via colloidal processing methods. SWNTs are used to develop 40 to 45 vol% solids composites, containing 1 to 6 vol% SWNTs, ready for robocasting at pH ~6. The CTAB dispersed silicon nitride suspension was predispersed at pH 4 before the addition of dispersed SWNTs. Colloidal processing techniques are used to evenly distribute multicomponents when they possess similar surface properties, and they obtain a viscosity suitable for robocasting based on a combination of electrostatic and electrosteric stabilization mechanisms. A paint shaker was used in between pH adjustments in order to aid in the dispersion of the Si_3N_4-SWNT slurries. The Si_3N_4-SWNT slurry, with 1.0 vol% SWNTs, is extruded through a 10-cm^3 syringe with a 0.84-mm inner diameter tip. The table speed of the platform moves at 10 mm/sec, and the parts were built submerged in an oil bath. Each solid part ($25 \times 10 \times 5$ mm) is built in less than 6 minutes and follows a length-wise, layer-by-layer building pattern. Robocast parts are dried overnight before high-temperature sintering. Si_3N_4-SWNT green parts are packed into a graphite crucible within a powder bed of 50 to 50 wt.% blend of silicon nitride and boron nitride powders. The furnace temperature was held at 1800°C for 1 hour under ultrahigh purity nitrogen gas at a pressure slightly over 1 atm. Nanotubes will further enable the robocasting process, making it applicable to a broader range of applications because the nanotube will enhance strength and lend to a multifunctional composite. The robocasting process will also promote nanotube alignment, which will further enhance the NCC properties.

11.5.4.6　Ceramic-Coated MWNTs

MWNTs coated with Al_2O_3 ceramic particles via simple colloidal processing methods provide a significant enhancement to the mechanical properties of the monolithic ceramic [58,59]. Coating the surface of MWNTs with alumina improves the homogeneous distribution of MWNTs in the ceramic matrix and enables binding between two phases to be tighter after sintering. MWNTs are treated in NH_3 at 600°C for 3 hours to change their surface properties. Treated MWNTs are put into a solution containing polyethyleneamine (PEI), a cationic polymer used as a dispersant. Alumina is dispersed into deionized water, and polyacrylic acid (PAA) was added into this very dilute alumina suspension. Sodium hydroxide is used to adjust the pH. The prepared dilute alumina suspension with PAA is added into the as-prepared carbon nanotube suspension with PEI, then both suspensions were ultrasonicated. The coated carbon nanotubes collected from the mixed suspension are subsequently added into a concentrated alumina suspension of about 50 wt.% in ethanol. Finally, the content of MWNTs is only 0.1 wt.% of alumina amount. Further drying and grinding result in MWNT-alumina composite powder that is sintered by SPS in a graphite die at 1300°C with a pressure of 50 MPa for 5 minutes in an Ar atmosphere.

MWNTs are heated to 600°C in a flow of ultrahigh purity N_2 for 3 to 6 hours then cooled to room temperature. The same treatment was repeated in NH_3 in order to obtain another sample. The coating process of MWNTs with ceramic particles is carried out by heterocoagulation; an electrostatically stabilized suspension of the particles to be coated (MWNTs) must be mixed with an electrostatically stabilized suspension of the smaller adsorbing particles (ceramic particles) under conditions where the two types of particles have surface charges of opposite sign.

A colloidal processing route is an effective way to improve the mechanical properties of MWNT-alumina composites. By adjusting the surface properties of the alumina powder and those of MWNTs, it is feasible to make them bind together with attractive electrostatic forces producing strong cohesion between two phases after sintering. During the sintering processing, the growth of alumina particles is believed to wrap MWNTs, which should have an increase in the effectiveness of the reinforcement. The addition of only 0.1 wt.% MWNTs in alumina composites increases the fracture toughness from 3.7 to 4.9 MPa $m^{1/2}$, an improvement of 32% compared with that of the single-phase alumina.

11.5.4.7　Ceramic-Coated SWNTs

Individual SWNTs coated with SiO_2 can be used to develop highly sensitive sensing device structures due to the unique properties of SWNTs [69]. The selective etching of silica-coated SWNTs either as small ropes or as individual tubes provides a route to site-selective chemical functionalization as well as the spontaneous generation of tube-to-tube interconnects. The individual SWNTs may be coated in solution and isolated as a solid mat. Either the end or the center of the SiO-SWNT may be etched and exposed. The exposure of the central section of SWNTs has potential for sensor and device structures. Processing of the coated SWNTs takes place during an *in situ* chemical reaction between fumed silica and SWNT-surfactants structures in suspension. The choice of surfactant type is important in determining whether individual SWNTs or small ropes are coated. It has been found that anionic surfactants will result in the formation of coated ropes, where cationic surfactants will results in individual coated nanotube ropes. It has been proposed that this effect is a consequence of the pH stability of the surfactant-SWNT interaction.

11.5.4.8　Y-Ba-Cu-O/MWNT Composites

Copper nanowires prepared using surface-modified MWNTs as templates are used to create MWNT-encapsulated copper nanowires as effective reinforcements for the brittle Y-Ba-Cu-O superconductor without sacrificing its superconductivity [70]. Through the protection of metal oxides, the MWNTs are prevented from decomposition during sintering at 930°C under air for 24 hours. Reinforcement effects are demonstrated to increase by 20% with the addition of 5 wt.% MWNTs-containing Cu nanowires as compared with the pristine superconductor material.

11.5.4.9 Alumina Nanocomposites Processed *In Situ*

SWNT/MWNT-metal-oxide powders (Al_2O_3, $MgAl_2O_4$) can be prepared by reduction in H_2-CH_4 of the corresponding oxide solid solutions [57]. The composite powders contain a high amount of SWNTs and MWNTs homogeneously dispersed using this nonconventional processing technique. The MWNTs and SWNTs are long and tend to form small bundles. Using solid solutions in the form of porous foams, as opposed to powders, allows for a fourfold increase of the quantity of SWNTs/MWNTs [71]. The SWNT/ MWNT-metal-oxide composite materials prepared by hot pressing are electrical conductors (0.2 to 4 S.cm^{-1}) whereas the corresponding oxide and metal-oxide materials are insulating [72]. Alignment of the SWNTs/MWNTs by hot extrusion results in an anisotropy of the electrical conductivity [73,74]. The densification of the SWNT/MWNT-containing composites is relatively poor (less than 90%); therefore, no mechanical reinforcement is observed. However, SWNTs/MWNTs mechanically blocked between the matrix grains or those entrapped inside the matrix grains could contribute to the reinforcement on a microscopic scale and later for further development to produce strengthening on a larger scale.

11.5.4.10 Multifunctional MWNT Films

A piezo-silica sandwich beam with an embedded MWNT-film sublayer can be used to increase the baseline structural damping by 200% and increase baseline stiffness by 30% due to MWNT reinforcement [75]. The multiscale system exhibits superior damping properties particularly at elevated temperatures. Potential applications of this technology range from microturbine blades to mesoscale systems such as machine elements and manufacturing tools to macrosystems such as airframes and automobiles.

11.5.4.11 Conducting Ceramics

SWNTs can be used to convert insulating nanoceramics to metallically conductive composites. Using the nonconventional consolidation technique, SPS, and conventional powder-processing techniques, dense SWNT/Al_2O_3 nanocomposites can show increasing electrical conductivity with increasing SWNT content [76]. The conductivity of these composites increases to 3345 S/m (semiconducting range 10^3, metallic conducting 10^4) for a 15 vol% SWNT/Al_2O_3 nanocomposite, at room temperature, which is a 13-fold increase in magnitude over pure alumina. In addition, these nanocomposites show a significant enhancement to fracture toughness over the pure alumina: a 194% increase in fracture toughness over pure alumina, up to 9.7 MPa m$^{1/2}$ in the 10 vol% SWNT/Al_8O_3 nanocomposite [60].

11.6 Summary and Outlook

Opportunities abound in the field of composite materials development based on carbon nanotubes. The broad property base of the nanotubes is enabling composites for electrical, mechanical, and many other applications. No doubt the true near-term outcomes will be related to the multifunctional materials that are being developed. Nanotube composite systems are being investigated in the fields of polymer, metal, and ceramic systems, so there are opportunities in a wide variety of new composites and also augmenting existing composite systems. Functionalization is further enabling the wide use of nanotubes for composites. Although a current interest has been on low nanotube concentration systems, hybrid and composite systems alike will emerge with high concentrations of nanotubes being used. These composites will likely enable properties that are not yet fully anticipated and will further open new directions for composite development. Also, composite systems produced from several nanomaterials are likely to emerge in the future because they will offer mutifunctionality to a higher scale. More smart materials will be a basis in nanotube composites, and the success of the field will strongly be linked to the continued development of nanotubes based on diameter and length control, extending their length, and providing for their separation. To this end, our education of the use of nanotubes for composite development will lead to a fuller use of nanotubes for composite development.

Acknowledgments

The authors would like to acknowledge Brian Mayeaux of NASA Johnson Space Center for sharing investigational thermal coating applications for nanotubes.

References

1. K. Jiang, Q. Li, and S. Fan, Nature, 419, 801 (2002).
1a. V.A. Davis et al. Macromolecules, 37(1), 154 (2004).
2. J. Fournier et al. Synthetic Metals, 84, 839 (1997).
3. Z. Ounaiesa et al. Composite Sci. Technol., 63, 1637 (2003).
4. E.V. Barrera, J. Mater., 52(38) (2000).
5. F.T. Fisher, R.D. Bradshaw, and L.C. Brinson, Appl. Phys. Lett., 80(24), 4647 (2002).
6. H.D. Wagner, Chem. Phys. Lett., 361(1–2), 57 (2002).
7. L.L. Yowell, Thermal Management in Ceramics: Synthesis and Characterization of a Zirconia Carbon Nanotube Composite, Rice University: Houston, 1 (2001).
8. O. Ley et al. in progress.
9. K.K. Chawla, *Composite Materials, Science and Engineering.* 2nd ed. Springer, 303 (1998).
10. E.T. Mickelson et al. Chem. Phys. Lett., 296, 188 (1998).
11. Jiang Zhu et al. Nano Lett., 3(8), 1107 (2003).
12. S. Kirkpatrick, Rev. Mod. Phys., 45, 574 (1973).
13. J. Sandler et al. Polymer, 40, 5967 (1999).
14. C. Park et al. Chem. Phys. Lett., 34, 303 (2002).
15. J.-M. Benoit et al. Synth. Metals, 121, 1215 (2001).
16. R. Safdi, R. Andrews, and E.A. Grulke, J. Appl. Polymer Sci., 84, 2660 (2002).
17. B.E. Kilbride et al. J. Appl. Phys., 92, 4024 (2002).
18. M.S.P. Shaffer and A.H. Windle, Advanced Mat., 11(11), 938 (1999).
19. S.H. Foulger, J. Polymer Sci. B, 37, 1899 (1999).
20. M. Cochet et al. Chem. Comm., 16, 1450 (2001).
21. Q. Xiao and X. Zhou, Electrochimica Acta, 48, 575 (2003).
22. Z. Jia et al. Mat. Sci. Eng. A, 271, 395 (1999).
23. C.A. Cooper et al. Composite Sci. Technol., 62, 1105 (2001).
24. R. Haggenmueller et al. J. Nanosci. Nanotechnol., 3, 1105 (2003).
25. R. Andrews et al. Macromol. Mat. Eng., 287, 395 (2002).
26. M.L. Shofner et al. Composites Part A, 34, 1207 (2003).
27. R. Andres et al. Appl. Phys. Lett., 75, 1329 (1999).
28. R. Haggenmueller et al. Chem. Phys. Lett., 330, 219 (2000).
29. A.B. Dalton et al. Nature, 423, 703 (2003).
30. S. Kumar et al. Macromolecules, 35(24), 9030 (2002).
31. J.J. Beaman et al. *Solid Freeform Fabrication: A New Direction in Manufacturing*, Kluwer Academic Publishers: Boston (1997).
32. E.T. Thostenson et al. J. Appl. Phys., 91, 6034 (2002).
33. J.D. Kim et al. in *SAMPE Conf. Proc.* ISBN 0-938994-95-6, Oct. (2003).
34. Q. Zhao et al., Comp. Sci. Technol., 61, 2139 (2001).
35. Q. Zhao et al., Polymers Adv. Technol., 13, 759 (2002).
36. P. Dharap et al. J. Nanotechnol., 15, 379 (2004).
37. Z. Li et al. J. Adv. Materials, 16(7), 640 (2004).
38. K. Rege et al. Nano Lett., 3, 829 (2003).
39. N. Zhang et al. Smart Mater. Struct., 12, 260 (2003).
40. E. Kymakis and G.A.J. Amaratunga, Appl. Phys. Lett., 80, 112 (2002).
41. P. Fournet et al. J. Appl. Phys., 90, 969 (2001).

42. B. Landi et al. Nano Lett., 2, 1329 (2002).
43. Y. -C. Chen et al. Appl. Phys. Lett., 81, 75 (2002).
44. E. Dujardin et al. Science, 265, 1850 (1994).
45. T.W. Ebbsen, J. Phys. Chem. Solids, 57(6–8), 951 (1996).
46. A.N. Andriotis et al. Appl. Phys. Lett., 76(26), 3899 (2000).
47. X. Chen et al. Comp. Sci. Technol., 60, 301 (2000).
48. Y. Zhang and H. Dai, Appl. Phys. Lett., 77(19), 3015 (2000).
49. Y. Zhang, Chem. Phys. Lett., 331, 35 (2000).
50. A. Bezryadin et al. Nature, 404, 971 (2000).
51. A. Loiseau and H. Pascard, Chem. Phys. Lett., 256, 246 (1996).
52. X. Chen et al. Carbon, 41, 215 (2003).
53. T. Kuzumski et al. J. Mater. Res., 10(2), 247 (1995).
54. S.R. Dong et al. Mat. Sci. Eng. A, 313, 83 (2001).
55. R. Zong et al. Carbon, 41(CO1), 848 (2003).
56. K.K. Chawla, *Ceramic Matrix Composites.* 1st ed., Chapman & Hall: New York, 417 (1993).
57. E. Flahaut et al. *Carbon Nanotubes-Ceramic Composites.* Nanostructured Materials and Nanotechnology Proceedings at the 105th Annual Meeting of the American Ceramic Society (2003).
58. J. Sun and L. Gao, Carbon, 41, 1063 (2003).
59. J. Sun, L. Gao, and W. Li, Chem. Mater., 14, 5169 (2002).
60. G.-D. Zhan et al. Nat. Mater., 2, 38 (2003).
61. M. Omori, Mat. Sci. Eng. A, 287, 183 (2000).
62. R.Z. Ma et al. J. Mat. Sci., 33, 5243 (1998).
63. C.N.R. Rao et al., Chem. Commun., 16, 1581 (1997).
64. S. Alperine et al. *AGARD Report 823: Thermal Barrier Coatings: The Thermal Conductivity Challenge.* Advisory Group for Aerospace Research and Development, Aalborg, Denmark (1997).
65. F.O. Soechting, in *NASA Conference Publication 3312: Thermal Barrier Coating Workshop.* NASA Lewis Research Center, Cleveland, OH (1995).
66. L.L. Yowell et al., Mat. Res. Soc. Symp. Proc. 633, A17.41 (2001).
67. G.L. Hwang and K.C. Hwang, J. Mat. Chem., 11, 1722 (2001).
68. E.L. Corral et al. in Rapid Prototyping of Materials, TMS Pub. 53 (2002).
69. E.A. Whitsitt and A. Barron, Nano Lett., 0(0), A–D (2003).
70. G.L. Hwang et al. Chem. Mater., 15, 1353 (2003).
71. S. Rul et al. J. Eur. Cer. Soc., 23, 1233 (2003).
72. E. Flahaut et al. Acta Materialia, 48, 3803 (2000).
73. A. Peigney et al. Ceram. Int., 26, 677 (2000).
74. A. Peigney et al. Chem. Phys. Lett., 1–2, 20 (2002).
75. N.A. Koratkar, B. Wei, and P.M. Ajayan, Comp. Sci. Technol., 63, 1525 (2003).
76. G.-D. Zhan et al. Appl. Phys. Lett., 83(6), 1228 (2003).

12

Other Applications

M. Meyyappan
NASA Ames Research Center

Carbon nanotubes, due to their interesting properties, are being investigated for a variety of applications in addition to the major ones discussed in previous chapters, and here a brief summary of these areas is provided.

12.1 Applications in Integrated Circuit Manufacturing

Whereas carbon nanotube (CNT)-based electronics may be a long-term prospect awaiting further development as well as the end of silicon CMOS scaling according to Moore's law, there are areas where CNTs can provide possible solutions to anticipated problems in the future generations of Si CMOS integrated circuit (IC) manufacturing. One such area involves interconnects, which is currently dominated by copper damascene processing [1]. Copper replaced aluminum about 5 years ago due to its higher conductivity. The industry has developed successful processing techniques to integrate Cu interconnects because it is notoriously difficult to etch. The current problem with copper interconnects appears to be electromigration when current densities reach or exceed 10^6 A/cm^2 [2,3]. In contrast, CNTs do not suffer even at current densities of 10^7 to 10^9 A/cm^2 [4] and can offer a solution to the interconnect problem.

Taking the vertical interconnect for DRAM application as an example, the current technology involves etching deep vias (13:1 or even deeper) in SiO_2, void-free filling of Cu and chemical mechanical polishing (CMP). A simple solution here is to use CNTs instead of Cu in the second step above as demonstrated in Reference 5. However, as CNTs deposit like noodles in such trenches, the CMP step will pull a substantial amount of CNTs out of the trench. Li et al. [6] suggested a viable alternative wherein the interconnect vertical multiwall nanofibers (MWNFs) are deposited first on prespecified locations. The gap between MWNFs is next filled with SiO_2 using thermal CVD with tetraethoxy silane (TEOS) as feedstock. The CMP step is then used to planarize the top before metalization. The resistance of individual interconnects was shown to be around 40 Ω, which is higher than the industry expectation. However, further work on improving the material quality and interface resistance can help reduce the resistance.

It is well known that heat-dissipation issues are becoming critical with every new generation of computer chips, with current levels exceeding 50 W/cm^2. CNTs exhibit very high-thermal conductivities in the range of 1200 W to 3000 W/mK depending on single-wall nanotube (SWNT) or multiwall nanotube (MWNT) diameter, etc. [7,8], and can offer cooling solutions in IC manufacturing. The issues involve appropriate methods to deposit the nanotubes for effective cooling. The high-thermal conductivity of

CNTs is also useful in heat dissipation in several other industrial instruments, though at this writing no industrial applications have been demonstrated.

12.2   Catalyst Support and Adsorbents

Materials with large surface areas have always been of interest in industrial applications such as catalyst support and adsorbants. SWNTs possess a large surface area, and several studies have characterized the BET (Brunauer-Emmett-Teller) surface area of nanotubes [9,10]. The nanotubes need to be purified first to eliminate the unwanted metal and amorphous carbon impurities. Yang et al. [9] showed a surface area of 861 m^2/g for SWNTs produced from the HiPCo process discussed in Chapter 4. Improved purification techniques and a debundling step reported in Reference 10 and described in Section 4.4.2 yielded a record 1587 m^2/g for HiPCo-produced SWNTs. The use of SWNTs as catalyst support in NO conversion has been reported [11] where a 1 wt.% Rh loaded onto nanotubes yielded a 100% NO conversion at 450°C. Generation of nitrogen oxides in the combustion of coals and other fuels as well as automobile exhaust are major sources of pollution with undesirable effects such as acid rain and global warming. Hence, elimination of NO_x is an important industrial goal, and supported-catalyst approaches to reduce NO_x are popular [11,12].

Eswaramoorthy et al. [13] showed that SWNTs are good microporous materials capable of adsorbing benzene, methanol, and other molecules and suggested that this would allow carrying out catalytic reactions in nanoscale reactors. Santucci et al. [14] demonstrated NO_2 and CO gas adsorption on nanotubes whereas Cinke et al. [15] showed that SWNTs adsorb nearly twice the volume of CO_2 compared with activated carbon. The CO_2 heat of adsorption from their experimental data is 2303 J/mol (0.024 eV).

12.3   Storage/Intercalation of Metals

Filling of the nanotubes with various metals has been studied for producing wires and other applications. One of the most popular is the electrochemical intercalation of nanotubes with lithium for battery application [16–19]. A desirable attribute for the battery is the high-energy capacity, which is determined by the saturation lithium concentration of the electrode material. CNTs have been widely investigated for intercalating with Li with the hope that the cylindrical pores and intertube channels would all be available for metal uptake. Gao et al. [17] showed a reversible capacity of 1000 mAh/g for Li-intercalulated SWNTs. Zhang and Wang [20] filled nanotubes with copper using microwave-assisted plasma CVD. Although no conductivity data were reported, Cu-filled nanotubes may be useful in interconnect and heat-dissipation applications. Small spherical crystals and elongated single crystals of Ru have also been filled inside nanotubes using wet chemistry [21]. Kiang et al. [22] produced 1-nm diameter bismuth nanowires by capillary filling of SWNTs with Bi. Except in the case of Li, most studies concentrated on demonstrating an approach to fill the nanotubes without any particular application in mind. An early review of this subject can be found in Reference 23.

12.4   Membranes and Separation

The nanoscale size and the hollow core of CNTs have prompted research into applications such as separation membranes, nanofluidic channels, and drug and other molecule delivery systems. Most of the early and current studies have focused on the fluid transport through these nanochannels [24–29], primarily through computational analysis. For example, molecular dynamic simulations [25] show that methane/*n*-butane and metane/isobutene mixtures can be separated using nanotubes. Skoulidas et al. [29] performed atomistic simulations to compute diffusivities of light gases in nanotubes and zeolites of comparable pore size and concluded that the transport rates in nanotubes are orders of magnitude higher than those in zeolites.

References

1. International Technology Roadmap for Semiconductors (Semiconductor Industry Association, San Jose, CA 2001); http://public.itrs.net/
2. G. Steinlesberger et al., Microelectron. Eng., 64, 409 (2002).
3. T.S. Kuan et al., Mat. Res. Soc. Symp. Proc., 612, D.7.1.1 (2001).
4. B.Q. Wei, R. Vajtai, and P.M. Ajayan, Appl. Phys. Lett., 79, 1172 (2001).
5. F. Kreupl et al., Microelectron. Eng., 64, 399 (2002).
6. J. Li et al., Appl. Phys. Lett., 82, 2491 (2003).
7. M.A. Osman and D. Srivastava, Nanotechnology, 12, 21 (2001).
8. J. Hone et al., Appl. Phys. Lett., 77, 666 (2000).
9. C.M. Yang et al., Nano Lett., 2, 385 (2002).
10. M. Cinke et al., Chem. Phys. Lett., 365, 69 (2002).
11. J. Luo et al., Catal. Lett., 66, 91 (2000).
12. Y. Nishi et al., J. Phys. Chem. B, 101, 1938 (1997).
13. M. Eswaramoorthy, R. Sen, and C.N.R. Rao, Chem. Phys. Lett., 304, 207 (1999).
14. S. Santucci et al., J. Chem. Phys., 119, 10904 (2003).
15. M. Cinke et al., Chem. Phys. Lett., 376, 761 (2003).
16. P.M. Ajayan and O.Z. Zhou, Top. Appl. Phys., 80, 391 (2001).
17. B. Gao et al., Chem. Phys. Lett., 307, 153 (1999).
18. J.S. Sakamoto and B. Dunn, J. Electrochem. Soc., 149, A26 (2002).
19. I. Mukhopadhyay et al., J. Elecrochem. Soc., 149, A39 (2002).
20. G.Y. Zhang and E.G. Wang, Appl. Phys. Lett., 82, 1926 (2003).
21. J. Sloan et al., Chem. Commun., 347 (1998).
22. C.H. Kiang et. al., J. Phys. Chem. B, 103, 7449 (1999).
23. M. Terrones et al., MRS Bull., 43, August (1999).
24. Z. Mao and S.B. Sinnot, J. Phys. Chem. B, 104, 4618 (2000).
25. Z. Mao and S.B. Sinnot, J. Phys. Chem. B, 105, 6916 (2001).
26. Z. Mao and S.B. Sinnot, Phys. Rev. Lett., 89, 278301 (2002).
27. L. Sun and R.M. Crooks, J. Am. Chem. Soc., 122, 12340 (2000).
28. S.A. Miller et al., J. Am. Chem. Soc., 123, 12335 (2001).
29. A.I. Skoulidas et al., Phys. Rev. Lett., 89, 185901 (2002).

Index

Printed in the USA/Agawam, MA
October 15, 2010

554637.141